BIOLOGY

of the

LEAF MINERS

by

Prof. E. MARTIN HERING

Berlin.

UITGEVERIJ Dr W. JUNK

's-GRAVENHAGE

1951

ISBN 978-94-015-7198-2 ISBN 978-94-015-7196-8 (eBook)
DOI 10.1007/978-94-015-7196-8

SOFTCOVER REPRINT OF THE HARDCOVER 1ST EDITION 1951

N. V. DRUKKERIJ HOOIBERG, EPE

BIOLOGY OF THE LEAF MINERS

ERRATA:

Index 18.	for	occurence	read	occurrence
„ 19.	„	ennemies	„	enemies
p. 1, l.12	„	basic	„	basis
l.35	„	combatting	„	combating
p. 4, l.10	„	there are	„	they are
l.20	„	dependant	„	dependent
p. 10, l.6	„	lay	„	lie
p. 12, l.10	„	rate	„	rare
p. 13, Fig. 13	„	oven	„	over
„	„	leaft	„	leaf
p. 21, l.39	„	Colephora C.	„	Coleophora. C.
p. 23, l.5	„	its	„	their
l.7	„	Colephora	„	Coleophora
p. 30, l.5	„	examplified	„	exemplified
p. 33, l.15	„	occured	„	occurred
p. 43, l.44	„	preceeding	„	preceding
p. 44, l.5	„	with in	„	within
p. 47, Fig. 38	„	aizoon	„	*aizoon*
p. 53, l.2	„	like plug	„	like a plug
p. 56, l.39	„	where	„	to that on which
p. 57, l.14	„	though	„	thought
l.22	„	larva	„	larvae
p. 58, l.17	„	instears	„	instars
l.19	„	larvae	„	larval
l.40	„	into	„	onto
p. 59, l.5	„	in Tilia	„	on Tilia
p. 61, l.7	„	gallcauser	„	gall-causer
l.26	„	exsuding	„	exuding
p. 64, l.11	„	inter	„	intra
p. 65, l.5	„	as	„	at
p. 67, l.23	„	as pupa	„	as a pupa
l.29	„	pupate	„	pupates
l.31	„	larva	„	larvae
p. 69, l.14	„	emergences	„	emergence
p. 70, l.36	„	time	„	mine
p. 72, l.20	„	nutritous	„	nutritious
p. 74, l.39	„	plants	„	plant's
p. 79, l.28	„	earliest spring	„	earliest days of spring
p. 82, l.14	„	tissue	„	tissues
p. 83, l.25	„	larva	„	larvae
p. 85, l.16	„	pre-dominate	„	predominate
p. 88, l.26	„	by Dufrane, after breeding the imago, named	„	named by Dufrane, after breeding the imago

p. 92, l.24	for	dregree	read	degree
p. 93, Fig. 70	„	predominately	„	predominantly
p. 94, l.1	„	preceeding	„	proceding
l.28	„	spirale	„	spirals
p. 102, Fig. 75	„	read	„	head
p. 112, l.39	„	consistancy	„	consistency
p. 113, l.16	„	very	„	vary
p. 115, l.5	„	immistakably	„	unmistakably
p. 125, l.42	„	quarter	„	quarters
p. 142, l.18	„	predominently	„	predominantly
p. 159, l.19	„	monotpyic	„	monotypic
p. 168, l.20	„	the	„	that
p. 177, l.7	„	is large	„	is to a large
l.17	„	predominantely	„	predominantly
p. 182, l.15	„	these	„	this
p. 184, l.30	„	"wine-red" distinct	„	"wine-red", distinct
p. 187, l.38	„	follows	„	follow
p. 188, l.12	„	pollenization	„	pollination
l.41	„	need	„	needs
p. 191, l.11	„	pollenization	„	pollination
p. 200, l.12	„	„	„	„
l.30	„	transferance	„	transference
p. 216, l.28	„	representitives	„	representatives
p. 229, l.2	„	in	„	on
p. 238, l.15	„	destignation	„	designation
l.20	„	serpentine of	„	serpentine mines of
p. 244, l.6	„	arose	„	arise
p. 246, l.3	„	as	„	has
p. 249, l.6	„	formed Gertz	„	formed. Gertz
p. 253, l.17	„	noman's	„	no man's
p. 257, l.5	„	quarter	„	quarters
p. 259, l.9	„	year tissue	„	year, the tissue
l.12	„	laef	„	leaf
l.17	„	smal	„	small
p. 260, Fig. 162	„	larva eaten out from within the mine by bird's beak-cut	„	larva pecked out of the mine by a bird
Fig. 163	„	then	„	later
p. 261, l.9	„	feedings	„	feeding
p. 269, l.4	„	time	„	times
p. 273, l.18	„	Agrymyzid	„	Agromyzid
p. 276, l.14	„	mature mature	„	mature
l.33	„	prejudice, each	„	prejudice each
p. 277, l.20	„	mid-rib, of	„	mid-rib of
p. 285, l.42	„	higher the temperature	„	higher temperature
p. 285, l.33	„	Cicades	„	Cicadas
p. 290, l.20	„	wheather	„	weather
p. 291, l.27	„	rule	„	role
l.28	„	foodstoffs	„	foodstuffs
p. 298, l.3	„	polyphetic	„	polyphyletic
p. 299, l.14	„	a	„	the
p. 301, l.28	„	which	„	which
p. 306, l.38	„	a produces	„	a procedure produces
l.44	„	hour's	„	hours
p. 309, l.10	„	attack	„	attach
p. 314, l.37	„	graduating	„	grading
p. 315, l.10	„	Card	„	card
p. 320, l.20	„	which in	„	which it
p. 332, l.16	„	fresh await	„	fresh mines await

CONTENTS

1.

INTRODUCTION

The development of specialised feeding habits during the course of time by human beings is paralleled in the majority of animals, and insect larvae in particular have developed special peculiarities, which in most cases are quite characteristic of the species concerned. This applies especially to phytophagous insect larvae, and anyone with the requisite experience can say with a fair degree of certainty which insect larva is responsible for any damage to be found on a plant. It leaves behind a definite "feeding pattern" which might be compared to a "visiting card" on which the genus and species are marked in runic characters. Whoever has learned to read the runes can readily determine who has been feeding on the affected spot, solely on the basic of the "visiting card" left behind. From the known factors — the name of the plant and the type of feeding pattern — and after some study of the various types of plant infestation, both the genus and species of the larva producing the feeding pattern can be worked out without difficulty.

The importance of "feeding pattern investigation" has now far outstripped the successes to be obtained by normal collecting. Previously, when wishing to list the species of insects present in any given locality they were caught with the net, by sugaring and other methods. This always resulted in a very defective "list" of the insects in fact existing in the locality concerned. If an expert collector of micro-lepidoptera is asked how many species of the genus *Nepticula* he has taken as perfect insects in the vicinity of his home, one will be surprised at the very limited number by comparison with those he has obtained by breeding. Moreover, it is much easier to identify species existing in a locality by means of the feeding patterns. Collecting is normally restricted to the limited time when the insect concerned is on the wing, while it is very often possible to determine at a much later period who has been enjoying hospitality in the area on the basis of the "visiting card" left behind. The investigation of the feeding pattern is thus indispensible both for the compilation of local lists of all species of insect present in a given locality, and on the other hand also for applied entomology, which is concerned with combatting injurious species. The conspicuous feeding patterns immediately reveal the presence of any pests, even though the insect itself, which in many cases is more elusive, cannot yet be detected. Steps can in this way be taken in good time to forestall the occurrence of plagues.

Two groups in particular amongst the many different feeding patterns of insects have already for centuries past attracted the special attention of both professional and amateur entomologists — these are galls and mines. In both cases the existence of the insect larva in question is endophagous; in other words it lives inside the plant tissue and does not merely gnaw at it from the outside. The great interest shown in these feeding patterns derives on the one hand from the special pleasure which was found in the frequently highly decorative and in other cases quite remarkable structures and also from the fact that the uniformity of the feeding pattern is so particularly pronounced. This has meant that in almost all cases it was easier to identify the genus and species of the insect responsible from its feeding pattern than from the frequently minute and subtle characteristics, which differentiate the imagines of the species concerned. Finally, the species living in galls or mines consist for the most part of highly specialised forms showing evidence of a variety of special adaptations to their mining or gall-producing existence, so that both anatomical and physiological investigations, as well as the question of the relationship between plants and animals inevitably gave promise of fruitful results, of interest to an extensive public.

Cecidology, the study of galls, was studied systematically considerably earlier than its sister science, hyponomology, or the study of mines, which only fully developed in the twentieth century. This explains why many more fresh discoveries can be expected from the study of mines than from the study of galls; there are, for example, still large areas in Europe in which leaf mines have not been specifically collected. However, even in the regions where many entomologists have already explored the different orders of insects, new species of miners still remain to be discovered.

2.

DEFINITION, SHAPE AND CLASSIFICATION OF MINES

Mines are feeding channels caused by insect larvae inside the parenchyma or epidermis tissues of plants, in which the epidermis or at least its outer wall remains undamaged, thus shutting off the mine cavity from outside. The mine accordingly provides both living and feeding quarters for the larva.

Normally, the mine cavity is extended inside the green parenchyma of the leaf but in the same way feeding channels, which can be designated as mines, may be established inside the parenchyma of fruits, stems or roots. The channels established in the different type of pith tissue of the stem, root and fruit and which lie below the green parenchyma layer of the organ in question do not fall within the same category.

The construction of a mine represents a peculiar specialisation, which can be clearly differentiated from the other feeding habits of insect larvae but which on superficial observation can often give rise to mistakes. There are a number of distinct types of feeding pattern. First, the one which results in the disappearance of the complete leaf, can cause no confusion; leaf perforation, resulting in holes being eaten throughout the thickness of the leaf, can easily be recognised as such. The three following types, however, are more frequently and more easily confused with mining, particularly as they occur predominantly in a genus having a whole series of mining species, as is the case with *Coleophora*. First, there is the type in which one epidermis and parts of the parenchyma below are eaten away; a second type results in the eating of the whole parenchyma tissue, merely leaving behind the one epidermis, which is then analogous to a glassy window covering the hole which has been made and which has been termed "window feeding"; thirdly, there is skeletonization, in which the whole thickness of the leaf is eaten away including both the upper and lower epidermis, but in contrast to the previous type, the leaf veins, even the flimsiest, remain intact, so that the skeleton of the leaf is preserved. Mining is distinguished from all these feeding forms by the fact that b o t h the upper and lower epidermis remain intact. In many cases the epidermis itself is devoured, while its hairy appendages (in very hairy plants) remain in position, thus creating the appearance of a mine. This is particularly striking in the case of the Noctuid *Eublemma noctualis* Hb. feeding on Helichrysum.

As appears from the definition of a mine given earlier, the cavity in the leaf parenchyma can only be considered as a mine, if it has been produced by an insect larva. Perfect insects have never yet been known to produce leaf mines. Many Rhyncophoridae insert their proboscis deep into the leaf parenchyma and so produce structures resembling blotch mines; there are, however, always characterised by their small size and do not accord with the definition given above, since they are indeed the feeding but not the living quarters of the insect concerned. All forms of attack by fungi are furthermore not considered as mines, although fungi can frequently cause blemishes, which in appearance closely resemble mines. However, in such cases there is no well-defined cavity, which in order to satisfy our definition must be produced by the mining insect.

In this connection it should briefly be mentioned that the concept of a gall envisages a quite definite reaction by the plant to the presence of the intruder, who in turn is specifically dependant on the tissue of the plant, which has been produced as this reaction. There are, however, certain intermediate stages between leaf galls and leaf mines which justify us in considering leaf mines as the most primitive form of leaf galls and conversely leaf galls as highly specialised leaf mines. In a subsequent chapter this relationship will be dealt with in somewhat greater detail.

In order to identify the mine of any given species of insect larva, information is required on three points — the tissue in which the larva has been feeding, the horizontal and vertical extension within this tissue and the direction of the feeding. By linking up these different possibilities with the species of the food-plant, a surprisingly large number of combinations is arrived at which make it possible to identify the producer merely on the basis of the mine. It is therefore necessary to agree on the definition of certain forms in order to be able to differentiate between the many different types of mines.

The mine quite generally is called the hyponomium or more shortly hyponome (from Greek ὑπόνομος — underground passage) and the terms for the different types of mines are formed with the final syllable of the short form — nome. The mine may be established in the parenchyma of various parts of the plant and this enables us to distinguish the stem mine or caulonome, the flower mine or anthonome, the fruit mine or carponome and the leaf or petiole mine or phyllonome. Intermediate forms may occur between these different types, as the same species may also mine first one, then another part of the plant.

In describing the extension of a mine, first of all its depth within the leaf must be investigated, in other words the vertical extension in the leaf. In the most primitive case the larva will consume the whole green-coloured parenchyma lying between the upper and lower epidermis of the leaf. A cross-section of a leaf (Fig. 1) starting from the

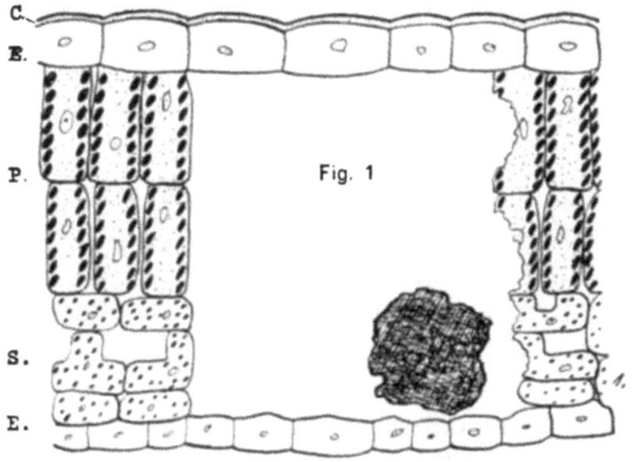

Fig. 1. Cross-section of a full-depth mine (C-cuticle, E-epidermis, P-palisade parenchyma, S-spongy parenchyma). In the centre a heap of frass.

upperside shows, beneath the layer of skin cells (epidermis), first, one or more layers of tightly-packed roughly cylindrical cells, which are called "palisade parenchyma" after their shape, or "assimilation parenchyma" after their function. On the underside of the leaf the epidermis is adjoined by a rather differently constructed parenchyma, which is looser and serves largely to carry away the products built up by the assimilation parenchyma. It is accordingly known as the spongy or conducting parenchyma. The spongy and palisade parenchyma are together known as the mesophyll of the leaf. If the larva has eaten up all the mesophyll, the mine will look glassily transparent when held against the light, since it is enclosed only by the epidermis, which of course contains no chlorophyll. Such a mine is termed full-depth. On the other hand, if the mine is constructed only in the palisade parenchyma, leaving the spongy parenchyma intact, it becomes more obvious from the upperside than from the underside of the leaf and against the light it appears greenish owing to the chlorophyll in the remaining tissue; this is known as an upper surface mine (Fig. 2). In the opposite case the larva consumes the spongy parenchyma and leaves the palisade parenchyma untouched; such a mine (Fig. 3) is more easily visible from the underside than from the upperside, and, similarly, looks green when held against the light; this is the lower surface mine. Generally speaking each species has its own specific characteristics concerning the vertical extension of the mine, which, however, may change during the course of the larva's life. There are only a few species which, in the same instar may be found either in an upper surface or in a lower surface mine, such as *Nepticula trimaculella* Haw. on Populus. We find the most specialised taste in those species which live exclusively inside the cells of the epidermis. In these cases the mine cavity is enclosed solely by the outer wall of the epidermis cells (with perhaps an outside cuticle) (Fig. 4). This type we call an epidermal mine, as opposed to the parenchymal mines mentioned above. Epidermal mining continues throughout the larval life only in a relatively small group of species; most frequently one finds that first an epidermal, and later a parenchymal, mine is constructed, particularly in the Gracilariidae and above all in the genus *Lithocolletis*. The epidermal mine can easily be recognised, since, when held to the light, it in no way stands out from its surroundings; the whole leaf appears uniformly green because no cells containing chlorophyll have been consumed by the larva. In due course air enters the empty cells of the epidermis giving the track of the mine a characteristic silvery lustre if it is on the upperside of the leaf, while it otherwise always assumes a conspicuous whitish coloration. Frequently the impression is caused of a snail having crawled across the leaf, with its dry track corresponding to the channel of the mine. Parenchymal mines are sometimes, although not often found, which are neither upper nor lower surface mines, since they can be seen equally well from both sides. Their course follows the most centrally situated layers of tissue in the leaf, in a lower layer of the palisade parenchyma or an upper layer of the spongy parenchyma, in any case in layers with no direct contact with the epidermis (Fig. 5). Such a mine is termed "inter-parenchymal"; it is usually distinguished by a characteristic yellowish-green coloration, owing to a still intact layer of parenchyma which lies between it and the outer world. For the same reason it is more difficult to see than the simple upper surface or lower surface mines. Mines of some flies

Fig. 2. Cross-section of an upper surface mine. Abbreviations as in Fig. 1.

Fig. 3. Cross-section of a lower surface mine. Abbreviations as in Fig. 1.

Fig. 4. Cross-section of an epidermal mine of a *Phyllocnistis*. Mine only in the upper epidermis.

Fig. 5. Cross-section of an inter-parenchymal mine (*Phytomyza affinis* Fall. on Cirsium). The undamaged cells of the uppermost palisade parenchyma are of a ligther green.

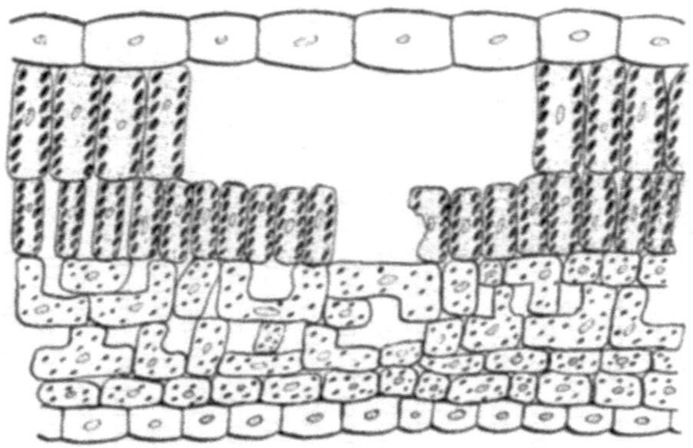

Fig. 2

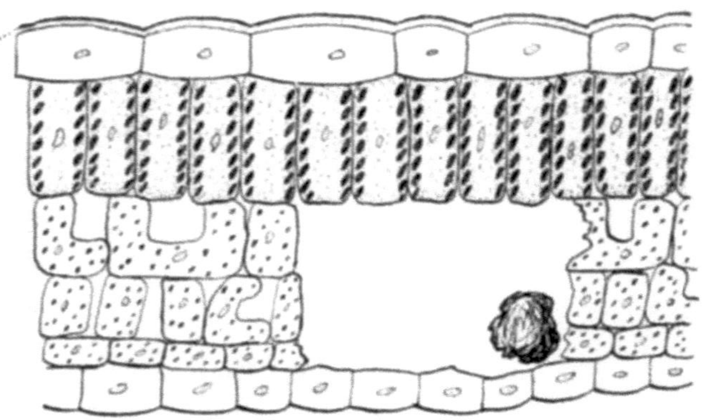

Fig. 3

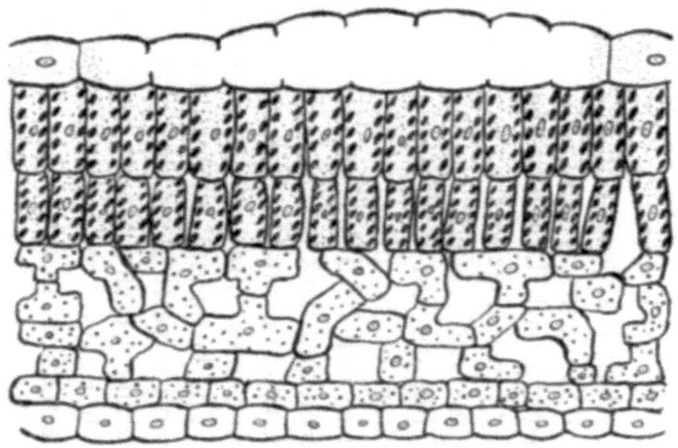

Fig. 4

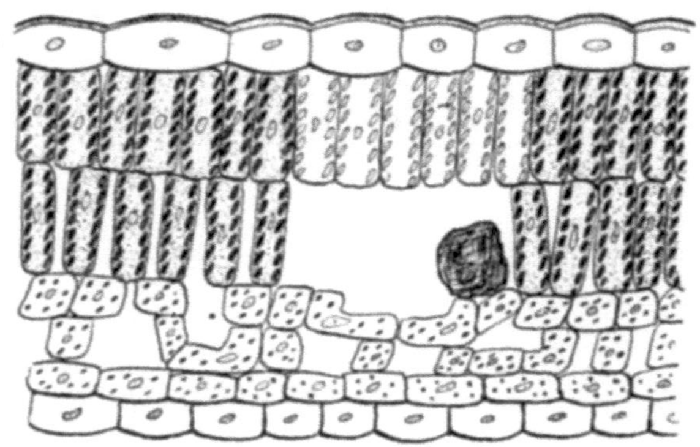

Fig. 5

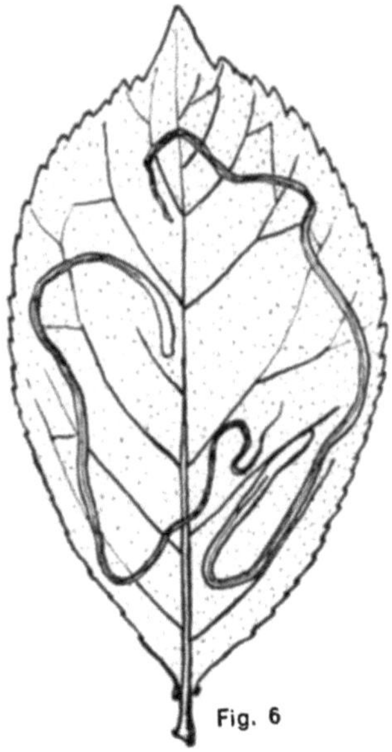

Fig. 6. Prunus cerasus: ophionome of *Lyonetia clerkella L.*

such as *Phytomyza heringiana* Hend. on Malus may be mentioned as an example. In this case, as with other leaf mines the larvae frequently gnaw deeper in a vertical direction in certain spots than in the immediate surroundings; these deeper holes in the mesophyll then appear as lighter patches and the mine is then "m o t t l e d". In interparenchymal mines this mottling is particularly well marked with Umbelliferae parasites such as *Phytomyza heracleana* Her. and in general it is most clearly marked in the species of the genus *Lithocolletis*.

The vertical extension inside the leaf tissue largely determines the colour of the mine, in so far as anthocyanine bodies do not arise; these will be discussed later. The colours from white to yellowish-green and dark green depend solely on the extension of the mine in the thickness of the leaf, while the brown to black coloration which is sometimes found can be ascribed to changes in the cell membranes.

The horizontal extension of the mine has a greater influence even than the vertical extension in producing a characteristic appearance, making it easily recognisable and particularly conspicuous. The type of horizontal extension of the mine cavity is largely dependent on the manner in which the larva eats its way through the plant tissue and in this, considerable differences can be noted.

If the larva moves only in one direction, in other words works its way through the food only forwards, the l i n e a r m i n e, t h e ophionome, is produced (Fig. 6). If, however, the larva eats away in several directions, a blotch mine or stigmatonome is formed. The blotch mine for its part may arise in two ways. If the larva eats around in all directions without any system, without maintaining a particular direction while eating, an orthogenous blotch mine is formed (Fig. 7). In contrast to this many larvae, particularly of Diptera, are accustomed in the first place to construct a linear mine and after a short while turn round and construct the next track flush with the one they have just completed, so that no leaf tissue remains between the two tracks. This process is repeated a number of times and the portions of track thus lying side by side together form a secondary blotch, which is termed an ophiogenous blotch mine. This difference in origin can almost always be detected in the completed mine, since the orthogenous blotch mine can only contain primary feeding tracks, while the ophiogenous mine always contains secondary feeding tracks, which together with the primary ones very frequently form the conspicuous "herring-bone" pattern (Fig. 8). The section on the differentiation of

Fig. 7

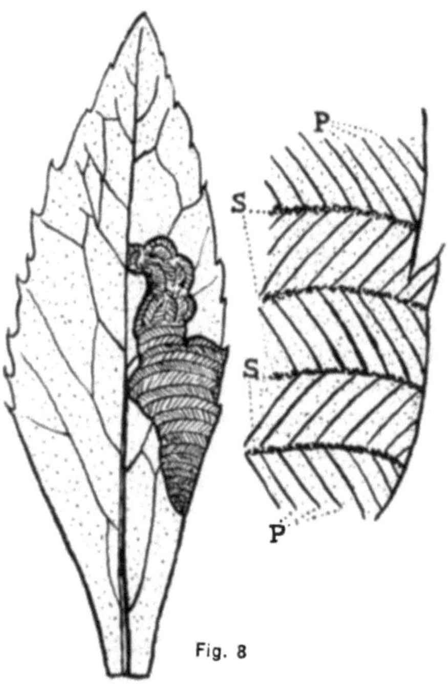

Fig. 8

Fig. 7. Betula: orthogenous stigmatonome of *Nepticula argentipedella* Z.

Fig. 8. Solidago: ophiogenous stigmatonome of *Dizygomyza posticata* Mg., on the right a part with greater magnification shows the primary (P) and secondary (S) feeding lines, the „herring-bone" pattern.

mines according to the Order of the producer will deal more fully with the formation of these primary and secondary feeding tracks. The two chief types of mine, the ophionome and stigmatonome, show in their turn a whole number of modifications, the most important of which will now be mentioned.

In many cases the bends of a mine track lay around each other in a spiral, so that the whole course of the mine is limited to a small patch (Fig. 9). In this way the serpentine mine, the heliconome, is formed, which is especially found in the early stages of very many mines. Its formation is explained by the fact that the young larva is unable to cross the stronger leaf veins, with their many hard strengthening elements. This form of existence is very often discarded only when the larva has attained a more substantial size. Such serpentine mines are found amongst Lepidoptera (*Nepticula acetosae* Stt. on Rumex, *Bucculatrix frangulella* Goeze on Rhamnus), as well as among Diptera (*Liriomyza eupatorii* Kaltb. on Eupatorium and Galeopsis, *L. cannabis* Hend. on Cannabis). In other cases the serpentine tracks are not made around each other in a spiral but, like the coils of an intestine, close beside each other. In this way the "intestinally-coiled" mine is found, the visceronome. Visceronomes are also found amongst other orders of insects; they can be ascribed to the same principle as heliconomes: the larva wishes to avoid as far as possible having to cross the stronger leaf veins. This habit of the larva was responsible for the name of the species *Nepticula viscerella* Stt. on Ulmus (Fig. 10). Finally, there are a whole series of mines in which the larva, setting out from a central patch, drives short tracks or galleries into the leaf in all directions, forming in this way a star-shaped pattern. These are therefore known as digitate or star mines or asteronomes and particularly fine examples are represented by *Phytomyza xylostei* Kaltb. on Lonicera and Symphoricarpos (Fig. 11) and by *Phytomyza gentianae* Hend. on Gentiana. In these cases also the desire of the larva to postpone as long as possible crossing the stronger leaf veins must be considered responsible for the development of this type of feeding. That this is the reason for the peculiar mine construction is shown by the fact that mines of the same species can assume quite different shapes, according to where the egg was laid. If the egg of *Nepticula atricollis* Stt. is laid of the edge of the leaf, the mine track is completely straight, following the serrations along the edge of the leaf. If, however, in the same species the egg is laid in the centre of the leaf, the initial track of this species represents a distinct visceronome, as can be seen particularly often on Crataegus.

While many species attempt to avoid the leaf veins in this way, others seem to be especially attracted by the leaf veins when con-

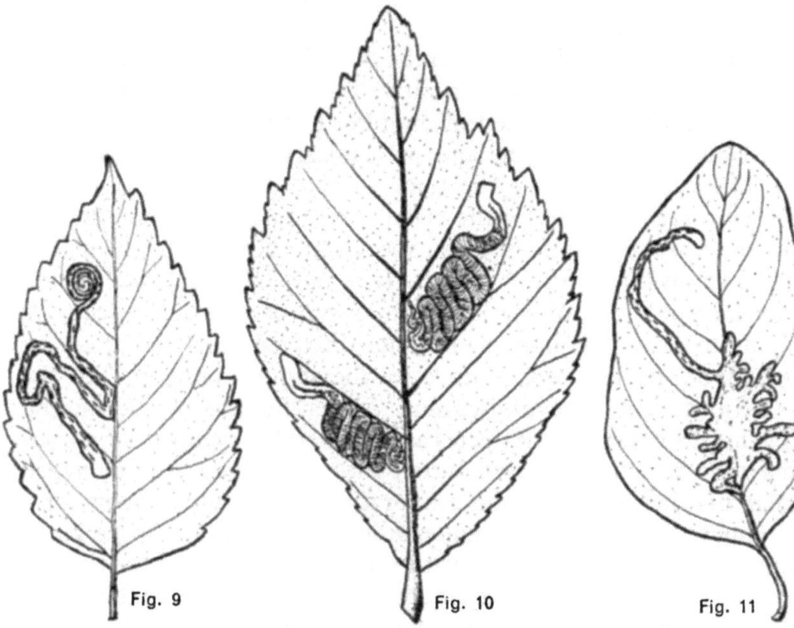

Fig. 9. Galeopsis: Heliconome of *Liriomyza eupatorii* Kaltenb.

Fig. 10. Ulmus: Visceronome of *Nepticula viscerella* Stt.

Fig. 11. Lonicera: Asteronome of *Phytomyza (Napomyza) xylostei* Kaltenb.

structing their mine. In such cases the mine runs directly above the vein and forks at the point where a lateral vein branches off. In this way the mine becomes pinnately or palmately-branched. During pauses in feeding, the larva generally seeks out a spot above a vein, where it is less visible than in the other parts of the leaf surface. Such forked tracks are very often found on the most diverse plants, usually produced by the polyphagous *Liriomyza strigata* Mg. (Fig. 12). In other cases the larva lives predominantly in the mid-rib, only driving out short tunnels to the sides; this is particularly conspicuous with *Phytomyza ramosa* Hend. on Knautia and here also the species has been named after the characteristic position of its mine. Mines in this position can be conveniently considered as modifications of asteronomes, even though they might have a different cause from genuine asteronomes. In this case a special association with the vein is sought after and the larva has a preference for mining in the parenchyma of the vein. On the one hand it gains increased protection through this position of the mine, but it probably also enjoys better nutriment than in other parts of the leaf through this close connection with the veins, in which naturally there is a particularly active movement of nutritious products.

The stigmatonome, the blotch mine, like the linear mine, has similarly many forms. In its most primitive form the blotch-mine represents a blotch or patch in the leaf, more or less uniformly developed, or eaten out in all directions and with no other distinguishing features. The type for this is the mine of the majority of species of *Coleophora*, recognisable by the circular opening in the one epidermis of the leaf; the habits of these species are dealt with in detail in the chapter on case-bearing miners. It is not, however, rate to find that in one epidermis of the leaf shrinking processes set in, causing the epidermis over the mine to become wrinkled; in other cases one epidermis of the leaf arches up over the mine, giving rise to the b l i s t e r m i n e, the physonome. The influence of the mining larva is in this case only indirect: the considerable quantity of parenchyma cells which have been bitten into or perhaps the damaged vascular bundles result in the evaporation in the mine cavity of a large amount of liquid. Even under normal circumstances the air in the mine is fairly well saturated with water vapour and where there is such an excess the mine is expanded, causing the one epidermis to swell up, giving the mine the appearance of a blister.

This blister-like expansion of the mine can be found in many species, such as in the *Philophylla* mines in Umbelliferae, in conjunction with special conditions in the food-plant, or it may be a typical characteristic of the species, as with many Gracilariidae on Compositae and Labiatae. Another modification of the blotch-mine is the t e n t i f o r m m i n e, the p t y c h o n o m e. In the tentiform mine the epidermis of one side of the leaf contracts over the mine cavity forming folds and the opposite side of the leaf shows in this spot a correspondingly convex arching. The commonest examples of such tentiform mines are found amongst the Palaearctic species of the genus *Lithocolletis* (Fig. 13). In most species the young larva forms first of all an epidermal mine, usually enlarged from a track to a blotch (Fig. 14, top left), and lives during this time purely as a sap-feeder. Later, (after a change in the mouth parts has taken place) it also attacks the parenchyma and now becomes a tissue-feeder. The outer walls of the epidermis cells are all that remain above the original epidermal mine. Silken threads are now drawn over this remnant of the epidermis, each one passing over one or more spots on the epidermis without being attached, so that spots left free of thread lie in a line one behind the other.

When drying, this silken secretion contracts somewhat, causing a contraction of the outer wall of the epidermis above the mine; the spots left free of thread then form the "wrinkles" of the mine,

Fig. 12

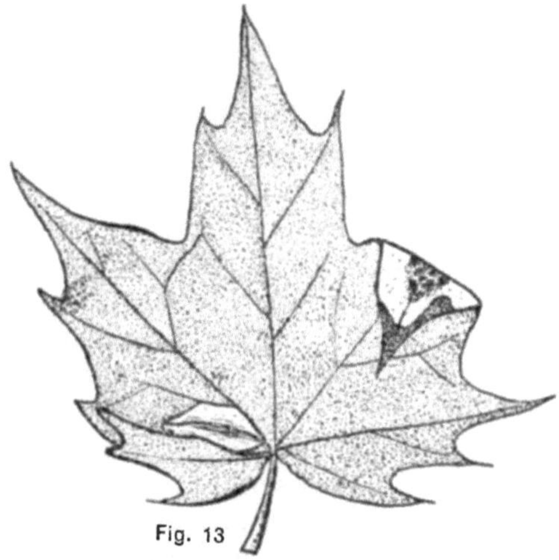

Fig. 13

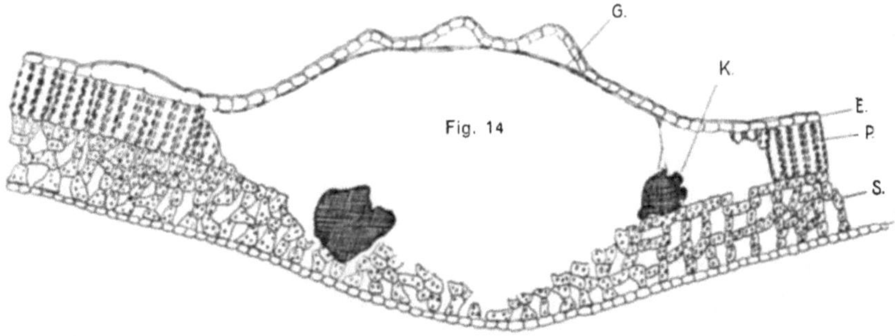

Fig. 12. Taraxacum: Asteronome of *Liriomyza strigata* Mg. along the leaf veins.

Fig. 13. Acer platanoides: ptychonomes of *Lithocolletis platanoidella* de Joan. Left, in the centre of the leaf; right, within a folded oven corner of the leaft.

Fig. 14. Section of an upper surface mine of *Lithocolletis* (E - epidermis, P - palisade parenchyma, S - spongy parenchyma, K - frass, G - Silk). Top left, *an early epidermal mine.*

which may appear in differing forms. There may be one distinct wrinkle or fold or a number which are less pronounced. In many cases folds are scarcely visible at all, as with *Lithocolletis tremulae* Z. on Populus tremula. The degree of development of the folds frequently enables the species to be identified, when a number of *Lithocolletis* species are feeding on the same plant, as on Quercus,

Alnus, etc. In other cases the quantity of web attached to the mine is so slight that the mine easily bursts open with the slightest stretching; this is, for example, typical of the former genus *Ornix*, now *Callisto* and *Parornix*, in contrast to the mines of *Lithocolletis*, which are difficult to break open. Ptychonomes can only arise in insect larvae able to spin; they are therefore not found at all with Diptera and amongst Lepidoptera occur chiefly with the family Gracilariidae. The use of silk in the mine is prevalent among many species, especially of Lepidoptera, without, however, always resulting in the formation of tentiform mines. Mines padded with silk in this way are always somewhat more opaque when held to the light and not transparent like glass; this fact is useful for identification purposes in a number of species.

There are a whole series of intergradations between the two main types of mine we have described, so that one form may develop from another within the life of a single larva. Very many species commence their mine with a track, only to enlarge this into a blotch after a certain time, usually after a specific moult. In this way the linear-blotch mine, the ophistigmatonome, consisting of a combination of track and blotch, is formed. This type of mine is extremely common amongst insects of all the mining Orders (Fig. 15). The ophiptychonome is produced in similar fashion, the early (epidermal) track giving rise subsequently to the (parenchymal) tentiform mine (as occurs with most species of *Lithocolletis*). The linear-blister mine, the ophiphysonome is formed in the same way. On the other hand many species first mine in a blotch, from which they finally drive out a channel of varying length before pupating. Transitions between linear and blotch mines develop particularly with ophiogenous blotch mines and are often very characteristic of the species.

Many mines cannot be classified in any of these groups. There are types in which the narrowness of the part of the plant attacked precludes the development of any special mine form; in such cases the larva is forced to mine the whole width of the area concerned, making the formation of any particular mine pattern impossible. This applies especially to the leaves of many grasses and also to the finely-divided leaves of many Umbelliferae (e.g. Anethum); in such cases the larva has to devour the complete parenchyma. Even the initially more characteristic track of the young larva is later obscured by the extension of the track of the older larva. A mine of this type is termed a pantonome; it may be produced even by the polyphagous and easily distinguishable species *Phytomyza atricornis* Mg. where it mines in the very narrow young leaves of Linaria vulgaris. Pantonomes can be recognised by studying other features of the mine,

Fig. 15. Prunus domestica: ophistigmatonome of *Nepticula plagicolella* Stt.

such as the depth, the traces of frass and, if necessary, the characteristics of the larvae and pupae in the mine. This is the reason for the difficulty in identifying grass miners, of which — unfortunately for the taxonomist — there are particularly numerous species.

It is difficult to decide whether, from the phylogenetic point of view, the linear or the blotch mine should be considered as the more primitive form. No generalised answer seems possible. Many examples can be found even to-day where free living gives place to mining habits, as species which are not normally miners adopt a hyponomogenous existence; Schröder (1903) has shown experimentally how in the larva of the Chrysomelidae, mining arose from the type of feeding where the one epidermis is left like a window covering the hole which has been eaten. *Nothris verbascella* Z. and *Eublemma noctualis* Hbn. feed on plants with very hairy leaves, the former on Verbascum, the latter on Helichrysum. The larvae consume the whole leaf, including the whole epidermis, while leaving intact the hairy attachments of the epidermis. It is not a big step from such a feeding method to one in which the epidermis is left intact, thus providing greater protection for the larvae. In this way the origin of orthogenous blotch-mines can easily be envisaged. Ophiogenous blotch-mines, however, are certainly a secondary development from linear mines. One might thus well be inclined to consider the blotch-mine as the more primitive form but it is important not to ignore the fact that there are very primitive insects such as *Eriocrania sparrmannella* Bosck. and *E. sangi* Wood on Betula, which, before they develop their large blotch, feed in a distinct linear mine. It is thus possible to consider the linear mine as the most primitive type of mining activity, and it is easy to envisage linear mines having arisen from channels inside the stem, in the cortex of the stem or in the cortex of the twig. Even to-day we can see a combination of phyllonome and caulonome in *Ophiomyia proboscidea* Strbl., which begins with a short track at the base of the leaf and then continues in a very long mine, winding around the stem of Hieracium and Aster. On the other hand species of *Lixus* on Umbelliferae begin their mine as a channel in the leaf, which continues through the mid-rib and petiole and finally ends up as a channel in the pith of the stem leading down to the root; this can no longer be considered as a mine.

It thus seems that no definite answer can be given concerning the origin of mining and it can be assumed that its development was polyphyletic and that in some cases the linear mine, in others the blotch mine must be considered as the more primitive form. The modifications of both these two main types must then in either case be considered as derived forms.

3.

LEAF MINES AND MINES IN OTHER PARTS OF THE PLANT

The physiologically essential aspects of insect leaf-mining activity i.e. feeding exclusively on parenchyma or epidermis cells and simultaneously being cut off from the outer world by a dividing barrier, are also to be found in a whole series of other feeding patterns occurring not in leaves but in other plant organs. In such cases there is a clear relationship to leaf-mining from the point of view not only of plant anatomy but also of insect classification, even though they are usually less conspicuous and diverge widely in form from leaf-mines. Essentially the same families and genera are concerned; in many instances even the same species, either through choice or compulsion during later stages of development, change their habits from leaf-mining to perhaps stem-boring. Frequently a virtue is made of necessity: if the leaf is very small and provides insufficient food for the larva, it will move on through the petiole into the stem. We can easily imagine how the phyletic development of stem-boring arose in a similar way. We can even to-day find a number of species with the early stages of their mine in a leaf but which later move on to the stem, where they form the main portion of the mine. However, there are differences of degree even in these cases.

Scaptomyzella incana Meig. mining in very small leaves, like the young leaves of Stellaria uliginosa, finds insufficient food to complete its development and is therefore forced to visit one or more fresh leaves. Like many other miners, it tries if possible to avoid having to bore into a new leaf; the epidermis cells, which are covered by a tougher and often stronger cuticle serving as a protective organ for the leaf, demand a greater output of strength during biting than the parenchyma cells. Therefore, when the larva has eaten out one leaf, it moves on into the next and maybe even another via the stem. These occasional movements through the stem are not yet essential for the larva but merely a makeshift solution. We see in such cases that the larva does not feed in the same way in the stem as in the leaves; the part of the mine in the stem has no or only very little frass. This suggests that while on the move the larva is more interested in removing the impeding cells blocking its way than in feeding, and for this reason it seeks to conclude this stage as quickly as possible and return to a leaf. In other species stem-mining represents a normal interlude between feeding in leaves, in which the same amount of nutriment is consumed as in the leaf itself. Thus with *Nepticula headleyella* Stt. which mines in Labiatae, we find in contrast to the first example that the traces of frass are equally developed in the petiole and stem mines as in the leaf. The development of caulonomes follows on from here in easy stages to those species, in which the most important part of the mine is constructed in the parenchyma of the stem cortex. *Ophiomyia proboscidea* Strbl. has already been mentioned, where the mine commences as a short little track in the leaf of Hieracium and Aster, so short that it scarcely remains visible and is usually only discovered if the track of the fly larva is traced backwards from the stem to its origin. The main portion of the mine descends the whole length of the stem in several coils and usually only comes to an end just above the root. We find exactly similar conditions in a number of species of *Elachista* in Gramineae and Cyperaceae; many species of this genus have entirely abandoned their mining habits in their later stages and then feed primarily in the pith cells of the grass stems. The mining activity of *Phyllocnistis saligna* Z. is remarkable (Fig. 16). The caterpillar starts with a short length of track in a leaf, which continues in an almost straight line to the cortex of the twig; here in the cortex is the main part of the mine. It is quite broad and long but is naturally not so conspicuous as the part of the mine in the leaf, owing to the greater concealment on the leaf-covered twig. When the caterpillar is almost full-grown, it leaves the cortex once more and transfers itself to a new leaf, where the final mine is made, containing at its end an area drawn together by threads or a curled up leaf edge, indicating the pupal chamber. *Phyllocnistis* live exclusively in the epidermis cells; it would seem that the content of these cells does not essentially differ whether in the stem or the leaf, as may well be the case with the parenchyma in the same organs. This would perhaps explain the fact that geographical sub-species of this species show considerable differences in the relative lengths of their mines in leaves and those in the cortex of the twig. *Marmara arbutiella* Busck lives in this way in N. America on Arbutus menziesii: first of all the leaves are completely mined and then the larvae go from them into the cortex of the twigs and from this sometimes back again into another leaf. A whole series of species have finally cut out entirely the first stage in a leaf; their mines now consist solely of caulonomes, with pupation often taking place in the cortical parenchyma of the stem. A number of species have been forced to adopt this habit, because the leaves of the plant are much

Fig. 16. Salix: ophionome of *Phyllocnistis saligna* Z., leaf — stem cortex — leaf. At the end, the edge of the leaf is curled over the pupal chamber.

Fig. 17. Silene: linear mine of *Liriomyza strigata* Mg. on the gall of *Gnorimoschema cauliginella* Schmid.

too minute to provide a medium-sized mining larva with adequate sustenance. The small, often frail little leaves of Sarothamnus do not preclude the possibility of mining, as the presence in them of *Agromyza johannae* de Meij. shows. However, they are not adequate to ensure the completion of the development of long-lived caterpillars such as *Leucoptera spartifoliella* Hb., *Trifurcula immundella* Z. and *Lithocolletis scopariella* Z. These species therefore pass their whole existence in the green stem cortex. In this plant the greater part of the photo-synthesis in any case takes place in the stem cortex and in this respect the leaves play a subsidiary role. The parenchyma of the stem cortex will accordingly have the greatest affinity to that of the leaves and will thus be especially suitable for miners. The choice of this area for feeding by mining species is favoured by the existence of a stem which is winged, as in Genista sagittalis, whose special characteristics are made use of by species of *Trifurcula* and by *Lithocolletis cerasinella* Rtti. In many plants the leaves have become completely atrophied and as a result the shoot and stem have undergone transformations, making them particularly suitable for photo-synthesis and they frequently assume the appearance of leaves; such modified shoots known to the botanist as phylloclades thus also provide a home for mining larvae. Examples can be adduced from the Americas, where in N. America *Marmara opuntiella* Busck, a species of *Melitara* (Tortricidae), and in S. America *Lauxaniidae* (Diptera) live in the almost leaf-like phylloclades of the well-known species of Opuntia.

The needle-shaped, leaf-like phylloclades of our asparagus have become too minute and therefore *Melanagromyza simplex* Loew., and in the Mediterranean area other species of this and the related genus *Ophiomyia*, live in the green cortex of the main shoot, which is just like a normal stem. The larva of *Ptochomyza asparagi* Hering, however, even manages to pass its entire existence in the fine, needle-shaped phylloclades of our asparagus the result being its modification to its present minute size; it is the smallest mining species, which has yet been discovered. There are, on the other hand, a whole series of caulonomes in which the stem-boring existence of the larva has not been such a forced development from the structure of the plant. Several species of *Melanagromyza* and *Ophiomyia* live in the stem cortices of Labiatae and Compositae, without this being essential since the leaves are fully developed. There is evidence of special adaptations in these cases, causing the larvae to rely solely on these tissues. The same thing applies to the Nepticulidae of the genera *Zimmermannia* in Europe and *Ectoedemia* in North America, which spend their whole lives in the cortical parenchyma in the twigs of deciduous trees, particularly in the cortex of branches of Fagaceae. An existence of this kind results in the exclusion of a genuine Gracilariina (*Dialectica simploniella* F. R.) from its sub-family. The caterpillars of *Lithocolletis* seem to be able to exert great strength by their spinning action, forming as they do in their mines in the shoots, folds in the epidermis, which are most characteristic of the genus; the tension in the epidermis tightly surrounding the stem must be much greater than in a leaf, where the surface is far more pliable and conducive to the formation of folds. Characteristic examples of this are *Lithocolletis scopariella* Z. and *L. ulicicolella* Stt., which mine in the cortex of Ulex.

The course of caulonomes shows a special characteristic, in that the track, as far as linear mines are concerned, practically never runs along the stem in a straight line but spirals round it. We noticed this already in *Ophiomyia proboscidea* Strbl., we can see it in *Trifurcula* and *Leucoptera* on Sarothamnus and the same thing is found in the tracks of the Cecidomyid *Clinodiplosis syringogenea* Hering on Heracleum. The course of linear mines in leaves is frequently determined by the larva seeking to avoid the stronger veins, which contain more strengthening elements, are less nutritious and more difficult to work through. This, however, does not apply to the structure of stems. If in such cases the larva winds around the stem, it is each time forced to break through strengthening elements in the corners of the stem — collenchyma or even sclerenchyma — which impede its progress. A satisfactory explanation of this peculiarity is not at the moment available; it is certain, however, that the collecting entomologist finds such mines much more difficult to see than if they ran along the stem in a straight line. It is not clear to what extent this fact might also be applied to animal enemies; naturally it has no bearing on parasites which do not find their victims visually, as is shown by the large number of parasitised larvae in *Ophiomyia proboscidea* Strbl., to which we have already referred.

A mine, which throughout its course is limited to the surface of a gall, must be considered as a peculiar exception. *Gnorimoschema cauliginella* Schmid causes a shoot gall on Silene cucubalus, resulting in a considerable thickening of the stem. The polyphagous *Liriomyza strigata* Mg. made a branched linear mine on this large gall surface, confining itself solely to the surface of the gall and leaving the neighbouring undeformed parts of the stem untouched. This species has so far only seldom been observed as a miner of stem cortex; presumably its choice fell on this special part of the plant owing to the enlarged surface of the cortex resulting from the gall-causing activity of the caterpillar. The fact that it confined its mine to the surface of the gall and did not attack any other part of the stem, despite the considerable length of track normally to be observed in leaves, seems to suggest that it was well satisfied. Presumably the gall tissue, characterised by more rapid cell division and a stronger flow of sap, facilitated particularly rapid assimilation of nutriment.

The flower mine, or *anthonome* is to be seen much less frequently than the stem mine. Petals are not very fruitful for mining activities, containing little parenchymatous tissue corresponding to that in the leaves and genuine chlorophyll is in particular almost always lacking (though this does not constitute the sole ingredient of the necessary food); furthermore, petals are often so frail that the whole development of the larva might be endangered. Genuine mines are thus found in petals only as an occasional 'abberation of taste' (Xenophagy) which might be anticipated especially in the miners possessing the greatest range of normal habitat, as in the genus *Colephora*, *C. fuscedinella* Z. was once found in flower mines in Caltha, which was growing beneath Alnus, the usual foodplant of the caterpillar. The caterpillar which had probably fallen from the tree, doubtless in this case started eating the pleasantest plant substance first found by its mandibles. Other parts of the flower such as the sepals are more suitable for providing food. On a number of occasions species which normally feed in the green leaves of a plant have also been observed in the sepals, especially when these are large and durable as in Agrostemma, in whose sepals the sizeable blotch mine of *Dizygomyza flavifrons* Mg. has been found. The larva of *Phytomyza hellebori* Kalt. acts in the same way in the green sepals of Helleborus viridis (these striking 'petals' are the genuine sepals of the plant, while in this case the genuine petals have been transformed into nectaries). The larva is able to complete its whole development in them, which it normally begins and ends in the parenchyma of the leaf.

Voigt (1932) termed this transfer from one organ of the plant to another 'Organoxeny'; it represents a distinct stage in the adaptation of insects to plants. Organoxeny is an interesting counterpart to 'Xenophagy' which we shall deal with later.

It seems more difficult to understand why the fruit-mine or *carponome* is found so relatively seldom. The majority of fruits possess green parenchyma in their skin, which would seem suitable food for leafminers. In addition there is a particularly strong flow of nutriment to the developing fruit; it is calculated that in our fruit trees the constant photo-synthesis of about 20 leaves is necessary for the formation of a single fruit. Finally, the fruit is generally green for sufficiently long to give the larva adequate time for its development. Despite these favourable factors, there are only very few species which pass their existence in the green parenchyma of the fruit. *Marmara pomoniella* Busck constructs dainty channels in the still green skin of apples in N. America, while here *Hoplocampa testudinea* Klug sometimes but not always mines in apple skins during its infancy, later burrowing into the flesh of the fruit. According to Costa Lima (1945) *Marmara isortha* Meyr. mines the fruits of cocoa-plants in S. America in similar, dainty, serpentine channels. Some species of *Nepticula* here, which mine in the fruits of maple are only found in the parenchyma of these fruits; the best-known of these are *N. sericopeza* Z. (Acer platanoides) (Fig. 18), *N. decentella* Z. (A. pseudoplatanus), and *N. sphendamni* Her. (A. campestre). They all agree in the choice of their mining site, first constructing a slender linear mine in the 'wing' of the fruit, which takes a fairly direct course towards the actual fruit. The main part of their development then occurs in the fruit itself between the ends of the two wings. However, it is strange that such polyphagous species as *Phytomyza atricornis* Mg. and *Liriomyza strigata* Mg. are not found more often in the green pods of the Leguminosae (the former has occasionally been seen in the pods of Pisum). Apparently the parenchyma of fruit parts is substantially different to that of the green leaves, causing the larvae to avoid them. As might be expected, fruit miners, repre-

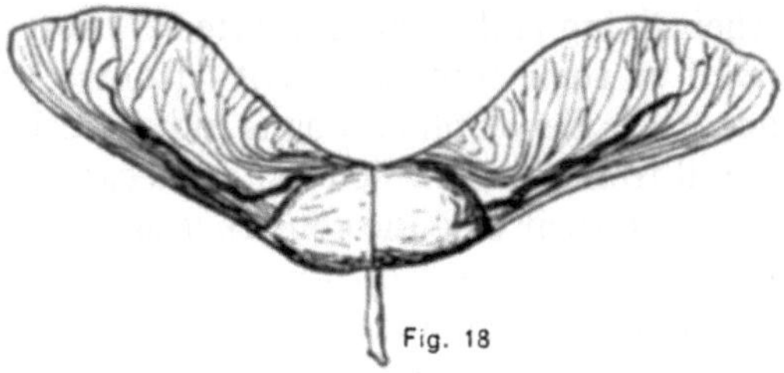

Fig. 18

Fig. 18. Acer platanoides: carponome of *Nepticula sericopeza* Z.

senting evidence of occasional organoxeny, are to be found most frequently in the genus with the widest range of habitat, in *Colephora. C. serratella* L., *hemerobiella* Sc. and *anatipennella* Z. are thus not infrequently found in fruits in orchards, particularly on Malus and Prunus. Contrary to its normal habits, the larvae mining in such cases start with a mine but then penetrate right inside the flesh of the fruit. Another species mines the characteristic *Colephora*-blotch in the green fruits of Atriplex, in addition to normal leaf mines on the same plant. In this group of Chenopodiaceae-feeders leaf mines merge with fruit feeding, as many species in their infancy live in a leaf mine, later exclusively feeding in fruits. In another group of the same genus one finds a reversed sequence in the forced organoxeny, combined also with xenophagy — the transference to a plant, which is not in any way related systematically. *C. lixella* Z. lives while young in the seeds of Thymus and the closely related *C. ornatipennella* Hbn. in the calices of Labiatae (Salvia), while after hibernation both live exclusively in large blotch mines in Gramineae, from which they also construct their final case, while the juvenile case used in their early stages was constructed from the mined seeds or calices of Labiatae.

A physiological specialisation of a quite different type is found in the mining larvae, which construct their feeding channels inside dead or dying leaves. The processes of autumnal disintegration of chlorophyll do not in general affect mining larvae. In as far as they are still feeding at all so late in the season, they do so in the 'green islands' which will be discussed in detail later, and in which they find the same living conditions as the larvae which lived earlier in the leaves which were still green. Larvae mining in dead leaves depend for their nourishment, however, on quite different products. *Lauxaniidae*, members of the genera *Sapromyza* and *Lauxania*, have in particular been found feeding as larvae on dead, decaying leaves. According to the definition given at the beginning, they can be considered as true miners, for they feed in the parenchyma (admittedly more or less disintegrated or decayed) and leave the epidermis intact; the cavity thus caused is their living and feeding quarters. These species, however, have arrived at their mining activity from a quite different type of existence; their development, as is indicated by their family name, has occurred from forms which fed on decaying matter and it can be assumed that in their case the construction of a mine is not an essential feature. This is a case of specialisation in a saprophytic rather than a phytophagous existence, as in the other mining larvae.

A well-known beetle larva, *Coenorrhinus pauxillus* Germ. provides a link between them and the majority of miners. It also, like the Lauxaniidae, only mines in the dying or dead leaf but its habits are derived from phytophagous relatives. The puncture made during oviposition, usually occurring at the junction of the petiole with the leaf blade, less frequently in the midrib in the centre of the leaf, injures the leaf in such a 'suitable' way, that it soon falls to the ground and normal weathering provides the larva with the food it needs. The regularity of this form of mining in the most varied Rosaceae (particularly frequently on Malus) suggests in this case a special adaptation to a half-saprophytic form of life.

4.

CHANGES IN THE TYPE OF MINE

PERMANENT AND TEMPORARY MINERS

The mining larva in many cases retains throughout its life the type of mining it has chosen and in this way feeds consistently only in upper surface or in lower surface mines. In a whole series of other cases, however, it changes the type of its mine on one or more occasions. Such changes in the feeding instinct frequently occur after completing a moult. They then often run parallel to changes in the structure of the mouth parts, which are now adapted to a different type of food. There are very many linear mining dipterous larvae which initially live on the under-side of the leaf in the spongy parenchyma, later continuing their channel in the palisade parenchyma on the upper-side, and then often transfer the last part of the channel back to the under-side, where they can pupate in a position less exposed to view. Such a double change of the side of the leaf infested can be frequently observed in the polyphagous *Phytomyza atricornis* Mg. The change from a linear-mining to a blotch mining existence can also often be found running parallel to moulting; ophiogenous blotch mines represent a transitional form between the two forms of mining. The connection between the change of skin and the change of instinct expressed in the different method of feeding can be seen particularly clearly in certain species of *Bucculatrix*. *B. noltei* Petry forms a narrow linear mine in the first larval instar in the leaves of Artemisia, resembling the mine of a *Nepticula* and leaving its frass in a continuous central line. It then abandons this linear mine and moults, as always in this genus, under a delicate moulting cocoon. When this process is completed, it continues feeding, but now no longer makes a linear mine like *Nepticula* but forms blotch mines, which are exactly similar to those of *Coleophora*. We find the same thing with many species of *Coleophora*, which first construct a mine and live inside the leaf, stacking away their grains of frass inside the mine. This first mine is usually a small blotch mine, but in other cases (*Coleophora arctostaphyli* Med.) a linear mine with a central line of frass similar to *Nepticula*. Only later does it cut a case out of the leaf, usually from the mined part, continuing its mining activity from within this case but now outside the mine, and the mines thus formed are always free of frass. (Fig. 19).

Fig. 19

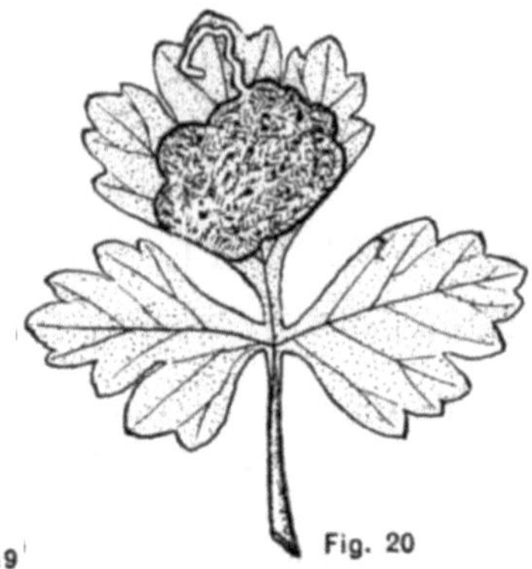

Fig. 20

Fig. 19. Arctostaphylos: *Coleophora arctostaphyli* Med. a) Initial mine b) Later mines and case cut-out.

Fig. 20. Apium: *Philophylla heraclei* L. (celery fly), initial channel full-depth and transparent, later blotch upper-surface, greenish.

The depth of the mine in the leaf tissue is also subject to remarkable variations during the course of a larva's life. In general the type of depth with which feeding commences is retained throughout. What starts as a full depth mine does not later become upper surface or under surface, and vice versa. However, a few exceptions exist here too. The initial channel of *Philophylla heraclei* L., for instance, is normally a full depth mine on the leaves of Umbelliferae, the subsequent blotch on the other hand is upper surface and greenish (Fig. 20). Many Anthomyidae behave in a similar way. Conversely, a mine may be first upper surface or lower surface, and later with the increased growth of the larva, becoming full depth; such a change in instinct is more understandable and is frequently though not invariably found in species of *Nepticula*, particularly *N. luteella* Stt. The majority of miners do not usually alter the depth of their feeding. The mines of the Gracilariidae represent another noteworthy exception, especially those of *Lithocolletis*. In its first instar the caterpillar lives as a sap-feeder and, as such, starts by forming quite a flat epidermal linear mine, which contrasts very little with the leaf and is quite invisible when held up to the light. This linear mine is soon extended to a blotch, which is still confined to the epidermis and thus no parenchymal cells are attacked. When the size of the future mine has been more or less decided, the caterpillar starts extending the mine to greater depths; only now does it become a tissue-feeder and at this stage also spinning begins to throw the outer walls of the leaf into the folds which are usually characteristic for each species. The origins of the final mine in the epidermis cannot readily be detected; microscopic examination of the cross section of the mine however shows that only the vertical walls and not complete cells of the epidermis were removed. Particularly extensive mines of this type can be found in *Lithocolletis platani* Stgr. on the underside of the leaves of Platanus. A similar process occurs with many other Gracilariidae. The moment when the caterpillar changes over from being a sap-feeder to a tissue-feeder may apparently vary. It has been observed that in *Callisto denticulella* Thbg. on apple leaves the initial consistently epidermal mine becomes transparent much more rapidly and extensively when it is on the underside of the leaf than when it is on the upperside. The same thing can be observed in *Caloptilia elongella* L. on the leaves of Alnus.

The miners we have mentioned so far live during all their stages in essentially the same way, that is to say they feed continuously on the mesophyll between the upper and lower epidermis. Mining larvae of this kind which, throughout their feeding right up to the time of pupation, live in mines, are called permanent miners. The majority of species from the Orders Diptera, Hymenoptera and Coleoptera belong to this group. As opposed to these are all the species which feed for only a limited period in mines, at other times, usually later, leaving a different feeding pattern. They are termed temporary miners. Such temporary miners are found predominantly amongst Lepidoptera. Mining activity usually occurs at the beginning of the caterpillar's development, while the different type of feeding that develops later is found in its more advanced stages. However, there are some species which form an exception to this rule. Certain species of *Coleophora* of the *ornatipennella* group for instance, spend their infancy in the calyx or seeds of Labiatae, adopting a mining existence only after hibernation but now feeding in Gramineae. Apart from these exceptions, temporary miners normally live in mines only while they are young and in their later stages feed on the plant externally. Many species change their food plant at the same time as they leave their mine. The minute young caterpillars are on the one hand better protected in their mine against attack from outside than if they were feeding in the open, while on the other hand they can make more rational use of their food, since they can digest far more of the mesophyll of the leaf than if they also had to eat the epidermis, which contains less food value and is in addition often covered by a particularly hard cuticle. When the caterpillar has become bigger, its mouth parts are stronger and the consumption of less nutritious food is no longer as important as in its earlier instars. In a large number of species, once the caterpillar has abandoned its mining existence, it no longer enjoys any external protection, as for example, in the species of the genus *Bucculatrix*, which, after

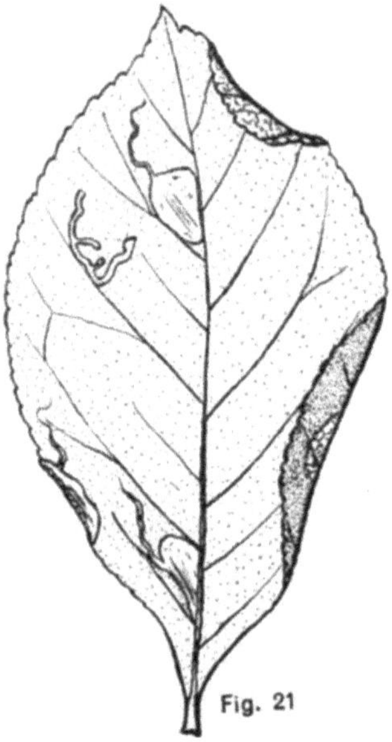
Fig. 21

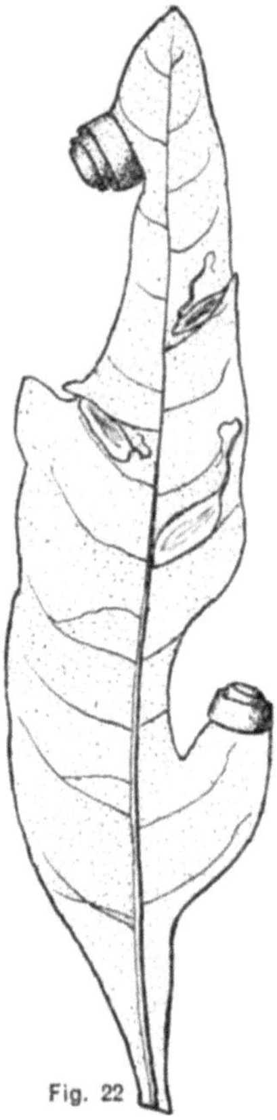
Fig. 22

Fig. 21. Malus: *Callisto denticulella* Thbg. First instar epidermal mines and folded over leaf-edges.

Fig. 22. Polygonum: *Euspilapteryx phasianipennella* Hb. with initial mines and later leaf cones.

leaving the mine, feed on the plant externally, first causing on the underside holes covered by one window-like epidermis (Fensterfrass), and later often perforating the leaves. When danger threatens, they readily let themselves fall and in this way escape many of their enemies. Yet other species seem to continue to require a greater need for protection. They therefore construct special living quarters, though not now inside the leaf. This is achieved in many different ways. In many of the Gracilariidae, as for instance *Caloptilia* and *Parornix*, the edge of the leaf is folded over after leaving the mine and within this fold the caterpillar gnaws away at the leaf until finally only the upper epidermis remains. In any case the need of the caterpillar for protection is satisfied, since many of its enemies are unable to attack it under the folded leaf edge, which is firmly webbed down at the edges. In the simplest case the edge of the leaf is forced to bend over downwards (less frequently upwards) by silken threads, as in many species of the genera *Callisto* (Fig. 21) and *Parornix* and many of the *Caloptilia*. Other species proceed more artistically and construct a so-called leaf cone. An incision is made into the margin of a leaf; the lobe of the leaf partially severed in this way is then wound round into the form of a cone and within this the caterpillar feeds. It is then able to consume several layers of leaf under this fold, while within the leaf edge which is simply folded over by *Parornix* and *Callisto* the caterpillar merely has at its disposition the two layers of leaf delimiting its living quarters (Fig. 22). After leaving the mine, the number of habitations constructed in this way by leaf-edge folds or leaf cones is probably constant for each species, and sometimes very similar species can be distinguished within their mine by the

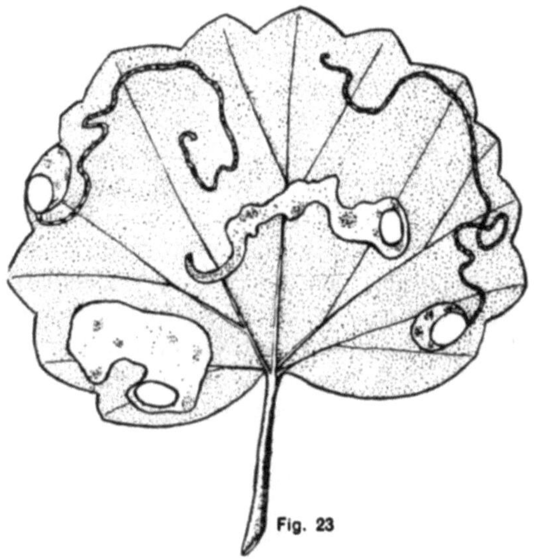

Fig. 23

Fig. 23. Saxifraga rotundifolia: Early linear mines, blotch mines, and areas cut out *(Incurvaria trimaculella* HS.).

number of leaf-edge folds, which is usually two or three. A number of species, especially from the genus *Caloptilia* are less artistic in constructing their homes; the leaf is rolled up by their spinning and the leaf-rolls so characteristic of the species are formed. An important point here is whether the caterpillar begins by attaching the thread to the tip or to the marginal edge of the leaf. In the first case the leaf is rolled over longways, in the second case, crossways. The differences

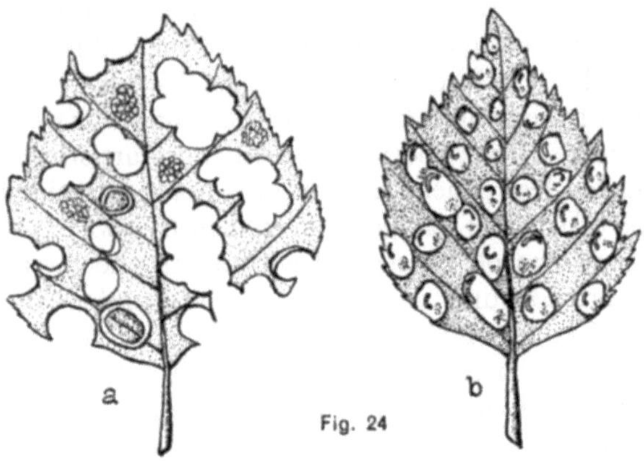

Fig. 24. Betula: *Incurvaria pectinea* Hw. a) Leaf after construction of cases b) mined leaf.

in these habits among individual species are constant and serve to distinguish otherwise identical mines.

The later leaf habitations constructed by the Tortrices, which mine while they are young, are far less artistic and consistent, as examplified especially by the species of *Cnephasia*. They occasionally draw parts of the leaf together like a tube, in other cases parts of the leaf or single leaflets are merely spun together irregularly and the caterpillar consumes beneath the shelter of this web construction either individual layers of the leaf or its full depth. In this group it is not possible to distinguish the species by the construction of the mine nor by the subsequent larval habitation; however, various characteristics of the caterpillar, as well as of the pupa, which is to be found between the spun-together leaf parts, facilitate identification. Even after the moth has emerged, the empty pupal case showing its distinguishing characteristics, which can be found in the neighbourhood of the old mine, makes it possible to identify the species concerned.

The mining Incurvariidae similarly form a mine only during their first instar (Fig. 23), and many can often be found together on one leaf. They can be seen in a wide range of deciduous trees; *Incurvaria pectinea* Haw., for instance, lives very frequently in the leaves of Betula (Fig. 24). The caterpillar later cuts out of the leaf a roundish case, with which it drops to the ground; it now no longer mines but continues feeding in green or drying leaves, occasionally also partaking of other plants. During this stage of feeding the case is successively enlarged by attaching to it new, larger portions cut out of a leaf, while the early case, which is now receding from view, is gradually consumed; simultaneously, however, holes are eaten in the leaf.

While the mines in the leaves are very easy to find, especially if they are very numerous on individual infested leaves, it is distinctly difficult to see the caterpillars when they are later living on the ground.

The identification of temporary leaf mines is in most cases dependent on a knowledge of the subsequent behaviour of the caterpillar. It is advisable therefore, when collecting such mines, to examine the nearby leaves where the later larval habitations can usually be found. With some practice we can tell from the mine itself whether it is a temporary type and one must then pay great attention to what larval habitations can be found in the proximity of the mine. They are occasionally on the same leaf but more usually on neighbouring leaves and should be kept together with the mine, or at least notes on the larval habitation found should be attached to the mine, to make possible its later identification. Caterpillars which are temporary miners in pine-needles, later very often live in a silken channel on the cortex of the twig, the structure of which should be examined.

The differentiation between permanent and temporary miners is usually specific and constant but there are, however, species in which there is no hard and fast dividing line. Edmundsbury (1937) noticed, for instance, that the caterpillar of *Parornix scoticella* Stt., which normally mines the leaves of Sorbus only during its infancy and later continues its feeding under a folded over leaf-edge, may continue mining right up until pupation. This represents an individual case of a temporary miner becoming permanent. Similar variations may be seen in *Teichobia verhuellella* Stt. The caterpillar mines in various Polypodiaceae; it later leaves its early mine and constructs a protective shelter by spinning a web amongst the Sporangia of the plant; it may go on mining under this shelter or it may cause holes leaving one epidermis intact (Fensterfrass). If a new leaf is attacked, the protective shelter is taken along too. The mining instinct in this case is not firm and probably depends on the nature of the food plant.

The species of the genus *Hyponomeuta* behave in a very strange way. The eggs are deposited during mid-summer in a pile on a twig, where they remain for the winter. In the following spring the caterpillars begin their mining, seeking out the freshly formed leaves in which they often tunnel jointly, producing a relatively small mine. They evacuate this after a short time, to band together and form "nests" reminiscent of "web worms", now destroying the leaves by "window-feeding", or later skeletonizing. An important contrast is seen here with the other miners, which deposit the egg directly on the leaf in which the subsequent mine is formed.

The change from one type of feeding to another, whether it be an alteration in the type of mine or, as in temporary miners, a complete change in the manner of feeding, is necessarily connected with a certain transformation of the mouth parts. This has already been shown in the first case by the classic investigations of *Trägårdh* (1913); analogous changes can be expected in the temporary miners' transfer from mining to a free-living existence, though this has not so far been investigated in detail. The morphological changes connected with the variation in mining habits will be amplified in a later chapter. The corresponding situation in temporary miners offers a fruitful field for future research.

It appears that in some cases mining may assume a variety of forms. For instance, it has been reported that *Cidaria incultaria* HS. which lives in typical mines in the leaves of Primula, may also live in the fruit capsules of the same plant. The extent to which this variation occurs within the same generation has not been discovered, though it seems very possible that the separate generations have different habits in this respect. We know that in *Stenoptilia pneumonanthes* Schleich, which lives on Gentiana, the young caterpillars of the hibernating generation feed on the shoot of the plant; in the following generation the eggs are laid on the leaves on which the caterpillars form their first mine before going into the flowers to continue feeding. The clearest example of the variations in feeding habits in the same species and at the same time is to be seen in *Acrolepia assectella* Zell., whose caterpillar feeds on species of Allium living either as a genuine miner inside the leaves, as in Allium porrum or eating a type of "Fensterfrass" in the pipe-like leaves of A. sativum and others, or finally, it may be found inside the shoot, where it may completely consume the undeveloped young leaves and the pith of the flower stalk. Chr. Schröder (1903) has also induced normally free living larvae to start mining. In his experiments he used *Phyllodecta vitellinae* L., whose larvae occur on poplar and flat-leaved willow, skeletonizing the leaves from below. He deprived these larvae of their normal food, giving them instead leaves of a species of Salix, which on the under-side were very woolly. The larvae continued to feed normally, pushing the severed wool of the under-side of the leaf, which was mixed with portions of epidermis, in front of them like a continuous disc and beneath which they were quite invisible. Some larvae, however, apparently felt impeded by all this wool they were pushing in front of them and eat into the inside of the leaf, thus producing a genuine mine $3\frac{1}{2}$ cm long. This case well illustrates how in many instances mining may have arisen.

No answer can be given to the question whether permanent or temporary mining should be considered the more primitive stage in the development of this special type of phytophagous living. Both forms of feeding have arisen polyphyletically and it may be broadly assumed that both developed from optional miners.

5.

CHANGING FROM ONE MINE TO ANOTHER

The great majority of miners consist of very small species; in such cases the substance of a single leaf is usually sufficient to provide all the food necessary for the larva from the time it hatches from the egg until it pupates. The organs of movement of leaf-mining larvae are also frequently so largely atrophied that walking from leaf to leaf presents real difficulty. These two factors bear a close connection with the practice of the majority of mining larvae of living throughout their development within the leaf — this applies, of course, to permanent miners. In general, this involves only a small part of the leaf substance being used up as nourishment for the larva. Meanwhile, temporary miners usually change from one leaf to another; they are usually larger species, for whom a single leaf would provide inadequate food. The possibility of forming a new mine depends also to a certain extent in many larvae on the morphological changes which have occured as a result of their mining habits. For example, in most caterpillars the head-capsule was initially o r t h o g n a t h o u s: the head is so arranged that the mouth parts are situated ventrally; the construction of their body at least enables such species to eat into a new spot in the leaf, if they have left their original mine either voluntarily or through some compulsion. In many mining caterpillars, however, the orthognathous arrangement of the head has been transformed into a prognathous arrangement, so that the mouth orifice is directed forwards. They are completely limited to eating in a forward direction within the leaf and if they once find themselves outside their original mine, they are not able to eat their way in at another place. A prognathous head arrangement of this type is found in very many species of mining Lepidoptera larvae, especially in the epidermal miners, the sap-feeders. In mining Diptera larvae the mouth orifice is invariably situated on the under-side of the head, so that if they have moved out of their mine they can eat their way in again at another place in the leaf; however, they are not often found to change mines and such changing is limited to certain families. Among the Trypetidae and Anthomyidae, for instance, there may be a change from one mine (or one leaf) to another, while the Agromyzidae, in general, never change in this way. If a larva is removed from the leaf it is not likely to eat its way in again elsewhere, and it normally dies. Therefore, in the Agromyzidae changing mines does not occur. The same applies to miners of the Order Hymenoptera and to many Coleoptera larvae, even when the structure of the mouth parts would not exclude such changes.

The wide variation among lepidopterous larvae results in the occurrence of species which never leave the mine throughout their lives, together with others which change mines and even leaves several times. Species with a prognathous head formation, such as *Phyllocnistis*, can never change their mines; the caterpillar removed from its mine always dies. The same thing applies to Eriocraniidae, Nepticulidae, and Heliozelidae. In many of the Gracilariidae no change is possible while the larva continues to be a sap-feeder; later however, it may change mine and leaf several times, when it has been transformed into a tissue feeder. The least difficulties in changing mines are encountered by members of Gelechiidae and Tortricidae. It is also frequently seen with the Elachistidae, Glyphipterygidae and Lyonetiidae. The larger size of the body makes a change of leaf essential for many Pyralidae and for the Zygaenidae and Geometridae, especially when the caterpillar has to continue feeding after hibernation, when the leaf containing the original mine has been destroyed, as for instance with *Procris* and many species of *Elachista*. Changing mines occurs most regularly in the species of *Coleophora*. There are, of course, a number of species here which make very large mines and apparently attempt, as far as possible, to avoid changing to another; the majority of the species, however, change both mine and leaf several times, tunnelling in again elsewhere. This occurs particularly where a new case has been cut out. Some of these species mine only very small blotches and then immediately seek out a fresh part of the leaf or a new leaf and produce in this way a large number of small blotch mines. This can be seen very clearly in *Coleophora fuscocuprella* HS. in the leaves of Corylus, which as a result of this infestation often look as though they have been lightly spotted.

The species of *Coleophora* which are especially predestined to frequent change of domicile by their case-bearing habits derive the advantage from these frequent changes that enemies on their track have far greater difficulty in finding the living larva, owing to the majority of mined blotches being empty. The micro-lepidopterist is particularly familiar with this difficulty, when he has to search each leaf in order to take a few or perhaps only one specimen of the many species of *Coleophora* which produce numerous blotch-mines. The Psychids of the genus *Apterona* behave in the same way, living in a case and producing only small blotch-mines, which become conspicuous owing to their large numbers. In all these instances it is not therefore shortage of food which causes the larva to change its mine. If these same larvae are given only one leaf to feed on in the breeding cage, the change of mine occur less frequently than under natural conditions, without the development of the larva being in any way impaired. In other cases the change of mine is dependent on the nature of the food plant; species with small or narrow leaves which do not suffice for the larva to feed up on, are infested by the greatest number of mine-changing larvae, as for instance, Coniferae and Glumiflorae, (Gramineae and Cyperaceae). This explains the almost total absence in such families of plants of the families which are incapable of changing their mines for morphological reasons.

No species of Hymenoptera have yet been discovered which could change their mine or even the leaf. Among Coleoptera the majority of species do not leave the mine they have begun and, as experiments with *Rhynchaenus* have shown, they are not able to eat their way into the leaf at any other place. There are, on the other hand, a number of species which are capable of changing mines and do in fact on several occasions make a fresh mine, as for example *Hydronomus* among the Curculionidae, many Halticinae among the Chrysomelidae, and the genus *Dicladispa*. In the *Apteropeda* changing mines seems to be the rule; many mines can frequently be found among the members of this genus, feeding on Labiatae, Scrophulariaceae and Plantaginaceae, which were apparently abandoned without any compelling reason and the new channels formed by the larva can be found not far away.

It is necessary to make special mention of the species which change their mine and also often their leaf when they reach the end of their larval stage. Many of them but not all have changed mines already a number of times before. When the larva is preparing for pupation, however, it makes a new, very short, mine, which often shows differences from its earlier mine; it is now a definite "Pupation mine". The pupation mine is frequently limited to either the upper surface or the lower surface, while the previous mine was full depth, in other cases the frass normally found in mines of the corresponding species is missing. For one reason or other the mine is invariably less conspicuous and its construction is explained by the desire to provide the helpless pupa with as well protected a resting place as possible. Such pupation mines are found among species of *Dicladispa* (Coleoptera), among Lepidoptera in the Momphidae and particularly striking examples among Diptera in *Hydrellia griseola* Fall. In this last species the larva also departs from its normal habit of mining only in monocotyledonous plants and makes its pupation mine sometimes also in dicotyledons. If this fly is very common in any locality, it can be found pupating in the most varied plants, such as Lychnis, Trifolium and many others, which happen to be growing in the vicinity.

A change of mines may occur in many instances after a moult, which has resulted in a transformation of the mouth parts. A parallel change in the type of mine then appears. Changing mines is observed in *Lyonetia prunifoliella* Hb., especially when the caterpillar is changing over from linear-mining to blotch-mining. In this species the original channel is therefore very often separated from the subsequent blotch, the two not infrequently even being in different leaves. There is a parallel between mine-changing in this case and the transition from mining to free-living habits in temporary miners. These parallels are particularly striking when a temporary miner for some exceptional reason remains permanent, as Edmundsbury (1937) observed on *Parornix scoticella* Stt. This species normally lives on Sorbus initially in a lower surface tentiform mine, later feeding in the open under a folded over leaf edge or in a leaf-cone. Edmundsbury observed how the larva left the mine, did not however fold over the leaf-edge for its new habitation and commence external feeding but made a new mine, changing mines in this way a number of times, while it normally leaves its mine only once.

The female frequently deposits its egg on leaves which are not yet fully developed. It may then often happen that the leaf in which the larva is growing becomes stunted for some reason and does not grow to its full extent. Some species regularly cause such malformation of the leaf and this serves as a valuable indication for recognising and identifying the species of mine; this will be discussed further in a later chapter. What happens when the stunting of the leaf is so considerable that the miner is unable to obtain sufficient food? If it belongs to the species which are not able to form a second mine owing to the nature of their mouth parts making it impossible, it would seem that it must inevitably die. However these species are provided with a "back-door", which enables even them to make a new mine, if necessary in a fresh leaf. This occurs by the larva moving from its leaf into another without thereby appearing in the open at all; it leaves the leaf through the petiole and moving by way of the stem cortex and through a new petiole reaches another leaf. Throughout this time it does not have to interrupt or give up its mining existence. The type of feeding remains identical, even though the food itself must be somewhat different, owing to the miner having to consume quantities of unaccustomed products. The larvae of *Nepticula*, for instance, which under normal circumstances are not able to leave the infested leaf for a fresh one, are usually forced to adopt this method, when they occur in the usually very small leaves of Labiatae (e.g. *N. headleyella* Stt. on Prunella), if they wish to mine one or more new leaves. A continuous trace of frass which can be followed through the petiole and stem right into the next leaf shows that the feeding of these larvae is not interfered with or interrupted during their move. On the other hand when the larva of *Scaptomyzella incana* Mg. mines the minute upper apical leaves of Stellaria uliginosa and moves into fresh leaves via the stem, no piles of frass are to be found in the part of the mine running through the stem, an obvious sign that it has made a tunnel by biting through the cells but that little or nothing has been eaten in the process. Only when the new leaf has been reached are piles of frass to be found once more. A particularly striking example of this "roundabout way" of changing mines is to be seen on the species of *Phyllocnistis*, which commence their mine in one leaf and finish it in another and in between make a long mine in the parenchyma of the stem cortex, like *Ph. saligna* Z., in various species of Salix. A virtue is made of necessity and the change of mines becomes habitual. Whenever the leaves are too small the larvae of Agromyzidae, which are not normally able to move from one leaf to another, make use of this same method of completing their development in a normal way. *Liriomyza strigata* Mg, when feeding on small or finely-divided leaves, has often been observed to wander considerable distances on the stem before selecting another leaf. This species which makes the greater part of its mine in the mid-rib of leaves is, however, normally accustomed to living under comparable conditions, since the mid-rib more closely resembles the stem, with its vascular system and strengthening elements, than other parts of the leaf. However, even larvae, for whom it is not "necessary", often choose this way of reaching another leaf, as can frequently be seen in *Pegomyia silenes* Hering in the leaves of Silene although it can, like all Anthomyidae, move directly from leaf to leaf and straight away make new mines without difficulty. The shorter the time the larva spends outside the mine, the greater is the protection against many enemies which the mining habit represents.

Species which normally under no circumstances form a second mine can nevertheless be induced experimentally to change the leaf and even the food plant. This is done by the transplantation of the miner, which will be discussed in a later chapter. If a slit is made in the epidermis of another leaf and the larva is pushed inside, in most cases it will continue its mining in the usual way. If this does not succeed, a small portion can be cut out of the leaf with the larva inside it and introduced into the new leaf to make the transfer more easy for the larva. H. Buhr was the first to carry out such transplantation experiments on a large scale. The further development of the larva is then completed normally as in the first mine.

A change of mines is frequently accompanied by a change of

food-plant, although a preference is shown for closely related plants, if the normal food plant of the species is no longer available. A number of characteristic examples of this type are quoted in the chapter on monophagy and polyphagy. A change of mine in conjunction with a change of food plant is a normal occurrence in some species after hibernation, as in *Coleophora ornatipennella* Hb., which in autumn mines the calyx of Labiatae and in the spring is found in the leaves of various Gramineae.

6.

MINE PRODUCERS

Present information shows that genuine mines are produced both in Europe and elsewhere in the world, including the tropics, only by insect larvae. Many other types of feeding, such as that of adult weevils produce patterns resembling mines but which do not conform to our definition of a mine as both feeding and living quarters. As far as we know the larvae of only 4 insect Orders can be considered as producers of mines, of which Lepidoptera and Diptera comprise the greatest number of species, while Hymenoptera and Coleoptera have been identified as mining insects less frequently.

The part played by Thysanoptera in leaf mines, where they are sometimes encountered, has not yet been satisfactorily determined. Many factors suggest that they are merely inquilines, only occasionally visiting these cavities in plants, which have been formed by other insects. The possibility of members of this Order one day being discovered to be true miners is suggested by the fact that H. Buhr has found Thysanoptera in natural cavities of plants and their remnants of frass showed that some feeding had been taking place. Further investigations will have to determine whether at least in some cases true mining does occur in this Order.

The other four insect Orders produce mines as larvae, which in most cases are sufficiently characteristic to be able to determine from any particular mine the Order to which the miner belonged. Some experience is obviously necessary before this is possible: it is difficult to lay down characteristics which would suffice under all circumstances to identify the Order, and forms will always be found in which the classification of a certain mine within one of the four Orders appears doubtful. However, certain guiding principles can be laid down, which make it possible to decide in a large percentage of cases the Order of a larva responsible for any given mine. Whenever this is not possible, a superficial examination of the larva will soon settle any doubts. Long practice enables the expert on leaf-miners in most cases to make a rapid and accurate decision in this matter.

Mines are classified into four groups after the Orders of their producers: the moth mine is known as the lepidopteronome, the fly mine as the dipteronome, the saw-fly mine as the hymenopteronome and the beetle mine as the coleopteronome. The larvae and pupae

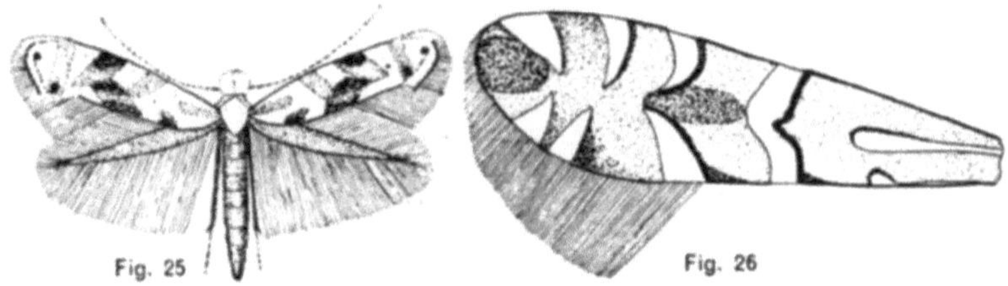

Fig. 25. Mining moth *(Bucculatrix)*

Fig. 26. Forewing of *Lithocolletis*

of these Orders differ just as do the imagines, and in some cases identification is easier by means of the first stages than from the shape of the mine. In a few rare cases, such as *Lithocolletis* on Salix and Quercus, the adult is also necessary for identifying the species but fortunately it can be bred out very easily.

A. Lepidoptera (Moths)

The perfect insect possesses four membranous wings, which are at least partially covered by broad scales (Fig. 25). The forewings, which during the state of rest cover the hindwings, are invariably more brightly coloured and possess very characteristic markings which are the most important factor in identifying the species (Fig. 26). The mouth parts are characterised by the possession of a proboscis, which is found in none of the other Orders and which may be atrophied even in Lepidoptera.

The moth larva is called a caterpillar. It always has a rather strongly chitinised and usually dark head capsule, which distinguishes it from most mining larvae of the Order Diptera. The head is composed of the mouth parts, minute antennae and also small eyes or stemmata each consisting of an ommatidium. The number of these eyes varies but there are usually 5 or 6, arranged in a curved row. Some of them are often atrophied: the lens and accumulation of pigment are the first to disappear, while a glassy-looking spot may continue to indicate the original position of the eye. Later, this also disappears and in many genera the number of eyes becomes reduced in this way; in others the arrangement changes and the curved row gives way to a straight line. *Eriocrania* and *Nepticula* alone now possess only a single eye, thus resembling the larvae of the mining Hymenoptera, the Tenthredinidae. On the upperside of the first (sometimes also the second) thoracic segment there are usually darker, more strongly chitinised structures, together known as the protho-

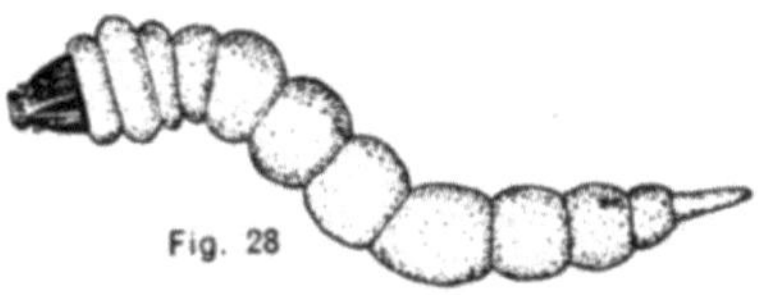

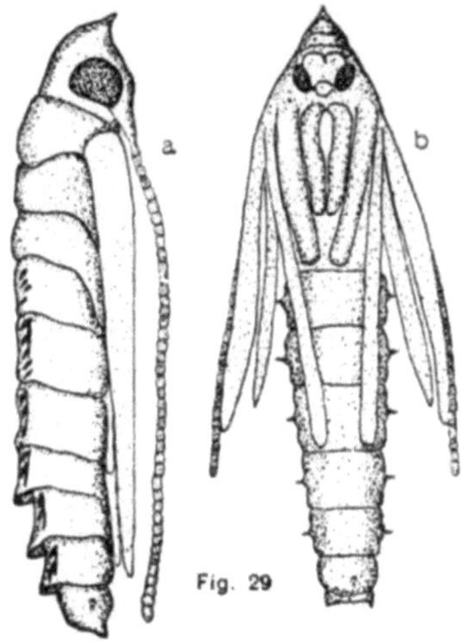

Fig. 27. Caterpillar of *Acrolepia* (Lepidoptera)

Fig. 28. Highly modified caterpillar *(Phyllocnistis)*

Fig. 29. Pupa of Lepidoptera *(Phyllocnistis)* a) Lateral view b) Ventral view.

racic shield or plate. Their shape and colour is often important for purposes of identification, especially with *Elachista.* On each of the first three (thoracic) segments behind the head the larva possesses a pair of jointed legs and also several pairs of abdominal feet (prolegs), usually on segments 6 to 9 and also on the terminal segment (Fig. 27). These feet on the abdominal segments consist only of unjointed, fleshy extensions of the body and end in a pad bearing a semicircular arrangement of chitinous hooks (crochets). Moths having larvae with this type of feet are classified as *Harmoncopoda* or Macrolepidoptera; if the arrangement of hooks on the feet is in 1 or 2 close rings (which may also be more or less atrophied or even entirely absent) the moths are known as *Stemmatoncopoda* or micro-lepidoptera. Mining Lepidoptera possess only prolegs on segments 6 to 9 and 13: the presence of legs or feet on other abdominal segments indicates that the larvae are not caterpillars but saw-fly larvae, the larvae of Tenthredinidae (which never have chitinous hooks on their abdominal feet). In a number of species feet are entirely absent (Fig. 28).

The caterpillar changes skin a varying number of times before it is full-grown: the number of moults may even vary in the same species, though it is normally 4 or 5. A change in the mode of mining can frequently be observed after a moult. A linear mine may develop

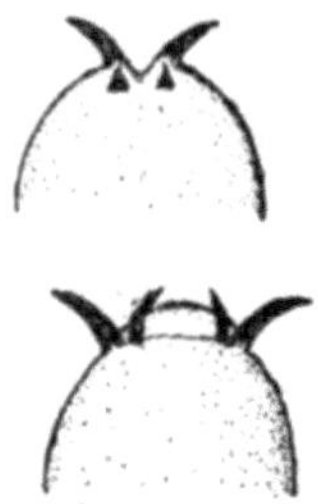

Fig. 30

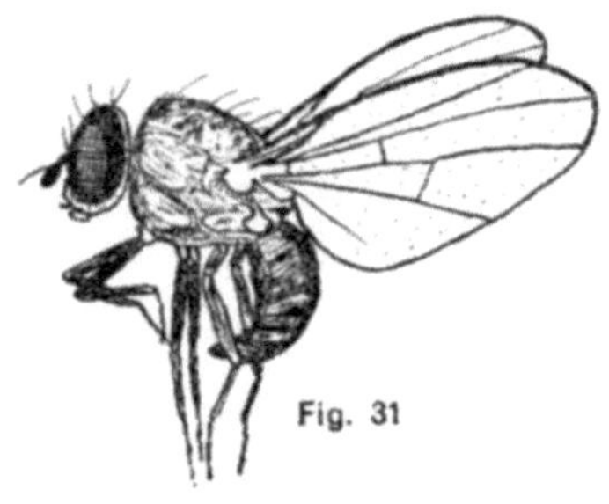

Fig. 31

Fig. 30. Cremaster of *Lithocolletis* on quercus. Above: *L. cramerella* F.; Below: *L. hortella* F.

Fig. 31. Mining fly *(Agromyza)*

a sudden change of direction or increase in width, or it is transformed into a blotch mine, or the method of feeding by mining may be given up altogether and the larva lives between spun-together leaf parts or on the plant in the open.

After its last change of skin the caterpillar is transformed into a pupa. The form of the pupa varies according to the family of the species in question and the sheaths of the various loose body parts may be free or largely fused to the body. The tail end of the pupa, which is often provided with hooks and a specially shaped chitinous structure, is particularly important for identification purposes: taken together it is known as the cremaster (Fig. 30). The shape of the cremaster frequently provides a certain means of differentiating the most similar species, such as the '*Wahlbomiana*' group of *Cnephasia* species and many of the oak-feeding *Lithocolletis*, which cannot be distinguished from the mine alone. In all these cases the empty pupal case can be examined either at the exit from the mine or between spun-up parts of the leaf near the mine and its structure provides a good means of distinguishing species which are otherwise difficult to differentiate from each other. The pupa lies either naked in the mine or it may be in a cocoon either in or outside the mine or in the ground.

Mine producers among Lepidoptera are found principally in the families of the Tineoidea, and especially in the Nepticulidae and Gracilariidae but there are also many in the Gelechiidae, Momphidae, Cosmopterygidae, Glyphipterygidae and others. The caterpillars of the Tortricidae are less frequently miners throughout their life but often in their early stages.

In Central Europe there are relatively few miners among the Pyralidae; the most striking of these, however, are *Phlyctaenodes pustulalis* Hbn. and *Cynaeda dentalis* Schiffm. whose mines on Boraginaceae form a globular swelling in which the larva spends the intervals between feeding and in which it later also pupates. In the Mediterranean area the number of mining Pyralidae is relatively greater. While the most primitive of the Hesperiidae, the *Megathymus* species of America, mine throughout their life, in Europe members of this family occur in mines only in their infant stages. Zygaenidae also only live in leaf mines in their early stages: *Procris* alone contains species which spend their whole life as miners. The Noctuids which are miners are chiefly those species living in the proximity of water *(Archanara, Sedina)* but in these cases mining tends to merge into pith feeding. We so far know only one mining species of Geometridae *(Cidaria incultaria* HS.*)* which feeds through out its life in the leaves of Primula.

It is usually possible to decide from the mine itself, whether it has been caused by Lepidoptera; hard and fast rules for this, however, are difficult to lay down, though long practice and experience enable one to distinguish at a glance lepidopterous mines from those of other Orders. A number of characteristics can be indicated, however, the presence of which show for certain that a mine has been made by a moth larva.

a) Purely epidermal mines throughout their whole length are made only by lepidopterous larvae (especially *Phyllocnistis*). In other Orders galleries are found in exceptional cases, which to begin with run only through the epidermis cells, but they soon expand to a greater depth and thereby include also the mesophyll of the leaf.

b) All mines in which frass is completely absent, without even traces of it, have been made by caterpillars (Lepidoptera). The only exception to this is represented by the larvae of Tendipedidae (Diptera) which in other respects also occupy a special position, which will be discussed further in the chapter on miners of aquatic plants.

c) If the mine is produced by a case-bearing larva, with the case attached to the outside of the leaf, then a species of Lepidoptera is involved. The larvae of the genus *Cryptocephalus* (Coleoptera), which live in similar cases, produce no mines.

d) The presence of silk thread in a mine invariably indicates that a moth caterpillar has been at work. Silk is also found in the mines of all the other Orders (except in Tendipedidae, living on aquatic plants), but only in conjunction with pupation: Hymenoptera and Coleoptera frequently pupate inside a silken cocoon and there are even Diptera, whose puparia are fastened to the mine with a few threads of silk, as for instance species of *Dizygomyza* in leaves of Gramineae. Spinning, however, occurs amongst the Lepidoptera in varying degrees even in the stages preceeding pupation. Full depth mines containing silk always appear more or less opaque against the light and not so clearly transparent as is usual with mines where the whole mesophyll has been eaten away. The silk threads are often found not in the mine but outside it. Thus many species of Lepidoptera mine from with in a channel of silk between spun-together leaves or from a passage in the ground lined with silk as for example, *Titanio normalis* Hbn. In other cases a web consisting of merely a few threads is attached to the outside of the leaf and the frass clinging to this web immediately identifies the mine as having been produced by a caterpillar. Tentiform mines can only be caused by moth larvae.

e) Mines in the epidermis having a circular opening less than 5 mm. in diameter have always been caused by Lepidoptera *(Colephora, Bucculatrix)*. Dipterous larvae leave their mine through an approximately semi-circular exit slit, Hymenoptera and Coleoptera sometimes also through a circular opening, but which is always very much larger.

f) Temporary miners, which adopt a different type of feeding after leaving their mine, in other words which change from being endophagous to ectophagous, are not found among the other three Orders of miners. If a larva leaves its initial mine to continue feeding between parts of the leaf which are spun together, it is a clear indication that it belongs to the Lepidoptera.

g) Blotch or blister mines, in which the frass lies in long threads are without exception caused by lepidopterous larvae. If the frass lies in a continuous central line in a linear mine, it is almost always a sign of Lepidoptera (exception *Phytomyza facialis* Kltb.): in linear mines caused by Diptera the grains of frass are deposited at the sides of the channel, in other words virtually in two lines. Double rows of frass are found in Lepidoptera only in certain species of *Nepticula*, which commence by living in the petiole or mid-rib, to which they regularly return during the intervals in their feeding in the leaf blade: in these species (*N. sericopeza* Z., *N. intimella* Z. and others) the frass does not lie in two rows of individual grains, but appears rather as two thick walls of the mine.

h) Mines in the needles of Conifers are almost without exception caused by larvae of Lepidoptera.

i) If the larva has a dark, chitinous head capsule, with at the same time more than one pair of abdominal feet having chitinous hooks on their pads, it is definitely a lepidopterous mine. These feet are absent in the larva of Tenthredinidae, and also of course in many lepidopterous larvae. If there are more than two pairs of pigmented stemmata (eyes) on the head, it can only be a lepidopterous mine.

2. Diptera (flies and midges)

The imago of this Order is characterised by having only two wings, which are membranous; the hind wings are modified to halteres. On the head below the frontal triangle are the antennae which are thread-like with many segments in the Nematocera (midges), but have only three segments in the Brachycera (flies); the remaining segments are modified to a terminal appendage known as the arista.

The wings of the Nematocera have a large number of free nervures, while on the other hand the families of the Brachycera which interest us here have normally only five clearly free nervures and one or two cross-veins, of which the upper or anterior is known as the small, and the posterior (when it is present) as the large or posterior nervure. The Brachycera comprise the majority of miners; among them the Calyptrata are distinguished by a cleft in the outer edge of the second antennal segment (and usually large hairy squamulae at the base of the wings), which is absent in the Acalyptrata (usually having small squamulae). The Calyptrata comprise Muscidae with many mining species and the Scatophagidae with *Hydromyza*, *Norellia*, *Clidogastra* and *Chylizosoma*. Anthomyidae are now usually included with the relatives of the common house-fly as Muscidae. The majority of mining Diptera belong to the Acalyptrata, and the families of Agromyzidae are in particular primarily miners; in addition miners are found in the families Trypetidae, Lauxaniidae, Psilidae, Drosophilidae and Ephydridae. Calyptrata and Acalyptrata are often further grouped together as Muscoidae; they may be contrasted with the mining forms of the Syrphidae which can easily be recognised by the large number of veins and cross-veins even on the outer margin (genus *Chilosia*). Of the Nematocera the following are producers of mines; Lycoriidae (Sciaridae), Itoniidae (Cecidomyidae) or Gall Midges (only some species) and Tendipedidae (Chironomidae) or Harlequin midges (only species living in water).

In most cases it is immediately possible to distinguish a dipterous mine from its shape in conjunction with the food plant but there are nevertheless some species in which the characteristics of the larva or pupa must also be considered in establishing the identification. We must therefore discuss more fully those characteristics of the dipterous larva and pupa which need to be referred to for identification purposes.

The mining larva of the Nematocera still possess a well developed head and clearly recognisable legs on the first and last segments of the body (Fig. 32 left). The larva of Brachycera on the other hand is

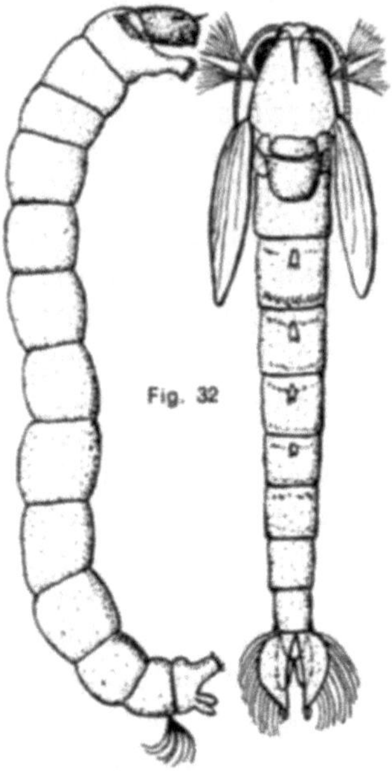

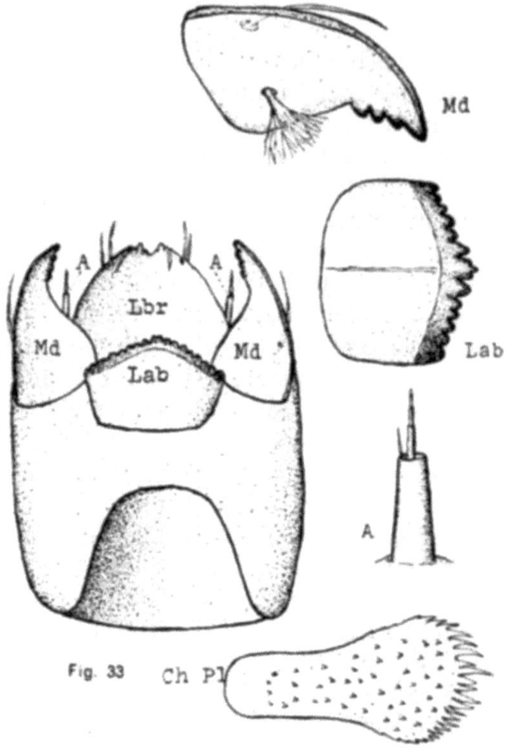

Fig. 32. Tendiped. Left - larva, right - pupa.

Fig. 33. Head parts of a Tendiped larva. (A - antenna, Lab. - labium, Lbr. - labrum, Md. - Mandible, Ch.Pl. - chitinous plates on the back of the abdomen of the pupa).

maggot-like; it no longer has a chitinised head capsule and the feet are completely atrophied. Among the Nematocera the mines of the Tendipedidae can only be distinguished from each other by the larva or pupa, which must therefore be given particular attention. They only live in water and further information on them is to be found in the chapter on miners in aquatic plants. The larva of Tendipedidae has a head with short antennae of several segments. (Fig. 33a). The mouth orifice is covered by the labrum or upper lip (Lbr) which may bear at its extremity a variety of different types of bristles. Beneath it are the mandibles (Md) and their type of serration and arrangement of bristles are important for identification. The maxillae which lie below are less easily visible and are not shown on the figure. On the lower side, the mouth orifice is closed by the labium (Lab) or lower lip. Its type of serration is of particular importance for identification.

There are normally two eye spots. The first thoracic segment bears a pair of prolegs and on the last segment there is a second pair with hooks, which can be useful in distinguishing the species. On the last segment there are also anal gills (mining species always possess two pairs). External breathing pores, known as spiracles are absent from Tendipedidae, which are apneustic and breathe through their skin. The sheaths of the wings and antennae are clearly visible on the pupa (Fig. 32, right) and on each side of the first thoracic segment there

Fig. 34

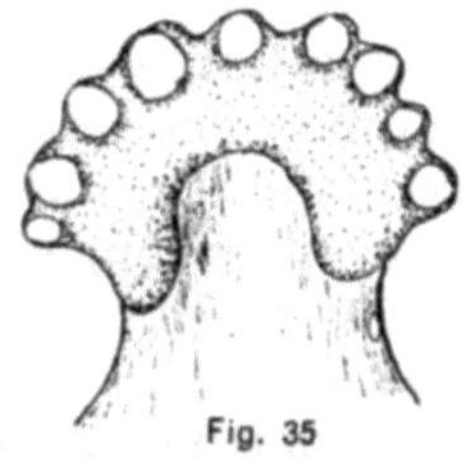
Fig. 35

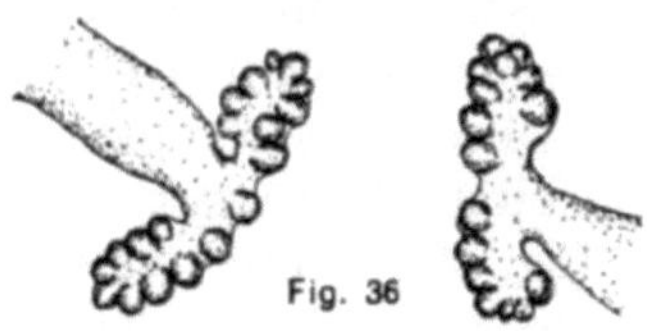
Fig. 36

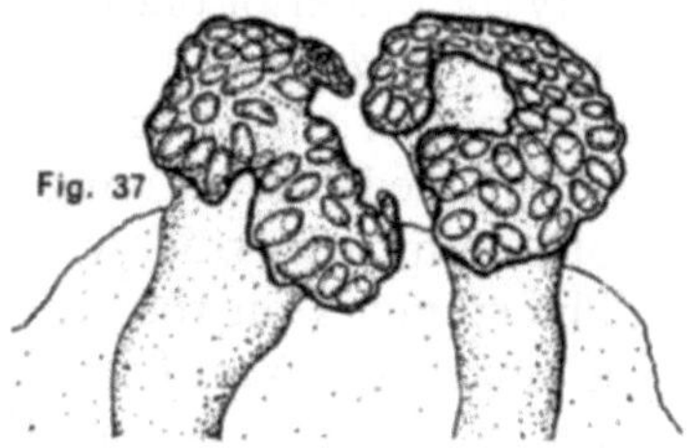
Fig. 37

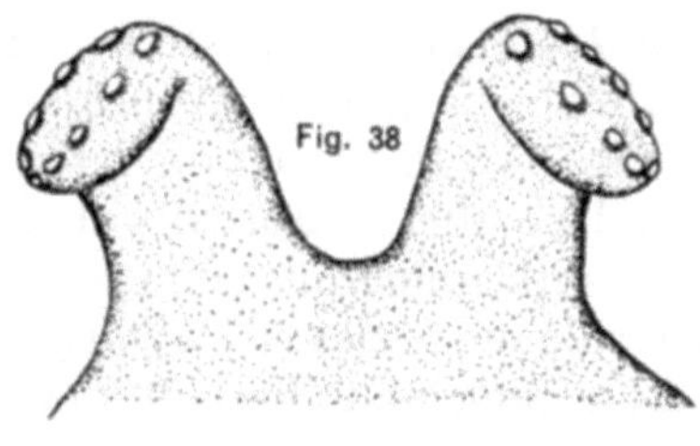
Fig. 38

Fig. 34. Larva of *Hydrellia*. Pointed posterior spiracles, mouth hooks with only one tooth.

Fig. 35. Posterior spiracular process of *Liriomyza mercurialis* Hering, spiracles arranged in an arc.

Fig. 36. "Two-horned" anterior spiracular processes of *Agromyza lathyri* Hendel.

Fig. 37. Posterior spiracular processes bearing numerous spiracles of *Agromyza lathyri* Hendel.

Fig. 38. Posterior spiracular processes of *Phytomyza* aizoon Hering.

is a simple or veined respiratory organ, the thorax horn; at the end it possesses the characteristic anal gill. In mining forms on the upper side of the second to sixth abdominal segments situated dorsally behind the centre of the anterior margin there are characteristic chitinous plates (in addition to the special thorny appearance), the presence and shape of which are valuable for classification. The necessary examination can best be undertaken on the empty pupal case, which shows all these features most clearly.

The larva of the Brachycera is maggot-like. It has neither legs nor a chitinised head-capsule (Fig. 34). Repiration takes place through special vents, called spiracles; the spiracles are usually grouped three or more on a respiratory process. The shape of the respiratory processes and the number and arrangement of the spiracles on them are always very important for classification. There may be either anterior or posterior respiratory processes (on the rear edge of the first and on the last abdominal segment, in which case the larva is termed amphipneustic (Agromyzidae, Trypetidae, Muscidae), or only the posterior ones are present, when the larva is metapneustic *(Hydrellia)*.

The structure of the respiratory processes and mouth parts can best be investigated, after leaving the larva for a while in a 4 % solution of Potassium hydroxide, when the detailed parts become readily visible. The spiracles on the anterior respiratory processes are frequently arranged in a semi-circular arc (fig. 35), in other cases they may appear in two groups and the processes are then called "two-horned" (Fig. 36).

In many species of *Hydrellia* the posterior processes bear only simple spiracles, which can be drawn up into a spike and inserted into the plant tissue (Fig. 34). In species of *Agromyza*, in Trypetidae and Muscidae there are usually three spiracles, though in *Agromyza* processes with numerous spiracles are also found (Fig. 37); this is the normal rule in species of *Phytomyza* (Fig. 38). Movement is effected by a creeping welt (Fig. 39 V) on the forward edge of most segments.

The mouth parts deserve special attention and are particularly striking in the light-coloured larva with their deep black coloration. Fig. 39 gives a sketch of the pharyngeal skeleton. The individual parts are almost all arranged in pairs and this should be carefully borne in mind in any investigation, as otherwise the number of mouth-hooks, for example, might be thought to be twice as large as in reality. In the very front of this apparatus are the movable mouth-hooks (M) which are considered by some to be modified mandibles, by others as modified maxillae or even as secondary new structures. Their function is to break up the plant tissue. Their rear extension is

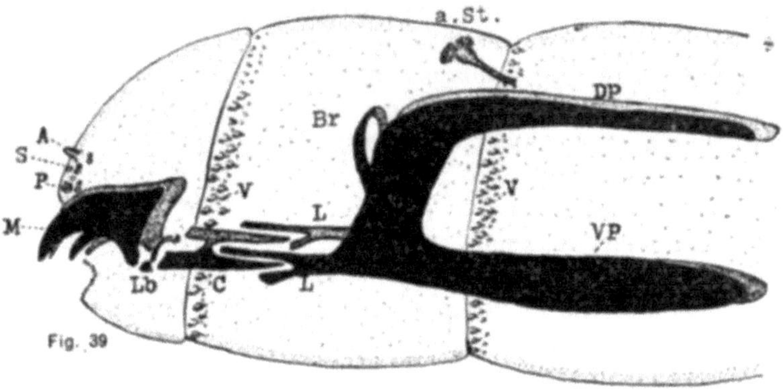

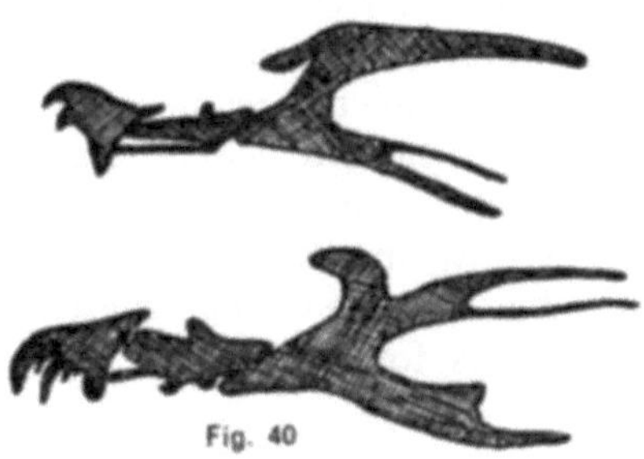

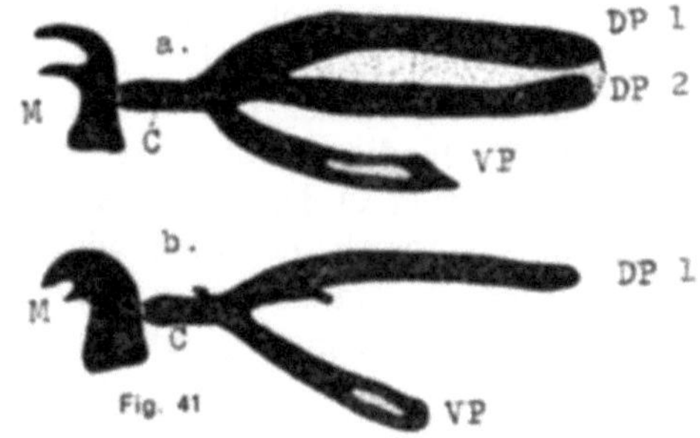

Fig. 39. Pharyngeal skeleton of the larva of Brachycera (A - Antenna, a.st. - Anterior spiracular process, Br. - Bridge between the vertical plates, C. - Collare, DP. - Dorsal part of the vertical plates. L. - Longitudinal sclerites, oral extension of the vertial plates, Lb. - Vestige of labium, M. -Mouthhooks, P. - Maxillary palpi, S. - Sensory organs of the head, V - Creeping welt (Verrucae), VP - Ventral portion of the vertical plates.

Fig. 40. Pharyngeal skeleton of: above, *Philophylla heraclei* L., below, *Pegomyia* sp.

Fig. 41. Pharyngeal skeleton of a. *Agromyza*, b. *Phytomyza* (Abbreviations as in Fig. 39, DP. 1. — upper extension, DP.2. — lower extension of the dorsal part of the vertical plate).

the rudimentary labium (Lb). They are adjoined by an H-shaped, horizontal collare (C), which is again linked up to the two highly developed vertical plates by two longitudinal sclerites (L). These two vertical plates are joined in front by a bridge (Br) and consist of a dorsal (DP) and a ventral part (VP). The dorsal part may again be divided into two extensions. Variations in the structure of this apparatus are illustrated in Fig. 40, which shows the pharyngeal skeleton of, above, a Trypetid and below, a Muscid. *Hydrellia* has only one tooth on each mouth-hook (Fig. 34), while in *Agromyza* there are usually two. In Agromyzidae there are differences in the structure of the vertical plate; in *Agromyza* and its closely related genera the dorsal part of the vertical plate is divided into an upper and lower extension, which may appear either free or more or less united, while in *Phytomyza*, and its related genera the lower extension is completely or largely atrophied (Fig. 41). Above the mouth-hooks there are rudimentary maxillary palpi (Fig. 39P), further dorsad the antennae (A), and close to them small sensory organs (S).

The use made of the mouth parts by mining fly larvae is fundamentally different from that of the larvae of other Orders. In the latter the larva lies either ventrally or dorsally to the centre of the leaf, while in the higher Diptera the larva lies on its side while eating, so that the mouth-hooks shear through the parenchyma cells like a sickle through grass. This results in two important characteristics of the dipterous mine. The larva causes in this way the primary and secondary feeding tracks, the "herring-bone" pattern, which is not found in other Orders, and frass is deposited not in central line but on the side; as the larva lies while feeding now on one side, now on the other, the grains of frass are deposited at the side of the channel, so that in principle a double row of frass is formed.

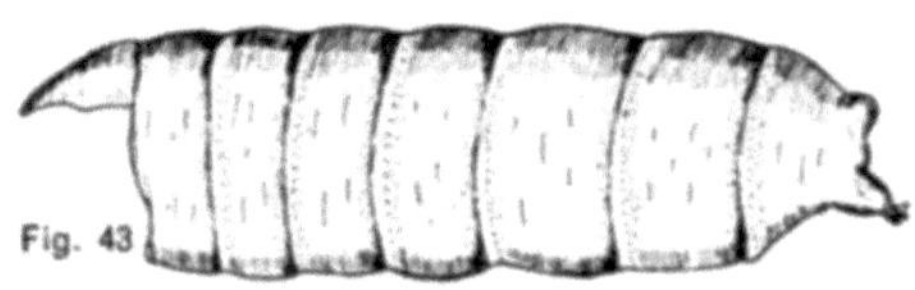

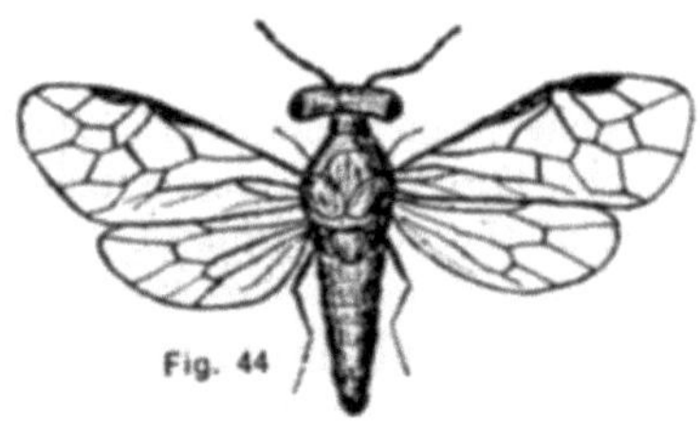

Fig. 42. *Phytomyza albimargo* Hering. a. Puparium, b. terminal segment with anal orifice and posterior spiracular processes, c. anterior spiracular processes.

Fig. 43. Empty puparium of an Agromyzid, showing in front the split-off section.

Fig. 44. Tenthredinid (Saw-fly).

In the Nematocera pupation occurs in the same way as in all the other relevant Orders; however, in the Brachycera and thus in the great majority of mining Diptera, pupation takes place in a quite different way. They do not shed the last larval skin, which, instead, hardens, darkens and itself forms the puparium, in which the actual pupation takes place. Naturally, the characteristics which serve to distinguish the different species of larvae, especially the structure of the mouth parts and the formation of the respiratory processes and their spiracles, may be observed and examined in this puparium, even though not perhaps with the same clarity as in the larva (Fig. 42). When the fly emerges from the pupa and the puparium, a section of the latter of varying size is split off (Fig. 43) and the imago gains its freedom.

Characteristics of mines of Diptera

1. All mines, in which primary and secondary feeding tracks are visible, are caused by larvae of Diptera.
2. If frass lies alternately on the two sides of a linear mine, the producer was a fly larva.
3. Completely straight linear mines, which have two similar round openings and are found on aquatic plants, have been caused by dipterous larvae (Tendipedidae). Silk is found in a mine throughout the larva's life only in this case; silk is absent from all other dipterous mines and small traces are only found during pupation as an exception.
4. A mine containing a larva without a brown head-capsule or a puparium without sheaths for the extremities, is always a dipterous mine.

3. Hymenoptera (Saw-flies)

Mining Hymenoptera always possess two pairs of membranous wings, which are never covered with hairs or scales (Fig. 44). The hindbody is broadly attached to the thorax and this serves to distinguish mining Hymenoptera from the parasitic Hymenoptera frequently obtained during breeding from the mines of other insect Orders, in which thorax and abdomen appear to be separated by a constriction, the "wasp waist". Normally the larvae of the mining Hymenoptera, the Tenthredinidae, possess a larger number of abdominal legs than caterpillars (Fig. 45). The abdominal legs are mere evaginations of the integument, on the pads of which the crochets so characteristic of the majority of Lepidoptera are *always* absent. The head

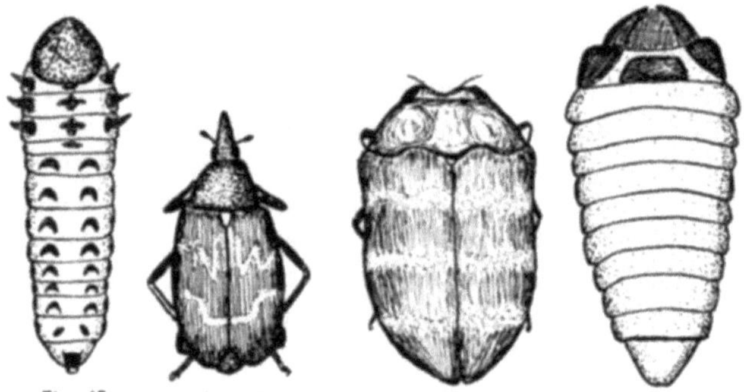

Fig. 45 Fig. 46 Fig. 47 Fig. 48

Fig. 45. Ventral view of mining saw-fly larva.

Fig. 46. Curculionidae: *Rhynchaenus salicis* L.

Fig. 47. Buprestid: *Trachys minuta* L.

Fig. 48. Considerably modified coleopterous larva *(Rhamphus)*.

capsule is always well developed and always has only one pair of eyespots (stemmata). The body is invariably very pale and the bright colours are lacking which characterise so many free-living saw-fly larvae and many caterpillars of mining Lepidoptera. The egg is always inserted inside the leaf and the saw-fly has a piercing apparatus for ovipositing, the sheath of which is often very important in classification. Pupation takes place inside the leaf only as an exception, when a circular cocoon is formed; in the great majority of cases the larva enters the ground when full-fed, where it spins a cocoon in which it usually passes a considerable time before actually pupating. A few species cut a circular cocoon out of the leaf or its upper epidermis, fall with it to the ground and there complete their pupation. There are occasionally dark, chitinous plates on the thoracic segments, which enable species with similar living habits to be distinguished by means of the larva. Miners are only found in the family Tenthredinidae.

Characteristics of hymenopterous mines are not so easy to establish as in the two previous Orders. They are mostly negative:

1. If the shell of an egg is found on the surface of the leaf, it can never be a hymenopterous mine.

2. Mines of Hymenoptera may possibly commence as linear mines, but are then rapidly extended into a blotch. Typical linear mines are not found in this Order.

3. Frass is deposited neither in two rows of pellets as in Diptera, nor in a continuous line as often in Lepidoptera. The grains of frass

are usually scattered irregularly about the mine cavity and sometimes piled up in the beginning of the mine like plug.

4. Hymenopterous mines never contain silk during the period when the larva is feeding; the only time when it may be attached to the mine is during pupation.

5. There are no temporary miners among Hymenoptera and no change of mine or leaf occurs. The only exceptions are the species of *Hoplocampa* mining fruit.

6. Larvae with numerous abdominal legs or with only one pair of eye spots on their head usually belong to the Tenthredinidae. A dark head capsule is always present. There are never chitinous hooks on the abdominal legs.

4. Coleoptera (beetles)

The mining forms of the perfect insect of this Order always possess fore-wings modified to elytra (wing-covers), which cover the membranous hind-wings while at rest. The part of the body lying behind the head is called the prothorax and behind it between the elytra is a small dorsal shield called the scutellum. The characteristic markings are usually on the prothorax and scutellum. In the Rhyncophoridae (Fig. 46) the head is elongated forwards into a beak. Mines are produced by members of the families Buprestidae (Fig. 47), Chrysomelidae (*Zeugophora*, *Galerucella*, many Halticinae and Hispinae) and Curculionidae.

The larva (Fig. 48) is usually very much degenerated and it may be completely devoid of legs. There may, however, be three pairs of thoracic legs but abdominal legs are never found. There is always a brown head capsule, which has only a single pair of eye spots. Frequently the fore-body of the larva is strongly thickened, which facilitates penetration into the plant tissue. Pupation takes place in the mine or in the ground.

It is impossible to lay down general characteristics for coleopterous mines. In many cases examination of the larva gives a clue.

1. A coleopterous mine is indictated if a spot of shiny secretion covers the egg at the commencement of the mine. Mines of Buprestidae can be recognised in this way.

2. Globular cocoons inside the mine are usually formed by Coleoptera, but are also found among Lepidoptera.

3. Carpetting of the mine cavity with silk only occurs in Coleoptera prior to pupation; there is therefore no silk or web during the larval stages.

7.

LIFE HISTORY OF THE MINING INSECT

a) Oviposition

After fertilization, the female lays its eggs (although ovipositing frequently takes place in Tenthredinidae and sometimes in other Orders without previous fertilization, in other words parthenogenesis occurs). Ovipositing assumes different forms according to the Order or family. In very many cases the egg is attached to the leaf on the outside of the plant tissue. It is cemented to the leaf surface with a sticky secretion and the empty egg-shell is still visible after the larva has hatched and is frequently a valuable guide to identification. In certain cases the egg is covered by a varnishlike substance which gives it added protection, especially in Buprestidae, whose mines can easily be recognised by the shiny secretion covering the egg (Fig. 49). The egg may be laid on the upper or lower surface of the leaf and the side of the leaf chosen is usually characteristic of the species concerned. Mines closely resembling each other, such as those of *Nepticula* on Rosaceae can often be differentiated by the side of the leaf on which the egg is to be found. In this genus, especially, the eggs are often laid in a well concealed position, frequently in the angles of the veins where they are difficult to find, especially when this part of the leaf is more hairy than elsewhere. In such species the mine track which is more easily visible should be followed back to its origin, where the shell of the egg can be found. A lens with a magnification of 20 always makes an examination of the egg possible without difficulty. There are some species which do not lay their eggs in the exact spot where the young caterpillar will later commence its mine. The females of *Hyponomeuta*, for instance, deposit their eggs in groups on the cortex of the stem or twig where they pass the winter and only in the spring do the young caterpillars seek out the leaves, which they burrow into. This is, however, an unusual exception and in almost all cases it will be possible to observe the type of oviposition at the commencement of the mine. When the larva has already changed its mine this is more difficult and the original mine must then be looked for in the vicinity of the secondary mine in order to establish the type of oviposition.

Fig. 49. Salix: Blotch mine of *Trachys minuta* L. with the patch of black secretion covering the egg shell at the beginning of the mine.

Many species of Lepidoptera, especially *Nepticula*, and also certain Diptera such as Cordyluridae and Muscidae attach their eggs to the outer surface of the leaf. These usually deposit a varying number of eggs lying close together and parallel with each other on the underside of the leaf. If some doubt arises whether a young dipterous larva belongs to Agromyzidae or Trypetidae, it is merely necessary to examine the beginning of the mine; in Agromyzidae and Trypetidae no egg shells will be found on the leaf.

Additional protection is afforded the eggs when the female first excavates a small hollow in the leaf, in which the eggs are deposited and then covered with a secretion, as occurs in *Zeugophora*.

In the great majority of mining species, however, the egg is inserted inside the plant tissue. The ovipositing female makes a puncture in the leaf and in each hole so made is pushed a single egg. The egg thereby gains increased protection and at the same time the young larva on hatching can more easily gain access to the part of the leaf it favours, without having to bite through the harder epidermis cells, which are often in addition covered with a tough cuticle. Depositing the egg inside the leaf assumes particular importance in saw-flies, since it is essential for the development of the egg of these insects to be able to absorb moisture from the plant tissue. All saw-flies, even non-mining species, therefore insert their eggs into the plant tissue where they often undergo a considerable increase in size during the development of the embryo. On hatching the larva then leaves the pocket in the leaf, and the hole that remains might well be the cause of it having been incorrectly considered as an early mine. Such cases will be referred to further in the chapter on pseudomines.

All mining Agromyzidae and Trypetidae also insert their eggs into the leaf tissue. The female possesses for this purpose a particularly well developed ovipositor and when not in use its individual parts are telescoped together. Cavities are sawn into the parenchyma with this apparatus and the eggs are deposited within. While ovipositing, the Agromyzidae "test" the leaf substance: the female creates a puncture with its ovipositor, then turns round and imbibes the exuding sap with its proboscis. It is not clear whether this is merely a sampling of the tissue or whether the fly is carrying out an additional act of feeding. Whatever the reason may be, many hundreds of such borings are not infrequently found in a leaf, of which only one or a few contain an egg. If the leaf is held to the light, the numerous punctures can be seen; a superficial examination may thus easily lead to a gross over-estimation of the number of eggs which have been laid, as occurred when Brischke claimed to have found up to 50 eggs of *Phytomyza crassiseta* Zett. in a Veronica leaf, of which only one developed. In reality he had counted only the borings of the female.

There are also a number of genera of Lepidoptera which are able to insert the egg into the parenchyma tissue by means of a special piercing ovipositor. Very closely related genera can sometimes be distinguished by the structure of this ovipositor, which suggests that there may be considerable differences in the act of oviposition, which have not yet been investigated. The ovipositing female of species mining submerged leaves of aquatic plants is forced to penetrate beneath the surface of the water in order to deposit its stock of eggs. This applies especially to species of *Hydrellia* and *Hydromyza*, where ovipositing of this type has been observed. The fly's thick covering of bristles retains air during this period so that the females which go under water are surrounded by a covering of air like water spiders, which enables them to breathe while ovipositing. Tendipedidae which mine aquatic plants lay their eggs spawn-like direct into the water.

In a number of cases the egg is inserted into the opposite side of the leaf to that of the subsequent mine; in many mines, which are otherwise confined to the upper surface, the egg is nevertheless deposited on the lower surface. This cannot be detected with lowpower magnification as the point of oviposition always appears as a light speck. As soon as the larva is somewhat developed, however, a short, often spiral length of mine can be found on the opposite side of the leaf, where the egg was laid, which was made before the larva wandered through to the upper surface. Such early mine tracks can be observed in many species of *Phytomyza;* the different species of *Phytomyza* on Clematis which have remarkably similar mines can be distinguished by the presence or absence of this short mine track on the lower surface of the leaf.

b) The Life of the Larva in the Mine.

In all cases where the egg was deposited inside the leaf tissue, the larva hatching from the egg finds its way directly to the leaf tissue. The shell of the egg appears first to be consumed in these cases, since no remains have ever been discovered at the beginning of the mine. If the egg has been deposited on the outside of the leaf, the young larva invariably chooses the shortest course to gnaw its way into the leaf by eating through the shell on the side of the egg adjoining the leaf and so directly into the leaf, and the entrance to the mine is thereby closed by the outer wall of the egg-shell. This sealing off of the mine channel is particularly important for larvae which live under water, as the penetration of water into the mine is prevented and larvae, such as those of *Hydrellia*, are able to live in the tissue surrounded by air (which enters the mine via the plant tissue). Portier (1918) though it could be assumed that all miners living in this way had achieved an existence where their mine channel was free from parasites and noxious micro-organisms and that similarly, within the larva itself, none of the bacteria, or other minute forms of organisms which are normally present, especially in the digestive organs of the larva, would be able to develop. It was later shown that there is no question of these mine channels being sterile and that an intestinal flora develops in these just as in free-living larva. This is, of course, quite understandable; the spot to which the egg is affixed was not previously sterilised and when the larva bores into the leaf any organisms which may be on the leaf epidermis are able to enter the mine cavity; furthermore, oviposition itself is not a sterile operation and when the underside of the egg-shell is consumed, micro-organisms are also able to enter the digestive tract of the larva.

Many species live while young, many others indeed throughout their lives, exclusively in the cells of the epidermis, which contain primarily liquid substances. The mine channels of such epidermal miners for this reason scarcely stand out against the light and can easily be overlooked, particularly as the cell contents which have been consumed are also primarily liquid and the larva therefore often leaves no obvious traces of frass. The majority of mining species, however, proceed directly into the chlorophyll containing parenchyma of the leaf and the mines now become clearly visible when held to the light, owing to the parts which have been eaten away appearing lighter than the rest of the leaf. The depth of the mine may remain the same from beginning to end or the vertical extension may increase during the course of the larva's life. In almost all species of Tenthredinidae the mine is full-depth from beginning to end; the whole parenchyma between the upper and lower epidermis is eaten away and the mines look particularly transparent. In some species, as for instance *Fenusa* the mine is only upper surface and a shallower part is evident at its beginning; later, the mine becomes deeper but remains upper-surface.

The larva in the mine always changes its skin at certain intervals in the same way as free-living larvae from the same Orders; fly larvae thus normally have three, lepidopterous larvae four to six instars, each of which is concluded by a moult. In certain genera of Lepidoptera the larva consumes no food during its last instar and is solely occupied in producing silk for its pupation *(Phyllocnistis)* and for this reason the instar is known as that of the pre-pupa. The first instar differs from the later ones in many respects, especially in the higher Diptera, in which an unpaired tooth occurs above the mouth hooks (explained as a vestigial labrum) but which is absent in the later instars. The arrangement of bristles is still very much simplified in first instar lepidopterous larvae and largely similar, in a number of families, while in later instears the number and arrangement of body bristles varies considerably according to the family and facilitates identification of families and genera. The cast larvae skin with its readily visible head-capsule which remains inside the mine, enables the place of moulting to be exactly determined and shows how a change in the type of mining frequently take places after a moult, resulting in an alteration in the vertical or horizontal extension of the mine. In other cases there may even be a change from one mine to another and in temporary miners the mining habit is given up altogether and replaced by external feeding. There is also frequently a change in the arrangement of frass after a moult and this will be discussed in greater detail in a subsequent chapter on account of its diagnostic importance. Many species also change the side of the leaf in which they mine, transforming their lower surface into an upper surface mine; a different attitude to the leaf veins can be noted and thus a number of instinct alterations can be observed after changes of skin.

c) Pupation

After the last change of skin the larva turns into a pupa. Pupation may take place either inside or outside the mine. The species which pupate outside the leaf leave the mine through a semicircular hole which is gnawed through the epidermis. This slit may be on the same side of the leaf as the mine or on the opposite side. The situation of the escape slit may in many cases be used for determining the species. The larva allows itself to fall into the ground and burrows into the soil or seeks some other suitable hiding-place; many do not fall but crawl from the leaf downwards along the twigs and transform in crevices in the bark or other safe places on the plant. Many lepidopterous larvae lower themselves on a silk thread which is blown against twigs or the stem, on which they can then gain a footing. This can frequently be observed in *Bucculatrix thoracella* Thbg. in Tilia, which usually pupates on the stem. Many fly larvae, especially in the genus *Liriomyza*, possess the ability to jump in their final instar and in this way cover larger distances than would otherwise be possible with their inefficient locomotory organs. A considerable number of species make themselves a case before pupating, either by cutting out from the leaf the upper epidermis and completing the case by spinning a covering wall, or a round or elliptical portion of the leaf is cut out, so that the cocoon consists on both sides of leaf epidermis. In the saw-fly *Phyllotoma aceris* McLchl. (Fig. 50) and the moth *Tischeria decidua* Wck. on Quercus (and also in many tropical Buprestidae) this pupal cocoon is only cut from the upper epidermis of the leaf. A piece of the whole thickness of the leaf is cut out by *Phyllotoma leucomela* Klug, among Lepidoptera by all members of the family Heliozelidae (Fig. 51), by one Nepticulid (Fig. 52), among Coleoptera by many species of *Rhynchaenus* (Fig. 53) but never by species of Diptera. The lenticular cocoon produced in this way enables spe-

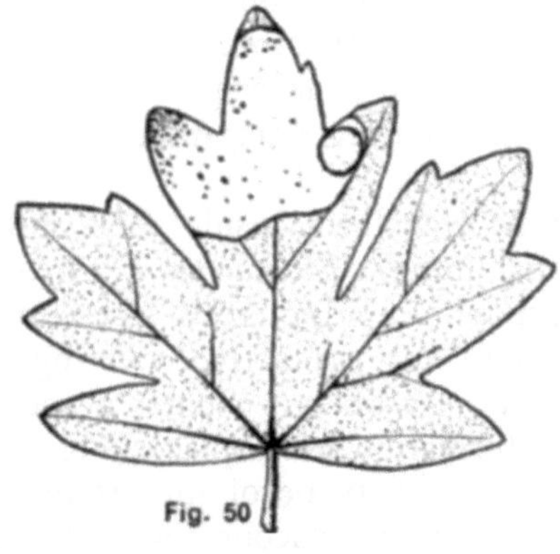

Fig. 50

Fig. 51

Fig. 50. Acer: Mine with a piece cut out of the upper epidermis of the leaf by *Phyllotoma aceris* McLchl.

Fig. 51. Cornus: Blotch mine with a piece of the whole leaf cut out by *Antispila pfeifferella* Hb.

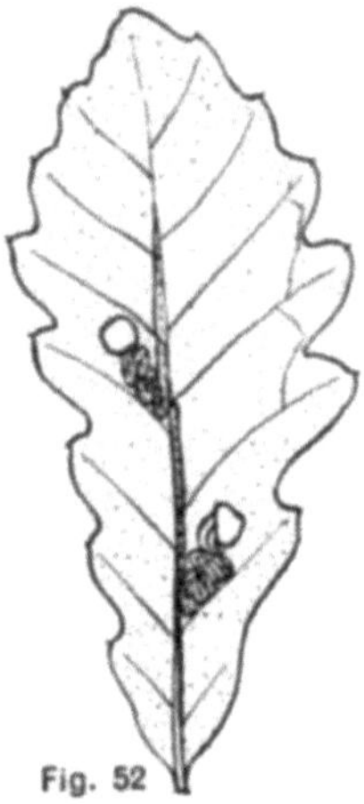

Fig. 52

Fig. 53

Fig. 54

Fig 52. Quercus cerris: Linear mine with a piece cut out of the leaf by *Nepticula phyllotomella* Klimesch.

Fig. 53. Salix: Linear mine on the margin of the leaf with piece cut out by *Rhynchaenus decoratus* Hrbst.

Fig. 54. Salix: Disc-shaped pupal cocoon in the blotch mine of *Phyllotoma ochropoda* Klug.

cies to continue their movement when on the ground by constantly changing position and so seek out particularly protected spots; for this reason, however hard one may search, it is never possible to find these creatures beneath their food plant, however common they may have been.

A large number of mining larvae transform inside their mine, frequently inside a new short "pupal mine" *(Dicladispa, Hydrellia)* constructed especially for this purpose. Lepidopterous larvae may pupate in this pupal chamber without any protective covering; in many cases, however, they and some saw-flies construct a special silken cocoon, in which their transformation takes place (Fig. 54). The colour and structure of the cocoon may be of importance in identification. Many *Lithocolletis* spin up this cocoon together with old pellets of frass, and even as larvae they can be seen to be constantly piling up and webbing down the frass on both sides of the most strongly dilated part of the mine, the future "pupal chamber"; they withdraw to this pupal chamber during pauses in feeding even before they are fully-fed. Species which transform in the mine "naked" build up a stronger silk lining to their pupal chamber before actually pupating. The Pyrales *Cynaeda dentalis* Schiff. and *Phlyctaenodes pustulalis* Hb. which mine Boraginaceae produce an almost globular pupal chamber of considerable toughness, which is used earlier by the caterpillar as a retreat during pauses in feeding. This structure is so striking that *C. dentalis* Schiff. was inaccurately considered on account of it as a gallcauser.

Mining Agromyzidae in many cases construct a "pupal blister". No semi-circular exit slit is made by the larva but when constructing the puparium the two anterior spiracles penetrate the epidermis and project in the open beyond, while the whole body of the pupa remains inside the leaf. In *Phytomyza* the pupa lies ventrally, in *Ophiomyia* and *Melanagromyza* dorsally to the outside of the leaf. In *Phytomyza* when the imago emerges, the ventral portion of the anterior end of the puparium springs open and the fly also has to burst open the outer wall of leaf epidermis before gaining its freedom. In *Ophiomyia* this process is similar, except that the dorsal portion of the puparium also splits along its central axis, so producing a narrow line through which the fly emerges into the open. The last stretch of the mine channel is usually without frass, and often shallow and therefore less readily visible. Not all the Agromyzidae which pupate in the mine, make such a pupal blister; others also pupate in the mine but lie unattached in the mine cavity and may roll about freely. Another group of species without a pupal blister attaches the puparium to the mine, either with a few threads of silk, like species of *Dizygomyza* in Gramineae, or cement it down with glue exsuding from the anal orifice, like species of *Dizygomyza* mining in Iris or *Dizygomyza humeralis* v. Ros. in Bellis, or like *Phytomyza cinerea* Hend. among the species of *Phytomyza* in its blotch mines in Centaurea.

In Agromyzidae which pupate in the mine, the mine cavity is sealed off from the outside and no exit is prepared in advance; the emerging fly bursts through the leaf epidermis. There are, however, not infrequently cases where the larva may pupate either in or outside the mine and here an exit slit is always prepared, through which the larva either leaves the mine in order to pupate on or in the ground or it may remain inside the leaf tissue. The presence of an exit slit therefore points to the fact that the species is one which does not pupate regularly and constantly in the mine, even though the puparium may still be found within; this is frequently seen in *Phytagromyza tridentata* Lw. in leaves of Salix and Populus. This is found particularly often in parasitised larvae, which on completing their feeding clearly have no longer sufficient strength to leave the leaf, as do healthy larvae of the same species. The presence of an exit slit at the end of the mine therefore shows that the species is one which normally pupates outside the mine. This does not only apply to Agromyzidae but also to other mining insects. A larva weakened by the attack of a parasite is frequently unable to leave the mine, and biting through the epidermis which is often covered by a cuticle, of course demands a greater output of strength than eating in the soft mesophyll of the leaf.

However, larvae are frequently found which remain in the leaf without having been parasitised, while the normal rule for the species is that they pupate outside the mine. This is especially the case when an unusually large number of larvae mine a single leaf. The obvious reason in such cases is that the weakening of individual larvae through undernourishment causes them contrary to their normal habits to remain in the leaf.

Various differences also occur in the same species according to the generation to which the larvae belong. A. Braun (1917) showed that the caterpillars of *Nepticula ulmella* Braun, a North American species, pupate partly in the mine, partly outside but that among the overwintering generation a far greater number remain in the mine than in the summer generation. The condition of the food-plant may well be the decisive factor for this variation in behaviour. The cell walls of the plant are much more woody in autumn than in summer and represent a far greater impediment to the caterpillar in gaining the open, so that only the strongest succeed in doing so. It is not impossible that local conditions also result in a variation in the condition of the food plant. For instance, in Central Europe, *Nepticula agrimoniella* HS. pupates exclusively inside the mine; in Bessarabia I nevertheless saw how some of the caterpillars left the mine to pupate, and transformed normally outside the leaf. Among the many hundreds of species of *Nepticula* practically all pupate outside the mine, though a few pupate in the leaf without any apparent compelling reason for this. In *Nepticula weaveri* Stt. which remains in the hard leaves of Vaccinium vitis-idaea L., this can be explained by the added protection which the pupa gains from the hard leaf; before pupating, the larva gnaws a silk-lined channel right up to the epidermis, so that the emerging imago has no further difficulties in getting into the open. It is not, however, clear why the same species remains in Agrimonia, nor why *N. septembrella* Stt. should remain in the relatively tender leaves of Hypericum, which can afford them no special protection.

Larvae normally mining in the cortex have varying habits in the selection of their pupation site. All species which mine the cortex of woody parts of the stalk or stem normally leave the mine to pupate. They would be in danger if they remained in the mine of being subjected to pressure and thereby being damaged by the increase in thickness which develops on the inside of the cortex. On the other hand when the larva mines in herbaceous stems, such as those of Campanulaceae, Compositae, Cruciferae, etc., pupation occurs predominately in the stem below the epidermis, as for instance in the many species of *Ophiomyia*, which live as stem miners.

The species of *Phyllocnistis* which only live in epidermis cells make special provision for protection before commencing to pupate. If they transformed in their mine, there would be too great a danger of the thin epidermis cells collapsing under the influence of the weather during the pupal quiescence, leaving the pupa exposed and unprotected on the leaf. On completion of their mining activity they therefore proceed to the margin of the leaf and make a web which causes the leaf-edge to fold over (by the contraction of the silk as it dries). Pupation then takes place under this folded over leaf-edge. If they do not succeed in reaching the edge of the leaf — this happens especially when several caterpillars are mining the same leaf — they then draw the leaf together in its centre with their silk threads, in the same way as the caterpillars of Gracilariidae *(Lithocolletis, Parornix,* etc.*)*. Parallel with this spinning activity is the fact that in the last larval instar this genus is able to consume no further food; the whole instar is devoted to spinning, the mouth parts are largely atrophied and the spinneret is strikingly enlarged.

In general, the mining larva which pupates outside the mine, first moves a greater or lesser distance away from the site of its mining activity. Certain species form an exception to this. The dipterous larvae mining aquatic plants have no legs and they are frequently, therefore, unable to move far on leaving their mine; if they simply let themselves fall, as they do from plants on land, they would fall into the water. Miners in aquatic plants especially, therefore, have considerably modified posterior spiracular processes, the bases of which appear to be thickened and lengthened like crutches while their extensions are broadly forked. With these extensions the puparia remain hanging in the tissue of the plant and the larva and puparium are thus prevented from falling into the water. This can be seen in *Liriomyza virgo* Zett., which mines in very damp places in the green stem cortex of Equisetum and in *Phytomyza cicutae* Hend. in Cicuta on the edges of ponds and lakes (Fig. 55). This connection is emphasised particularly strikingly in *Phytomyza calthivora* Hend. on Caltha palustris. Three species of the genus live on this plant, of which *Ph. calthae* Hering and *Ph. calthophila* Hering make their mines in leaves growing on drier ground. *Ph. calthivora* Hend., however, is only found where the leaves lie on quite wet or marshy ground and as a result of this their posterior spiracular processes are

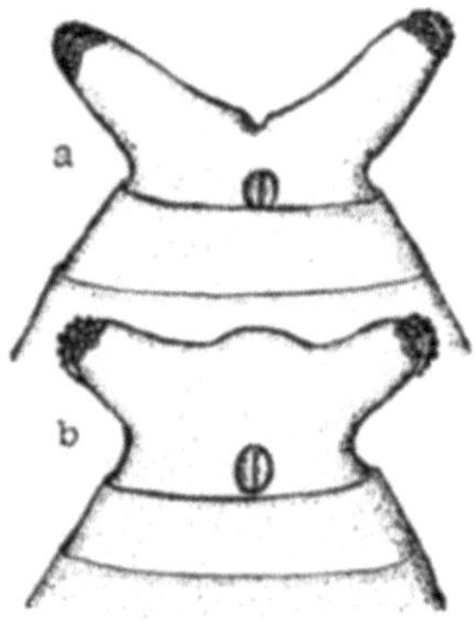

Fig. 55. Extensions of the posterior end of the puparium for anchoring in the plant tissue: a) *Phytomyza cicutae* Hend. (Cicuta), and b) *Liriomyza virgo* Zett. (Equisetum). At the extremity the posterior spiracular processes. Although belonging to different genera, great similarity in structures used for the same function.

developed in the way described. In these species the puparia are affixed so firmly to the plant tissue that they often collapse on any attempt to remove them. It can therefore be assumed that, like certain species of *Dizygomyza*, they are in addition fastened to the tissue by a secretion of glue. Many species of *Hydrellia*, which often live under water, achieve the same result in a different way. The posterior spiracular processes in this genus are extended into two long points (Fig. 34), with which they bore into the stem or leaves of aquatic plants. In this case they not only affix the puparium to the plant but also ensure an adequate supply of air for the respiration of the pupa, by absorbing it with their long spiracles from the air-filled inter-cellular channels, which invariably occur in aquatic plants.

The larva of miners which live in Glumiflorae (Gramineae and Cyperaceae) which has mined throughout its life in the leaf blade is frequently found to move over to the sheath prior to pupation, in order to complete its transformation there. This habit can be observed in several species of *Dizygomyza*. The interesting thing about this is that in the last part of the mine in the sheath of the leaf no further food is consumed and the larva merely clears its way to its pupation site; often feeding already ceases in the lower portion of the leaf blade, where the mine appears to come to an end; no mine channel can be observed against the light. Naturally the pupa enjoys greater security in the sheath. In such cases therefore when collecting or investigating mines particular attention should be given to the sheath of the leaf.

When pupation takes place in the mine, the larva frequently makes provision to facilitate the escape of the subsequent imago from the leaf. This usually takes the form of preparing an escape hole. The larva often constructs a long channel right up to the lower surface of the epidermis, so that as this point the mine encloses the way out only by a thin covering or perhaps merely by the upper wall of the epidermis, which can easily be broken through by the emerging imago. A channel of this kind with a prefabricated exit hole can be seen in *Nepticula weaveri* Stt. in Vaccinium vitis-idaeae L., in *Hydrellia butomi* Hering in the leaves of Butomus, in *Hydromyza livens* Fall. in the leaves of Nuphar. Such holes have not been found in other species of *Hydrellia* which mine leaves growing in water, as for instance in *H. hydrocharitis* Her. in Hydrocharis, and no special channel and prepared exit hole are found in another form of *Hydromyza livens* Fall., which emerges only after hibernation. In both cases there is the same connection: the leaves decay in the water and the emerging fly in any case has no difficulty in gaining access to the open.

The construction of the cocoon takes the same form as in other species of the insect Orders concerned. The Tenthredinidae transform in a cocoon woven together with grains of earth. The same thing occurs in many species of Lepidoptera. When the cocoon lies fairly freely, as on the stem, between fallen leaves, etc., its colour usually tones with its surroundings, making it not easily visible; it is thus usually ochrous to brown or blackish. In cases where the species of *Nepticula* spin a cocoon inside the mine, it is sometimes very brightly coloured, since it cannot be so easily seen in the mine cavity; for instance in *Nepticula agrimoniella* HS. in leaves of Agrimonia it is a beautiful glittering violet. The cocoons of the group to which belongs *N. rubivora* Wocke and which are formed outside the mine are often of a striking dark green colour but do not appear so conspicuous in their natural surroundings on the plant as in the breeding cage. The employment of pellets of frass to cover up the cocoon is a common practice, as may be observed in many of the *Lithocolletis*. The pupa has increased security under such a covering. It is remarkable that these masses of frass webbed up into the cocoon do not develop mould as long as the pupa is inside the cocoon. Presumably the frass pellets are in some way sterilised by the spinning which has taken place. Many species make a particularly robust cocoon even inside the mine, as may be seen very clearly in *Lithocolletis emberizaepennella* Bché. in the leaves of Lonicera. The construction of the ribbed cocoon in the majority of *Bucculatrix* is particularly striking. It has already been mentioned that the young caterpillar of this genus completes each moult under a frail, flat cocoon. Before pupating, however, they spin a boat-shaped, ribbed cocoon; the construction of the longitudinal ribs is only made possible by their spinning first one, then the other half and finally joining the two halves so neatly that frequently no trace of the seam in the ribs of the cocoon can be seen.

Many mining species of Diptera secrete a glue-like substance shortly before they transform into the puparium, with which the puparium is then fastened down inside the mine. It consists of the contents of the malpighian tubes, mixed with the last excrement voided and produces a characteristic black or dark-green mass, which rapidly hardens and often cements the puparium so firmly to the leaf tissue that it cannot be removed without breaking it open. Warm water, however, successfully dissolves this excretion, the chief matter dissolved presumably being the elements deriving from the frass.

The fact that many miners leave their protecting mine and pupate outside without a cocoon should not pass without mention. Many species of the genus *Elachista* which mine Glumiflorae are the principle examples of this habit. Some species of the genus still make a light web in which they pupate, many others, however, are attached to the plant unprotected with only one or more threads around their body, similar to the silken girth found with the pupae of many of the butterflies. Pupation in the open also occurs in many of the mining Agromyzidae, whose puparium is often found naked on some spot on the leaf which had previously been mined, frequently attached to the surface with glue or a few threads.

d. Emergence of the Imago.

In our latitudes mining species generally have one or two generations and only in the Mediterranean area are some species found with more generations in the year. The first larval generation, which develops from eggs laid in the spring is usually very brief; a few days suffice for the development from egg to pupa and the pupal stage is also very short. It is quite different with the succeeding generation. The life of the larva in the mine is now much longer and after spinning up, the larva frequently does not transform, pupating only in the following spring. The slower development of the second larval generation seems to be connected with the condition of the food plant. From early spring until summer the parts of the plant are very soft and full of sap and offer the mining larva both abundant food, only slight resistance as it feeds and also only a small percentage of indigestible food; on the other hand in autumn the plant is stronger and tougher, with in parts a thickening of the cell walls and woody tissue, so that the larva inevitably has to consume a much larger proportion of indigestible elements, thus requiring more time before the requisite amount of food has passed through its digestive tract. Furthermore, the resistance encountered by the larva in the tougher plant parts is much greater. It seems very possible that these differences are also reflected in an altered construction of the mouth parts of the feeding larva. This can only be confirmed by a comparative examination of the mouth parts of the larvae of the same species in different generations, which has unfortunately not so far been undertaken.

In primitive forms the time spent as a pupa in comparison to the time spent by the larva in the cocoon prior to pupation can be seen to be very short. Not infrequently the larvae spend the greater part of the year in their cocoon without transforming and then produce the imago a few days after having pupated. This occurs in the relatively primitive saw-flies and is seen in its extreme form in the Eriocraniidae, which mine in early spring and immediately burrow into the soil and spin up, completing their pupation only the following spring followed shortly after by the emergence of the moth. These, of course, belong to the most primitive forms of mining Lepidoptera. The more an insect is developed, — phylogenetically speaking — the shorter is the time it spends without transforming in its cocoon; a large number of miners for instance, pupate already in autumn, overwinter as pupa and the moth emerges in the following spring. While the primitive forms have only one generation in the year, the more specialised have two or more. Within a genus, closely related species can be found which behave differently in this respect. For example, of the species of the genus *Lithocolletis* living on Prunus, *L. cerasicolella* HS. and *spinicolella* Z. pupate only after overwintering, while *L. padella* Glitz, on the other hand, pupate already in the autumn.

In the species whose larva pupate in the mine, the emergence of the imago may occur inside the mine itself or outside. In the first case the perfect insect has to force its way out of the mine through the epidermis of the leaf. It would thus be necessary to bite through the epidermis, which flies and moths are not able to do, as the mining families no longer have functional mandibles. They are still present in Hymenoptera and Coleoptera and the emerging beetles or saw-flies can bite through the leaf and so get out into the open. It is frequently seen in Coleoptera that the beetle which has emerged remains in the mine for some time until the chitin has hardened up and only then leaves the leaf. Diptera, which have pupated in the mine, have a double task to perform: they have to burst open the puparium and then break through the leaf epidermis. They manage this through the use of their ptilinum, which projects from a fold in the frons which is filled with a liquid, and by forcing in yet more liquid can exert considerable pressure. Prepared lines of thinner chitin can be observed in the puparium and a portion of the anterior end of the puparium bursts away along these lines. On the forward edge of the first abdominal segment a line of this kind encircles the coarctate pupa; with only one line a cap would split off in front. There is, however, a second prepared line which, as distinct from the vertical one already mentioned, runs horizontally, thus causing the cap to break up into a dorsal and ventral part. Accordingly, in some species a ventral, in others a dorsal section splits off (Fig. 43). In some species this is again divided longitudinally into a right and left portion, so that only a small slit remains as the opening through which the fly leaves the puparium. In species of *Hydrellia* a section does not spring off but the puparium merely splits along the prepared horizontal lines; the fly them emerges through the slit thus caused. By pressing with the ptilinum, a gap is then also forced in the leaf epidermis, the mine is opened up and the fly escapes.

The problems of emerging are greater in mining Lepidoptera than in free-living species, as the latter merely have to break through their cocoon, while miners also have to make a hole in the leaf. In view of the frailty of moths which emerge from the mine, it is not surprising that the more difficult task of breaking through the leaf epidermis becomes the prior responsibility of the pupa. There are in fact a large number of species among the mining moths in which the pupa breaks through the epidermis shortly before the emergence of the moth and remains projecting outside the mine. This action is facilitated by certain structures, the most important part being played by the covering of bristles on the sides of the posterior segments. These bristles function in this instance as organs of movement and with their assistance the pupa succeeds by a series of wriggling movements in working its way out of the mine. The arrangement of bristles on the segments of the hind-body is characteristic for practically all Lepidoptera which pupate inside the mine; the subsidiary group of the Tineoidea possessing them is named *Acanthopleona.* There is also frequently a horn-like or cutting extension on the head of the pupa, which facilitates breaking through both web and epidermis; this is strikingly developed in *Phyllocnistis* (Fig. 56). Remarkable variations are found in this "cocoon piercer" in different species; presumably these variations developed in connection with differences in the consistency of the leaves.

It is surprising that in spite of the difficulty to be overcome by the pupa in working its way into the open, the shortest route is not always chosen. For instance, there are species of *Lithocolletis* which make lower surface mines and only eat isolated spots of the parenchy-

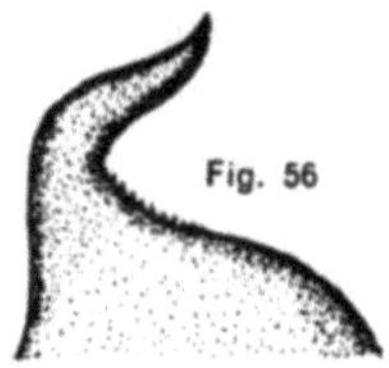

Fig. 56. Populus: Anterior extremity of the pupa of *Phyllocnistis suffusella* Z. with "cocoon piercer", with which the pupa works its way free from silk and mine.

ma adjoining the upper surface of the leaf, so that from the upperside the mine appears mottled. The pupa does not always choose the simplest way by breaking through the lower epidermis but in a whole number of cases it moves through the mottled part of the upper surface and is thereby forced to a far greater exertion of strength. In this respect species of *Lithocolletis* which otherwise have largely identical life histories behave quite differently, with the pupa of one species emerging through the upperside, with another through the underside of the leaf. This point should be considered when identifying mines; the empty pupal case remains for some time projecting through the opening of the mine and can still be found after emergence.

In *Nepticula weaveri* Stt., which forms large upper surface linear-blotch mines in the leaves of Vaccinium vitis-idaea L., emergences does not take place on the upper surface where the mine is found (the larva pupates in the mine) but on the lower surface of the leaf. Neither the moths nor the particularly delicate pupae of *Nepticula* would be able to force their way through the layers of parenchyma down to the underside of the leaf. Therefore, shortly before pupation, the caterpillar constructs a long silk-lined channel which leads right down to an opening in the underside of the leaf. The silk lining of this channel prevents the formation of a callus which may develop with time in many mine channels, and which might have blocked this narrow passage when the moth wished to emerge.

Larvae of Lepidoptera, of most Tenthredinidae and of many Coleoptera which transform inside cocoons in the soil make no special arrangements for the subsequent emergence of the imago. It merely has to bite through the silk and to work its way out of the ground. There are, however, species in which the task of working their way out is assigned to the pupa. The Eriocraniidae, for instance, have a pupa with moveable mandibles, which, however, play no part in providing food but have a quite different function — to facilitate the pupa working its way out with spade-like movements. While beetles often remain for some time after emergence in their cocoon and only seek their way into the open when they have fully hardened up, perfect insects of the other Orders are unable to remain for any length of time in their cocoon, since they must unfold their wings before hardening.

Abnormal duration of pupal quiescence is quite often encountered in miners, the imago does not emerge at the normal time but remains in the pupa for one or more further generations before emerging. The explanation of this lying-over is clear; in this way a number of specimens of the species concerned can be saved in an unfavourable season or an unfavourable year. We know of non-mining insects which may remain for several years in the pupal stage. In miners this lying-over has so far only been observed in species which transform outside the mine. When pupation takes place in the mine cavity, the perfect insects always appear at the correct time. This need cause no surprise; remaining in the pupa inside the leaf would assist the species in no way, since the leaves would later be blown far away, perhaps to places where the new generation would find inadequate means of subsistence (the perfect insects of miners are almost all weak on the wing), the leaf tissues soon decompose and the pupa inside would be exposed to numerous dangers.

Case-bearing miners *(Coleophora, Apterona)* firmly attach their case with silk from their spinneret to some protected spot prior to pupation. The larva then turns round in its narrow case, so that its head is towards the opposite end of the case. In both genera the pupa does not force its way out of the case but remains inside, so that on finding a case it is not possible to tell whether the moth has already emerged or can still be expected. In *Apterona crenulella* Brd. the female which has emerged from the pupa remains inside the case in the parthenogenetic race of Central Europe (*helix* Sieb.) and also lays its eggs there, while the males of the form with normal reproductive habits leave their case and seek out the females which have remained inside theirs. Some species of *Coleophora* leave their case before pupation and pupate in the ground.

Pupation takes place inside the time in very many Lepidoptera, in particular in practically all *Lithocolletis* (*L. helianthemella* HS. pupates outside the mine in a cocoon), in all *Phyllocnistis*, in many *Tischeria* and many others but rarely in *Nepticula*. Of the Diptera many species of *Dizygomyza* and especially of *Phytomyza* transform inside the mine, as well as the majority of *Ophiomyia* and many *Melanagromyza* but never members of *Liriomyza*. Among beetles the Buprestidae *(Trachys)* pupate inside the mine and also many Curculionidae, the former as a pupa loose in the mine, the latter often in a cocoon. Of the Tenthredinidae only members of the genus *Phyllotoma* transform in the mine; species are also found here which cut a piece out of the leaf (either from the upper surface or the whole thickness of the leaf) with which they drop to the ground and pupate inside it.

8.

DURATION OF MINING AND CONNECTION BETWEEN MINE AND SEASONS OF THE YEAR

A comparison of the time required by a mining larva for its development with the length of the larval stage of free-living insects shows that in the majority of cases mining insects have a shorter larval development than external feeders. Although details of extremely short larval stages may be based on inaccurate observation (for instance, it has been reported that *Nepticula malella* Stt. required only 36 hours from the commencement of hatching from the egg until spinning its cocoon), owing to the minute, hair-like early channel being frequently overlooked, it is nevertheless a fact that the larval stage, especially in the first larval generation, is often extraordinarily short. Webster and Parks (1913) recorded that in an American species of *Liriomyza* the duration of the larval stage was four days and that the shortest time of development from oviposition until the emergence of the imago was 18 days. Similarly short periods of development are found in many miners, especially in the genus *Nepticula*.

Two principal reasons may be adduced to explain the shortening of the larval stage by comparison with free-feeding forms. The larva of mining insects which lives inside the mine and feeds exclusively on mesophyll, consumes qualitatively highly nutritous food but apparently quantitatively less than free-living larvae. It is not forced to eat quantities of indigestible food of low nutritive value like the free-feeding larva, which in addition to the green leaf tissue with its high protein content is also compelled to eat the epidermis cells, which are often thickened by a cuticle, as well as many strengthening elements in the leaf, such as are found in the vicinity of the vascular bundles. The mining larva can thus develop with a far smaller quantity of food; as the larval stage is essentially devoted to the consumption of food, it is naturally shortened by its abundant supply. That this is the decisive factor is shown by the fact that even in its living quarters inside the leaf the larva seeks to avoid as far as possible having to consume the more indigestible substances. We know many examples of the larva avoiding the leaf veins, with their numerous strengthening elements (thick-walled cell parts of low nutritive value). When the larva wishes to reach a fresh part of the leaf, it crosses over the leaf veins only where they begin to flatten out (towards the margin) and contain a smaller proportion of non-nutritious tissue. Others follow the edge of the leaf with great precision (*Nepticula marginicolella* Stt. on Ulmus and other species) and only turn into the centre part of the leaf shortly before the end of their larval stage. The same species which normally mine along the edge of the leaf form a coiled linear mine if the egg has been laid towards the centre of the leaf (*Nepticula atricollis* Stt. on Malus and Crataegus); this also enables them to avoid as long as possible having to cross the veins of the leaf. Species of *Nepticula* virtually never cross the mid-rib; whenever this does occur it is a noteworthy exception, which would immediately become conspicuous. *Lyonetia clerkella* L. on Rosaceae and Betula, which has a far longer larval stage than the species of *Nepticula* just mentioned, not infrequently directs its channel across the mid-rib even in its infancy. A parallel process is found in some species of *Nepticula* (the poplar miners related to *Nept. argyropeza* Z.) which first eat out a part of the petiole before continuing into the leaf blade to make their mine. This peculiarity is accompanied by a particularly long larval stage. The same thing is found in *Heliozela*, which to begin with feed in the centre of the twig or in the mid-rib.

A second reason for the shortness of the larval stage of miners is the fact that the larva which is inside a leaf enjoys optimum living conditions. Experiments on the most varying insects have shown that their development is favoured by a higher temperature and a relatively high humidity of the air. In both respects this represents an advantage for the mining larva over its free-living relatives. It is warmer inside the mine than outside, for one reason because the seclusion of the mine cavity prevents the warmth inside escaping and secondly, the lens-shaped epidermis cells covering the mine have the effect of raising the temperature in the mine, especially when, as can so often be seen in the epidermis cells, their outer wall is convexly arched, thus increasing their "magnifying-glass" effect. The most violent storm causes not the slightest movement of air around the larva in the mine, which might lower the temperature through increased evaporation. The larva lives in the mine as in a glass-covered hot-house and just as plants can develop more rapidly with an increased temperature, so is the development of the larva accelerated in the hot-house it has itself constructed inside the leaf. One knows that the decomposition and fermentation of plant substances produce heat; in this connection it is also possible that the frass which remains in the mine may play a part in increasing the temperature of the mine cavity, even though certain observations suggest that in a number of cases the stored frass is made "inactive", as will be shown in the chapter on frass disposal.

The relatively high humidity of the air in the mine cavity is readily understandable. The lack of circulation of the air and the seclusion from outside almost entirely prevent evaporation. On the other hand, much moisture penetrates the cavity, owing to so many of the parenchyma cells conveying sap being consumed or at least cut into. All moisture so released remains inside the mine cavity and gives it its high water content, which is so conducive to the rapid development of the larva. Proof of the abnormally high relative humidity in the mine can be seen in the fact that in many cases special new growths develop inside the mine, which we know can only thrive in especially damp air; parenchyma cells may hypertrophy, or completely new tissues may arise, like the hyperhydric callus cells, which often grow so vigorously that their presence can be detected from outside without a microscope from the deformation of the leaf. We will discuss these growths in greater detail in the chapter on mines and galls. All moisture from the damaged cells, from the body of the larva and finally from the stored frass which evaporates, remains in the mine channel and saturates the air. Oxygen is given off from the parenchyma cells into the mine, and this is of special importance. The effect of oxygen as an agent speeding up development is well-known: all life functions of animal organisms become more rapid with an increased supply of oxygen and so development is accelerated.

We thus see that the larva living inside the leaf develops more rapidly than free-living forms, due to the qualitatively better food and the optimum living conditions in the mine cavity; however, other factors from outside play a part, so that the picture is by no means uniform. For instance, there is a very considerable difference in the various generations of the same species. Normally the larva of the first (spring) generation develops in an appreciably shorter time than the second (autumn) generation. This is a peculiarity which applies not only to miners but also to free-living larvae. Two causes appear responsible for this. Firstly, during the life of the second generation the temperature is considerably lower and as a result all activity of the larva is distinctly more sluggish; feeding is not carried on with the same rapidity as in spring. Secondly, the available food is qualitatively of less value than at the commencement of the plants period of growth, when the cells have a particularly high protein content and have become only very slightly woody. In autumn the larva finds the dietetic value of the cells much diminished; many more undigested elements have to be voided in their frass, so that a more lengthy period of feeding is necessary to complete their development.

This is clearly shown by the larger amount of excrement voided, and lines of frass are therefore broader in the second larval generation than in the first. Thus we find that short-lived forms like species of *Nepticula* develop into long-lived forms in the autumn generation; the period of mining may become fifty times as long as that of the spring generation. The time spent by the larva in the mine becomes especially long if it belongs to species which continue to feed throughout the winter, as far as the temperature makes this possible. We know a number of species which as soon as the temperature has risen slightly above freezing-point can be found feeding in winter, such as *Phytomyza ranunculi* Schrk. in Ranunculus, *Ph. ilicis* Curt. in Ilex, *Ph. hellebori* Kltb. in Helleborus, *Delia brunnescens* Zett. in Caryophyllaceae and others among Diptera, species of *Elachista* on Luzula, *Mompha propinquella* Stt. on Epilobium, species mining pine-needles and others among Lepidoptera.

However, not all differences between miners of the first and second generation can be ascribed to the change in season and the less favourable nature of the food; a largely unsolved problem, arising from the general difference in the two generations of the same species must also be considered. It has been found that despite the less favourable living conditions for the autumn generation of larvae, the imagines these produce in the genus *Nepticula* are normally larger than those which emerged in the first generation. Phylogenetic influences can clearly be detected in this connection. Single-brooded species must in any case be considered as more primitive than species with several generations. This view is supported by the fact that — among the miners, which particularly interest us here — the obviously more primitive species such as the majority of Tenthredinidae and some genera of Curculionidae still today have only a single generation in the course of a year. In the more specialised species, two generations have subsequently evolved and it is certainly not mistaken to assume that the generation of spring larvae should be considered as the derived form. Support is lent to this view by the extreme reduction in the time of larval development, by the smaller size of the body despite the better quality of the food (under identical living conditions primitive forms of insects are normally larger than specialised forms) and finally by the gall-like reactions of the plant in the spring generation of such mines. However, it would be false to conclude that species which mine in early spring should be considered as more specialised, even when they mine in only one generation. Having only a single brood may definitely be considered as a sign of primitiveness. For instance, the Eriocraniidae among Lepidoptera and the Cordyluridae among Diptera which mine in early spring are without doubt very primitive forms; parallel with this is the fact

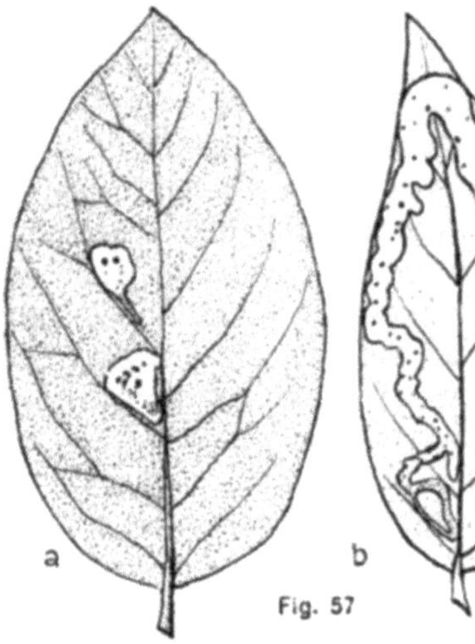

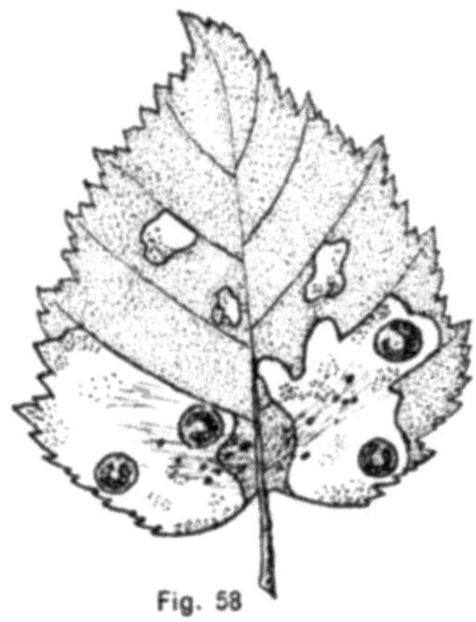

Fig. 57. Lonicera tatarica: Linear mine of *Phytagromyza lonicerae* R.D. a) Short mines in the thick, dark-green leaf. b) Long mine in the delicate, light-green leaf.

Fig. 58. Betula: *Atemelia torquatella* Z. Full depth blotch mines, containing hibernation cocoons, and outside thin webs with grains of frass.

that although their food plant provides every opportunity, they do not produce a second generation.

Finally, the nature of the food plant also plays a certain part. The length of the mine as well as the time spent on mining may be largely dependant on this. This can be well illustrated by species of plants, having both thin and thick leaves. If the mines of *Phytagromyza lonicerae* R.D., which is very common in early spring, are compared when they are in thin and when they are in thick leaves of Lonicera frequently side by side on the same bush, one sees that the mines in the thin leaves are often many times longer than those in the thicker leaves (Fig. 57). This can be seen even more clearly when different species of the same plant genus are involved, which have considerable variation in the consistency of their leaves. A. Braun (1917), for instance, observed that *Nepticula ostryaefoliella* Clem. produced a mine in the very thin leaves of Betula lenta which was twice as long as in the thicker leaves of other species of Betula, in which the larva can also live. The consistency of the leaf has an even greater influence on oligophagous species which attack widely varying genera of plants, as is clearly shown by the variation in the mines of *Phytomyza sonchi* R.D. in the various genera of Compositae.

A special influence on the time spent in mining and on the length of the mine is exerted by the choice of tissue eaten by the larvae. Species which live exclusively in the epidermis cells as sap-feeders, like the genus *Phyllocnistis*, make a very long mine in relation to the ultimate size of the imago. The explanation is that the epidermis cells provide only a small amount of food and, as the largely absent or almost liquid excrement shows, it is of relatively low nutritive content, consisting for the most part of cell sap. Differences can even be noted within the two types of parenchyma. It is generally found that in species which may mine either side of the leaf, the channels of the same species of miner in the same food plant are much larger when they are in the underside of the leaf than when they are in the upperside. One only has to investigate the behaviour in this respect of *Phytomyza atricornis* Mg. which is abundant on Sonchus to see clearly the variation in length according to which side of the leaf is mined. This is presumably explained by the fact that there is much more strengthening tissue in the (lower surface) spongy parenchyma than in the (upper surface) assimilation parenchyma. No mining insect larva is able to digest this strengthening tissue; they are nourished primarily by the protein content of the cells and this explains why they find the lower surface parenchyma less nutritious than that of the upper surface. This can be seen in a particularly striking form in the previously mentioned species group of *Phytomyza sonchi* R.D. on Compositae. The larva first makes a lower surface and only later an upper surface mine. Although the almost adult larva needs far more food than when it was young, the part of the mine in the upper surface is always much shorter than that in the lower surface, which shows that the upperside assimilation parenchyma provides the larva with much more food than the underside spongy parenchyma. This accounts for the fact that the side of the leaf mined is usually constant for each species; a few species, such as the polyphagous *Phytomyza atricornis* Mg. or *Phytagromyza populi* Kalt. on Populus form an exception in this respect, being found indiscriminately on either the upper or underside of the leaf.

The great majority of all miners have two generations, of which the first appears in spring or early summer, the second rather late in the autumn (or even throughout the winter). There is normally no mining activity during the driest and warmest months; the excessive heat during this time results in the premature drying-up and withering of the infested leaves, which might impede the normal development of mining larvae. It is thus rare to find inhabited mines between about the middle of July and the end of August, apart from certain exceptions which have specifically adapted themselves to this period. S t a r ý (1930) observed in *Anybia epilobiella* Roem. on

Epilobium that on hot days the mining larva is able to a certain extent to regulate its body temperature. On hot days the larvae leave the mine, remain outside the cavity and only mine with their forebody in a similar way to the case-bearing genera *Coleophora* and *Apterona* and as is the normal rule with some of the *Bucculatrix*. Not until September are conditions again suitable for miners to complete or at least to begin a second generation during the autumn and this development may be continued in some species throughout the winter and only be completed by pupation with the arrival of spring.

Single-brooded species may occur at the most varied times of the year. Some are found in the earliest days of spring, when the leaves have not yet fully unfolded from their buds. In these species immediately the egg is laid, an injury occurs to the leaf, resulting in a distortion in the growth of the leaf at the point of oviposition. Such distortion occurs all the more readily if the mid-rib, the bearer of food for the leaf, is also injured, as happens with many species of *Rhynchaenus* (*Rh. quercus* L. and *fagi* L.). In such a case the influence of oviposition and mine in the mid-rib invariably results in a deformity to the tip of the leaf. If the leaf was sufficiently young at the time of oviposition the uniform development of the leaf is also prejudiced in species which only mine the margin of the leaf; this can be clearly observed in the mines of *Eriocrania* on Betula, which occur in the earliest days of spring. Such growth disturbances are usually characteristic of the species and will therefore be dealt with in greater detail in a later chapter on account of their importance as a means of identification. Later on in the year others of these single-brooded species appear in accordance with their own particular habits. Later still in autumn more numerous species begin to appear, of which a part complete their larval development the same autumn, while the remainder, after a period of quiescence in the cold season, begin feeding again in early spring and pupate very soon after. Many species of *Nepticula* only appear in an autumn generation, such as *N. rubivora* Wocke on Rubus and the species group around *N. subbimaculella* Haw. on Quercus: the same thing is found among Coleoptera in many species of *Rhynchaenus*, such as *Rh. populi* F. on Populus and Salix, and among Diptera in *Delia brunnescens* Zett. on Caryophyllaceae. Among the Tenthredinidae and Agromyzidae so far not a single species is known which appears to be limited to the months of autumn. The species which occur so late in the year like those just mentioned are frequently accompanied by a chlorophyll conservation in the "green islands" (which will be discussed in detail in a subsequent chapter), which makes it possible for them to continue their mining even when the leaf has already fallen or assumed its autumnal discoloration; the vicinity of the mine is not affected by this discoloration but remains green and capable of assimilation.

The majority of species leave the leaf in late autumn on completion of feeding and pupate. Others transform in the mine, like almost all species of *Lithocolletis*, where the most closely related species may pupate either before or after hibernation. In the species which continue their feeding in the following spring, there is frequently a very distinct difference in the type of feeding and the location of the mine between the two seasons. Case-bearing species of the genus *Coleophora* frequently construct a quite different larval habitation in spring from the one they used in autumn. In autumn, *Nymphula stagnata* Don. makes a thread-like, full-depth mine in Sparganium, but in spring lives in a quite different way in a case on the same plant. The habits of *Atemelia torquatella* Z., which makes large blotch mines in the leaves of Betula (and Ulmus), are quite striking; it can easily be recognised by the web it spins outside the mine, which catches many of the grains of frass. Usually several larvae live together in such mines. In late autumn each one of them makes a circular, flat cocoon inside the mine, which serves solely as a habitation in which to overwinter (Fig. 58). The following spring the caterpillar leaves this overwintering cocoon and the mine; now, without mining any further, it spins an elongated, boat-shaped cocoon, in which pupation takes place. Many species, even before they have completed their life cycle, retire during the winter to the shoot of the food-plant and then leave this in the first days of spring to penetrate the new, highly nutritious young leaves. This is found particularly in cases where the food-plant has a shoot with a rosette formation or if the stem seems particularly tender with very little woody tissue. The larva may then re-commence mining in the earliest spring; *Phytomyza hieracii* Hendel is, for instance, found in February in the youngest leaves of Hieracium pilosella, which it has entered from the shoot, and even in the depth of winter if there is a mild spell of weather *Mompha propinquella* Stt. may resume its mining in the rosette-like leaves of Epilobium hirsutum.

Miners of the genus *Coleophora* which are least impeded in their freedom of movement often wander a considerable distance from their mines, in order to find a suitable place to hibernate. Apparently they are very particular in this respect; for instance, the larval cases of a certain species may be found in great numbers at some place, without there being any obvious reason to explain why it is so favoured. The slightest, scarcely perceptible differences in the amount of sun, dampness and location in respect of air currents may be decisive in influencing the choice of such overwintering areas. The sensitivity of the hibernating species of this genus is known to every breeder; it is common knowledge that even species which do no further mining in spring wander around apparently restlessly for a while, during which period a certain amount of sun is essential before they can pupate; without sun they merely continue walking until they finally die of exhaustion.

It is readily understandable that there are scarcely any insect larvae among the miners which have a life cycle extending over several years. There is normally no continuity of food, the substance of the leaf in spring and in autumn varies too widely, and the withering of the majority of food-plants in autumn have all militated against the development of a mining cycle spreading over several years. Certain species of *Coleophora*, which have a two-year life cycle represent an exception. We may mention as the best-known example *C. hemerobiella* Sc. which lives on every type of Rosaceae, especially fruit-trees, and whose first-year larval case is strikingly different from the case formed after the second hibernation.

9.

DISPOSAL OF FRASS IN MINES

Only a small part of the matter consumed by the mining larva while feeding is made use of to build up its body; the assimilated matter consists largely of protein products and while the remainder may undergo a change as a result of digestive action, it is not assimilated by the body. The chlorophyll from the plant is transformed into the dark green chlorophyllan while passing through the alimentary canal. The starch which comprises such an important part of plant cells is not attacked by the mining larva, according to the investigations of Haberlandt. Naturally enough, the tougher parts of the larva's diet, such as the walls of the cell with their high cellulose content and the thickened strengthening elements, (sclerenchyma fibres and cells), and also conducting vessels with their frequently thickened walls are excreted undigested. Since the larva eats almost uninterruptedly — apart from feeding pauses while changing skins — there is also a continuous excretion of frass, which contains undigested elements and matter changed by digestive processes.

The frass-line is of great importance in mining insects for identifying the species and demands special attention and often a very thorough investigation; it thus deserves more consideration than in free-living species. The manner of frass disposal in these is unproblematical. The free-living larva simply lets its pellets of frass drop, and pays them no further attention. Caterpillars which live in a web frequently cast the pellets away, often making use of a special structure on the abdomen, a sort of "pitch-fork", with which the grains of frass can be hurled out of their living quarters. If the frass pellets of a free-living larva are compared with those of an endophagous species, no apparent differences can be observed without a chemical analysis; both contain the same substances, except that in the pellets of the free-living larva there are elements which are naturally absent from those of the mining larva, such as epidermis cells and their hairy appendages; furthermore the pellets of the mining larva appear to be of a finer texture, owing to the differing nature of the mouth parts.

When considering this question, it is important not to forget that miners inevitably have to adopt a fundamentally different approach to their undigested substances from external feeders. When the larva has deposited the pellet of frass in the mine, the two are thereafter compelled to share the same habitation. We know that a large degree of decomposition takes place in the faeces; bacteria and other microorganisms represent a flourishing intestinal flora. It has already been pointed out that there is no question of the larva developing under sterilised conditions, however completely the mine channel may be isolated from outside influences. Processes of decomposition and fermentation of the type described represent a double danger for the mining larva. First, the larva may suffer a direct infection, as it comes constantly into contact with the piles of frass (especially in blotch mines, less so in linear mines). Secondly, the gases produced by the decomposition may contaminate the air in the mine cavity, which could represent a danger for the larva. Finally, there is the possibility of the tissue of the plant which are to provide the larva with food, also being affected by the decomposition, in the same way as a good apple is affected by a rotting one beside it on the tree or in the store-cupboard. While all these factors are of no significance for the larva which feeds externally, the mining larva is compelled to take precautions to obviate these dangers.

The larva has two possible ways of counter-acting the dangers arising from the decomposition of frass. It can either remove the frass from the mine or it must ensure that its condition is such as to cause it no danger. Removal of the frass is a safe method, which, however, is only applied by a single Order of insects — the caterpillars of Lepidoptera. It is surprising that this process has not found far wider adoption and there seems to be no good explanation, why, for example, the Tenthredinidae, which in every respect are so similar to lepidopterous larvae, do not remove their frass in the same way. The species which throw their frass right out of the mine are behaving in essentially the same way as external feeders: they subsequently have no further contact with their faeces.

Removal of frass from the mine in the species under consideration is achieved very simply. The rear segments of the caterpillar are protruded through an opening in the mine and the pellet of frass is ejected, whereupon the caterpillar returns inside the mine. A prerequisite for this method is that a hole is made for the ejection of the excrement. Rarely there are several such openings but usually the caterpillar always uses the same place, which was made specifically for this purpose. The habits of *Cosmopteryx orichalcea* Dru., which mines in the leaves of Hierochloa and Phragmites, represent a special exception. E. Hofmann (after Sorhagen, 1886) noticed that the caterpillar of this insect proceeds right outside its mine in order to void its excrement. In all species of *Cosmopteryx* the frass is ejected from the mine but in the other species this is not done in such a ceremonious way. When the caterpillar voids its faeces in the open, they simply drop down. In the breeding-cage, however, there is usually a greater humidity of the atmosphere, which affects the consistency of the frass, making it more damp and adhesive. It usually therefore collects around the ejection hole and this opening in the mine, which is normally difficult to see, can be found quite easily. This method of removing frass is not characteristic of special genera; for instance, all species of *Nepticula* leave their frass in the mine except one, *Nepticula subbimaculella* Haw. on Quercus, which makes a slit on the underside of the leaf through which some of the grains of frass are ejected. A whole group of closely related species make no such slit in the lower epidermis, species which are so closely related to the above-mentioned, that they were systematically for centuries past united with it, until E. G. R. Waters drew attention to this point of difference.

Caterpillars which spin leaves together and then mine them usually leave no frass inside the mine. They are in any case rather active caterpillars which frequently leave the leaf they are mining and in general do not make particularly large mines. Considerable differences again occur in closely related species: of the species of *Phthorimaea* which mine on Lycium, one leaves its frass in the mines, while the others do not. The frass is then often found entangled in the web which is spun between the mined leaves. This habit indicates a remarkable and not readily understandable peculiarity of many mining larva, of which *Atemelia torquatella* Stt. on Betula (and Ulmus) and *Bedellia somnulentella* Z. on Convolvulus deserve special mention. These species occasionally change their mines but they do not spin the leaves together. On the underside of the leaf beneath the mined areas, however, they attach a number of threads of silk (Fig. 58) and a certain amount of the frass which they eject from their mine remains caught up in this web. There seems to be no explanation for this habit, unless perhaps one assumes that they wish to make identification easy for the minologist; these species, whose mine is very similar to others, can indeed be easily recognised by this peculiarity. The pest of the olive tree *Prays oleellus* Fabr. which varies so widely in its different generations adopts a similar procedure in its blotch mines. One would think that if the caterpillar takes the trouble to eject the frass, it would wish it to be removed as far as possible from the mine; it is therefore all the more strange that silk is used (and for no other purpose) to retain the frass in its vicinity.

The removal of excrement from the mine involves a considerable expenditure of time and energy for the mining larva, in particular when, as in so many species (almost all Diptera), the legs are so rudimentary that the power of movement is strictly limited. The fact that this removal of frass is encountered only in a certain limited number of cases is very probably linked up with the degeneration of the organs of movement. Extensive atrophy of the prolegs is found in the genus *Coleophora* and precisely in this case the larger larvae never store their frass in the mine; *Coleophora* mines can be recognised by the complete absence of frass, together with the circular hole in the epidermis, the point of entry of the larva. The reason in this instance is that normally the larva penetrates inside the leaf only with its forebody while the hind-body remains in the case, through the end of which the frass is ejected. The same thing, however, occurs when the caterpillar penetrates further inside the leaf, leaving its case altogether. When excrement is to be voided, the larva returns to its case. This example is all the more interesting since a change of instinct occurs during the course of the larval stage: the frass is left in the primary mine, from which the infant larva cuts out its first case. The same thing is found in other species, which cut a case out of the leaf tissue, such as *Heliozela*, *Antispila*, *Phylloporia*, *Incurvaria* and others; the frass is never initially removed from the mine, although many of these species unlike *Coleophora* have no reduced abdominal legs.

The majority of all miners dispose of their frass inside the mine cavity. The way in which this is done varies enormously and in many cases makes possible the identification of the mine-producer. Sap-feeders occupy a special position in this respect. Naturally, in their case only very small quantities of undigested matter are excreted. Since they live inside the epidermis cells and consume largely only their liquid content, it is readily understandable that their remnants of frass are not well defined. They have to bite through the vertical walls of these cells as they drive their way through the leaf, which might lead one to suppose that solid parts would form a certain proportion of their excrement. The absence of any regular frass-line suggests, therefore, that these transverse walls are bitten through but are not swallowed. In some species an extremely fine frass-line does occur, consisting presumably of the dried up remains of the liquid frass. Closely related species show variations even in this, for instance in *Phyllocnistis labyrinthella* Bjerk. on Populus tremula there is a well-defined narrow frass-line in the mine channel; however, when as an exception, *Ph. suffusella* Z. feeds on this species (it normally mines Populus nigra), there is no visible frass-line in the same way as when it mines its normal food plant. In other ways certain food remnants can be detected in sap-feeders; if the mines of *Lithocolletis* are opened up and the smallest, discarded head capsules are examined, in many of them there will be found a single grain of frass, which was clearly excreted by the larva at the end of the sap-feeding stage.

In all tissue feeders the quantity of undigested and excreted matter is appreciably greater. We previously made mention of the composition of the excrement of mining larvae. What we call the frass-line consists not only of matter which has passed through the intestine of the larva but also parts of the destroyed plant cells which were not consumed at all and which under certain circumstances can influence the coloration of the frass-line. The transformation of chlorophyll into chlorophyllan results in the frass-line being at least a darker green than the surrounding parts of the leaf and thus causes the mine to stand out more strongly from the leaf. In the vast majority of cases this dark green coloration becomes almost pure black and in some other cases the green of the frass merges into red and reddish-brown, since the chlorophyll is not a pure substance but contains an admixture of red carotene; according to the specific type of digestive process either the red or the green components in the frass may pre-dominate. These differences also often prove useful for identifying species.

In numerous cases the mining larva appears to take no precautions against the dangers arising from the decomposition of its frass. It leaves its frass in the earlier parts of the mine just as it was excreted and no disadvantages appear to ensue from such action. This applies especially to linear mines. Presumably the narrowness of a slender mine channel prevents the spread of any products of decomposition which may arise. The frass is thus deposited continuously throughout the whole length of the mine channel. On the other hand in other cases there is a clear attempt by the larva to produce as finely divided grains of frass as possible, resulting in better ventilation and more rapid drying of the excrement. At this point it is necessary to emphasise again the essential difference between dipterous and lepidopterous larvae in the disposal of frass in linear mines. The lepidopterous caterpillar eats lying on its back or on its belly, in other words dorsally or ventrally in relation to the epidermis. The anal orifice is then situated in the centre of the channel and accordingly the frass-line is also found in the centre of the channel (Fig. 59). The dipterous larva behaves quite differently. The structure of its mouth parts demands that it lies on its side while feeding, with alternately either its left or its right side towards the epidermis and it usually lies obliquely across the channel. The grains of frass are therefore deposited on the side of the channel, now on one side, now on the other, according to its change of position. The frass is therefore found in two more or less distinct rows at the sides of the channel. Further important distinctions between closely related species occur according to whether the excrement is deposited in small grains or in more continuous threads. In the stems of Campanula, for instance, linear

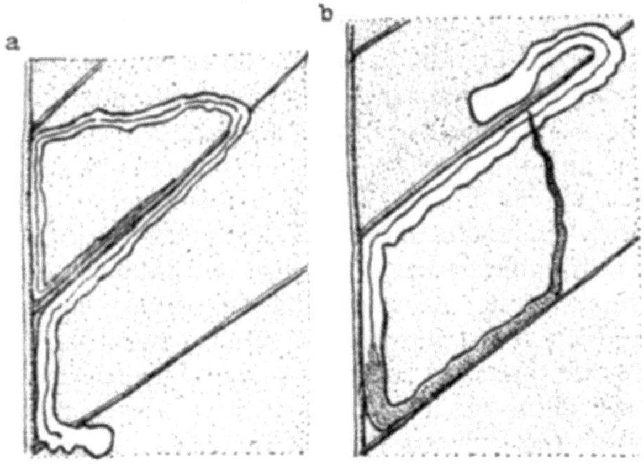

Fig. 59

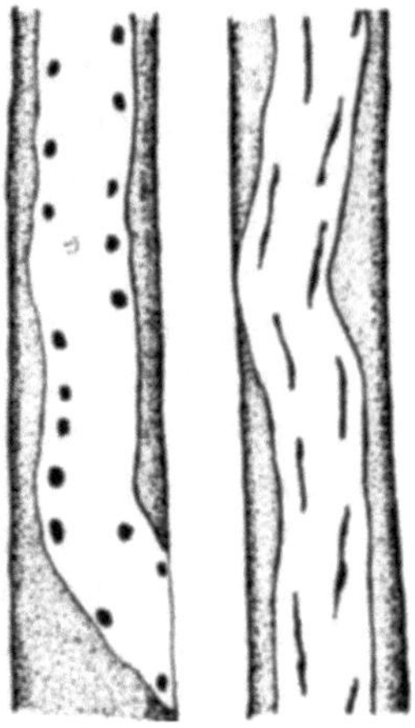

Fig. 60

Fig. 59. Betula: mine channels of *Nepticula*. a. *N. confusella* Wood, frass throughout as a thin, central line. b. *N. lapponica* Wocke, to begin with the frass fills the whole width of the mine.

Fig. 60. Campanula: stem mines of *Ophiomyia*. Left, *O. campanularum* Starý, frass in isolated grains. Right, *O. heringi* Starý, frass in threads.

mines of two different species of *Ophiomyia* are found, of which *O. campanularum* Starý deposits its frass in distinct, well-separated grains, while *O. heringi* Starý forms distinct, thin threads (Fig. 60). Frass is deposited as a central line in dipterous linear mines only as a rare exception, as for example in *Phytomyza facialis* Kalt. on Bupleurum (Fig. 61). In general, grains of frass are found in the genus *Phytomyza* and drawn-out threads frequently in *Liriomyza*, without however this being a specific characteristic of the two genera. Arrangement of frass in a double line occurs only rarely in Lepidoptera and is formed in quite a different way. This applies without exception to species which do not, as is the normal rule, leave the mined area behind then for good and all but constantly return to it during pauses in feeding. This is done, for instance, by species of *Nepticula*, which commence their mine inside the petiole or the leaf mid-rib, such as *N. argyropeza* Z. and its related group on Populus and *N. intimella* Z. on Salix. Each time they crawl back into the petiole or mid-rib the frass is pushed sideways by their body, thus forming two walls of frass which somewhat resemble the arrangement found in Diptera.

Two tendencies can be observed in the way the majority of miners arrange their excrement: the grains of frass are maintained as small as possible and they are deposited as far away from each other as possible. This ensures adequate ventilation and thus the rapid drying of the frass, and the processes of decomposition are strictly limited. In linear mines these two tendencies can be conformed to in two ways. In one case the very small grains of frass are distributed over the whole width of the mine channel, not infrequently in semi-circular arcs each separated from the other; they may fill up the whole or the greater part of the channel but in such a way that air can circulate between the individual grains. In another group of species a more or less compact, continuous central line is formed and drying is ensured by the free circulation of air along both its sides.

The type of arrangement is quite distinctive of the species and this provides a useful guide for distinguishing species with otherwise very similar mines. In certain cases the mine is completely filled up by the mass of fine-grained frass, leaving no narrow, more lightly-coloured edges. In such cases the mine is scarcely distinguishable from the surrounding parts of the leaf (Fig. 62), as occurs in *Nepticula continuella* Stt. on Betula, in *N. aceris* Frey on Acer and especially in *N. basiguttella* Hein. on Quercus. The larva deposits its frass in a

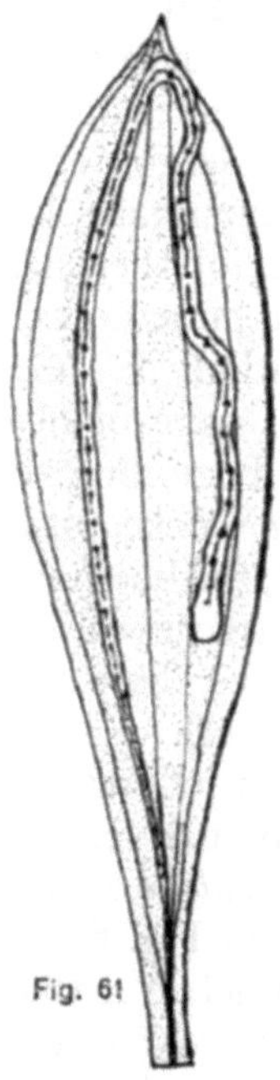

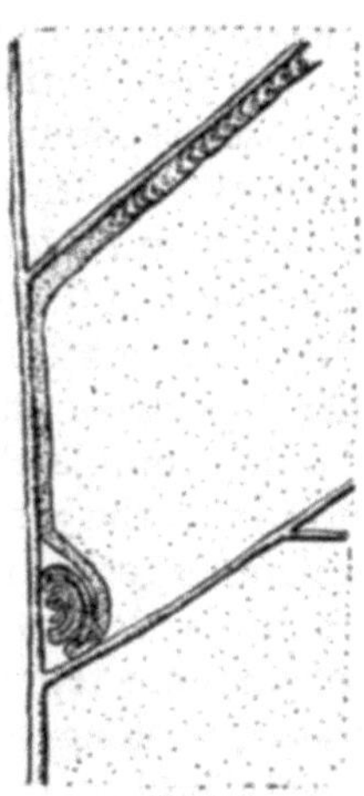

Fig. 61. Bupleurum: *Phytomyza facialis* Kalt., frass in a central line.

Fig. 62. Betula: Channel of *Nepticula continuella* Stt., with the channel filled by frass.

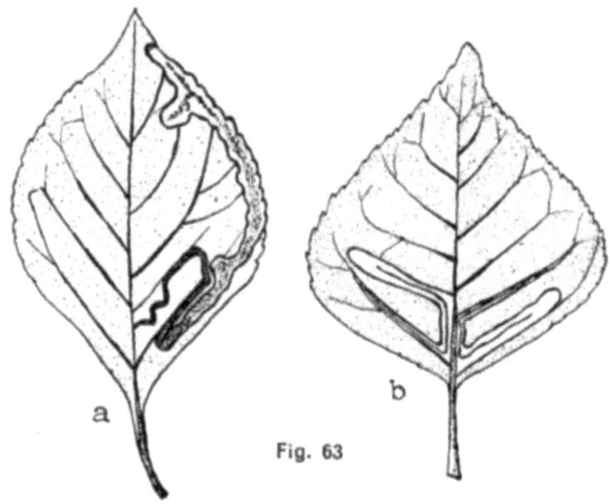

Fig. 63. Populus: *Nepticula trimaculella* Haw. a) Frass in a loose line. b) Frass in a thin central line. *(N. subtrimaculella* Dufr.*)*

central line only in the last part of the channel and only then does the mine become more visible. For this reason it is usually difficult to find inhabited mines in these species; the larva spends only a very short part of its life in the readily visible terminal portion of the channel. The arrangement of frass may undergo a change even earlier in the larva's life and it is necessary to investigate closely the type of frass arrangement in the very beginning of the channel. *Nepticula lusatica* Schütze has been described as a mixed species, because insufficient attention was paid to this variation; E. G. R. Waters showed that in fact two separate species were involved of which *N. confusella* Wood deposits its frass from beginning to end in a thin, black, central line, while in *N. lapponica* Wocke, which has a similar line over the greater length of its channel, the beginning is completely or very largely filled by light-green frass deposited in fine grains (Fig. 59). In other aspects of their mines these two species, both feeding on Betula, are completely identical.

A lack of uniformity in the arrangement of frass, without it being an indication of separate species, is found only very rarely. The best-known example of this is *Nepticula trimaculella* Haw. on Populus, of which the two extremes are represented in Fig. 63. In the more frequent case the frass lies along the channel in a thick, central line, which may be largely broken up into individual grains (a), or else there may be a hair-like central line, with the channel more distinctly adjoining the leaf veins (b). This difference induced Sorhagen to name this form, which could only be distinguished by its mine, as *N. populicola* Sorh. (by Dufrane, after breeding the imago, named as *N. subtrimaculella* Dufr.). In addition, in the one case the mine is more usually on the upper surface, while in the other it is generally on the lower surface. A similar lack of uniformity occurs in *N. tiliae* Frey on Tilia, where two species were suspected on the basis of the differing frass arrangements, although this has not been confirmed by breeding. These variations in the same species are such rare exceptions that they in no way prejudice the value of frass arrangement as a means of identifying species.

In certain cases it is found that the distribution of the grains of frass occurs with extremely wide intermediate spacing; the individual grains are then naturally not so small but appear distinctly enlarged. This is very striking in *Ophiomyia maura* Mg. which probably has the longest linear leaf mine in our geographical area (stem

Fig. 64

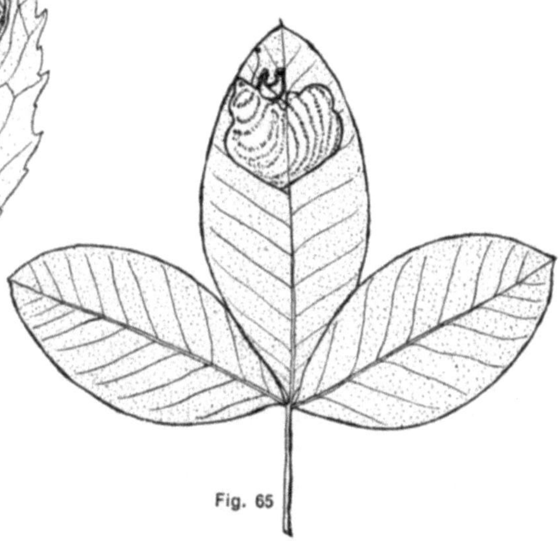

Fig. 65

Fig. 64. Solidago: Linear mine of *Ophiomyia maura* Mg. The longest linear leaf mine has reached 30 cm. Pellets of frass deposited at considerable intervals.

Fig. 65. Laburnum: *Leucoptera laburnella* Stt. Fine-grained arrangement of frass in a blotch mine.

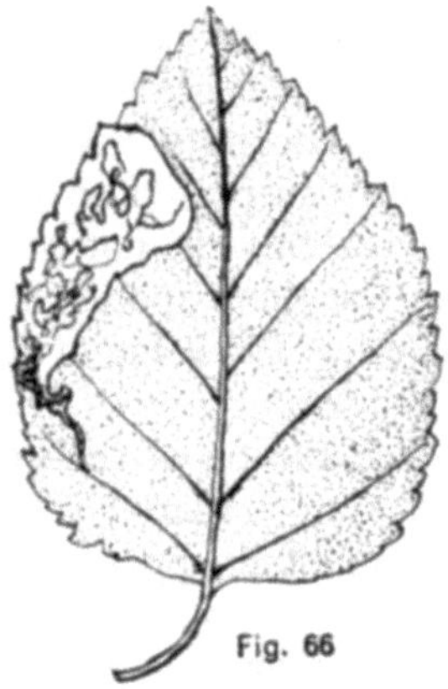

Fig. 66

Fig. 67

Fig. 66. Betula: Blotch mine of *Eriocrania.* Arrangement of frass in threads.

Fig. 67. Solidago: *Pegomyia nigrisquama* Stein. Central arrangement of frass.

mines are often much longer). In this case the frass is deposited in fairly large lumps, which are found at considerable distances from each other (Fig. 64). The extreme in this respect is represented by *Phytagromyza similis* Brischke on Dipsacaceae, where the mine appears almost free of frass despite its considerable length and width, and only at the last moment shortly before pupation is a substantial quantity of frass excreted.

The living space of the mining larva is much more airy in the blotch mine (and even more so in the blister and tentiform mine) than in the linear mine and it would seem that here the larva requires to pay less attention to drying its excrement. There are, however, species which arrange their pellets with great regularity throughout the mine cavity, especially the genus *Leucoptera* (Fig. 65). In other species they are deposited in the mine quite haphazardly, without any spots appearing to be more favoured than others. In a whole number of cases the pellets are piled together in a corner of the mine, since once they are dry they easily roll around and gather in the deepest part of the mine. This is found especially distinctly in many mines in grass blades, as in species of *Elachista* and *Dizygomyza,* but also in many tentiform mines of *Lithocolletis* in a wide variety of plants. This characteristic, for instance, distinguishes the large blotch mines of Tenthredinidae in Betula from those of Eriocraniidae in the same plant; the latter deposit their frass in very characteristic long threads, which always enables their mines to be easily recognised (Fig. 66).

In many species the larva proceeds regularly to the centre of the mine to deposit its excrement. Thus a sizeable patch of frass is formed in the centre of the mine — or in large mines there may be several such patches. By this means the frass is prevented from influencing the tissue on which the larva is actually feeding. Such mines with a central pile of frass are found very frequently in the Anthomyidae, for instance in *Pegomyia nigrisquama* Stein in Solidago (Fig. 67) or *P. genupuncta* Stein on Arctium. Application of the same principle can also be found in Lepidoptera; the best-known example is *Microsetia sexguttella* Thbg. (*stipella* Hb.) in the leaves of Chenopodium and Atriplex (Fig. 68), whose glossily transparent mine has the majority of the excrement in the centre of the blotch. *M. hermannella* Fbr., which is closely related to it, has a quite different type of mine and frass arrangement; it lives in a coiled linear mine, in which the frass is distributed in fine grains. Sometimes, there is the beginning of a localisation of the frass at the commencement of the mine, which is

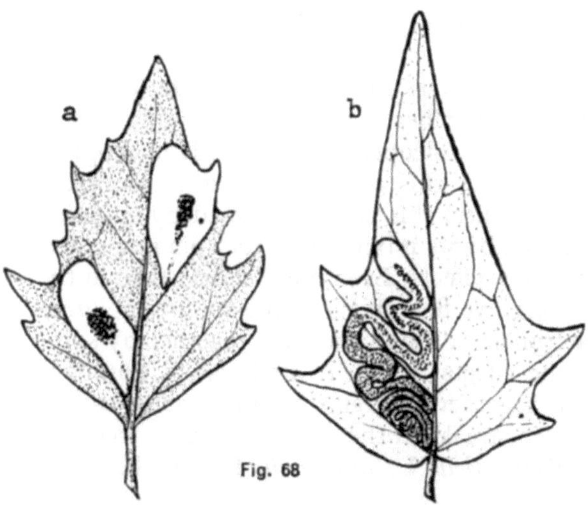

Fig. 68

Fig. 69

Fig. 68. Chenopodium: a) *Microsetia sexguttella* Thbg. (*stipella* Hb.). Central arrangement of frass. b) *M. hermannella* Fbr. Frass in fine grains which fill the channel.

Fig. 69. Betula: *Scolioneura nana* Klug with a pile of frass at the beginning of the mine.

subsequently characterised by a distinct wedge of frass, as for instance in the mines of *Scolioneura nana* Klug on Betula (Fig. 69). Some species again make short branches leading out from the blotch mine, which are employed like cess-pits and in which the main part of the frass is deposited. This is found especially with coleopterous larvae of the genus *Dibolia* and most frequently in *Scaptomyzella flava* Fall. on a wide variety of Cruciferae. Such frass tunnels keep the main cavity of the mine free and are made in places where the larva is in any case no longer feeding.

There are, however, species which do not seek to avoid the dumps of frass but on the contrary constantly revisit them. *Nepticula argentipedella* Z. in the leaves of Betula may be quoted as the best-known example of this. In this species the major portion of the excrement is deposited in the dark-coloured early part of the mine and the larva also returns here to rest, where it can be seen with difficulty and as a result obviously enjoys a certain protection from some of its enemies. Being in the immediate vicinity of its excrement seems to harm it in no way. A similar habit has been observed in many coleopterous mines, as for instance in species of *Rhynchaenus* on Ulmus, in *Apion flavofemoratum* Herbst on Genista and Cytisus, and others. Pupation usually takes place in the same spot. In these species it must be assumed either that a chemical change has occurred in the frass pellets rendering them less noxious to the larvae or the larvae have developed some dregree of immunity towards the harmful effects of the decomposing excrement.

The subsequent utilisation of frass for the construction of a rest area and ultimate pupation chamber is very widespread; this is found in many Lepidoptera and Coleoptera but not in Hymenoptera and Diptera. In many instances this resting-place is impregnated with frass in one way or another; this is accompanied by spinning and leads one to suppose that by its mixing with the spinning secretion the frass loses its noxious effects. Even more clearly than in the above-mentioned Coleoptera, we find this in some species of *Lithocolletis*, in which the caterpillar binds all the frass together by spinning, using it to cover the cocoon which is made later. Throughout its life the larva spins together its grains of grass, which can be seen against the light as two streaks in the mine. The most frequent example of this is *Lithocolletis rajella* L. (*alnifoliella* Dp.) in the leaves of Alnus.

The width of the frass-line is not always identical in the same species but may change as a result of various circumstances. If a species lives on various different plants, the width of the frass-line will vary according to the nature of the different food-plants. Variations may even occur on the same plant according to the state of development

Fig. 70

Fig. 70. Betula: *Nepticula argentipedella* Z. Predominately a central arrangement of frass.

of the infested leaves. In general, in the first generation the frass-line is more delicate, since less indigestible matter passes through the intestines; the larva's food is thus more nutritious than in the second generation, in which the larva finds far more woody and thickened elements mixed in with its food. A difference of this type can often be observed in the two generations in species of *Nepticula* and has also been found in *Pegomyia betae* Curtis. The differing width of the frass-line is particularly striking in species living primarily in the petiole or mid-rib, where the larva is forced to consume especially large quantities of indigestible substances. A cross-section of the mid-rib of an Alnus leaf containing a mine channel of *Heliozela resplendella* Stt. shows only slight traces of frass in the first generation; in the second, however, the channel is very largely blocked with frass. In species which are normally accustomed to distribute their frass regularly throughout the width of the mine, there are naturally no essential differences in the width of the frass-line between the two generations; variation in the nature of the leaf results in this case in a correspondingly shorter or longer mine.

Frass is not always excreted as a firm, well-formed pellet; gaseous and liquid excretions must also be considered. No detailed investigations have been made as to whether the mining larva excretes gaseous matter but there is little doubt that it does. One frequently finds that the upper surface of the mine shows a blister-like dilation, without any preceeding formation of folds as a result of spinning by the larva, of the type found in *Lithocolletis* and others. It is not yet known which gases are in evidence as the final products of the digestive processes but they occur in the mines of members of all insect Orders; in some cases such inflations of the mine are characteristic of the species concerned. It may well be that such a swollen epidermis prevents certain species of parasite attacking the larva owing to their ovipositors not being long enough to reach the larva.

We have already mentioned the liquid excretions of epidermal miners. Liquid as well as solid excreta also occur in other species which in some cases impregnate the cells adjoining the mine cavity, causing them to assume a characteristic discoloration. This is found very clearly in the mines of *Rhynchaenus populi* Fbr. on Populus and Salix. The mine appears almost blackish in contrast to the other species of *Rhynchaenus*, where there is no similar discoloration. This discoloration, however, cannot only be ascribed to the deposits of frass but there is also a certain tendency towards this in the food-plant itself (Salicaceae). It is known that in this family a blackening of the leaves can occur very easily, as happens to many species when they dry up and also as a result of various injuries. The two factors clearly operate together. The mines of *Rhamphus pulicarius* Herbst and of the species of *Zeugophora* on these plants are coloured blackish, while on the other hand many species mine Salicaceae without any blackening being observed. The influence of liquid excreta is much more apparent in *Bucculatrix frangulella* Goeze on Rhamnus. In this species there is a dark colouring in the early spiral mine; the walls separating the individual coils of the mine have assumed the colour of the frass. Immediately the spirale straighten out, the dark colouring disappears, which is moreover not found in the other species living on Rhamnus.

An even more far-reaching influence of the frass on its surroundings may be assumed from the investigations of G. Voigt (1945), who discovered that the formation of "green islands" in the discoloured leaves of autumn, which are found so frequently in conjunction with mining insects, occurred without any injury to the leaf as a result of the presence of bird droppings. Presumably, constantly recurring anthocyanine accretions, such as are often found in the vicinity of mines, are induced through influences arising from the frass. Both these phenomena will be dealt with more fully later.

The final defecation made by the larva before pupation must be considered as essentially different from all other deposits of frass. This is superficially apparent from its mere quantity. The final pellet of frass is especially large and frequently has characteristics not possessed by earlier ones. The reason is that the contents of the

Malpighian tubes are also excreted together with the last remnants of frass, and possibly also the contents of other parts of the digestive tract which are not normally voided. It has been shown that in larvae of Trypetidae symbiotic organisms living in special evaginations of the intestine are also excreted with the final voiding of frass and permeate this final frass. Probably similar processes occur in other miners; this has not so far been investigated. This last deposit of frass also has a certain biological importance: with Diptera which pupate in the mine, it serves in many cases to attach the puparium to its base. This is seen for instance in *Dizygomyza humeralis* v. Ros. in the leaves of Bellis. The puparium often adheres so firmly to the bottom of the mine that any attempt to remove it results in its destruction. This has probably been caused by the admixture from the Malpighian tubes. The content of this tract is not, however, always excreted with the frass but in some species it has been found that on one occasion it will be excreted, while on other occasions it will be incorporated in the puparium, without any apparent reason for such variable behaviour. In such cases there may or may not be distinct markings on the puparium, usually in the form of a dark longitudinal line, formed by deposits from the Malpighian tubes; this is of no assistance in classification (*Phytomyza xylostei* Kalt. on Caprifoliaceae).

After expulsion of the last remains of frass, pupation normally occurs; in certain cases a further short length of mine is made, which of course contains no additional frass. This can only be explained by the fact that the larva is now taking no more food and merely lays aside the parts of the plant cells it has bitten through, without eating anything. A clear example of this is *Dizygomyza luctuosa* Mg. mining in Carex. It normally deposits its last large lump of frass in the leaf blade but usually pupates in the sheath enclosing the stalk. It usually mines a considerable further distance without any additional traces of frass, so that it is clearly not eating.

The last section of a mine channel is invariably free from frass over a length equal to that of the larva itself; the length of this part of the mine in relation to its width serves as a guide in some cases to the species of the producer. The more elongated larva of *Lyonetia clerkella* L. (Fig. 6) leaves a longer frass-free final portion than in the mines of similar species of *Nepticula* in the same plants.

10.

CASE-BEARING MINERS

The species which in the larval stage bear a portable case externally on the plant, but are able to mine as they wander from leaf to leaf, represent a transition between permanent and temporary miners. In the biology of their feeding they do not differ from other mining insects; they likewise consume the parenchyma of the leaf and leave the epidermis little damaged. Apart from a few species the larva does not leave its case while feeding; it spins the mouth of the case to the leaf, and gnaws a hole in the epidermis through which it penetrates with its fore-body into the leaf and consumes the mesophyll; the end of its body remains in the case, to which the larva retires for rest and to void its excrement. The frass is ejected through the rear end of the case. Only a few species move right outside their spun-up case and penetrate for any considerable distance into the leaf; they return, however, to the case whenever they wish to move to another leaf. These species thus combine free-living on the plant, by which they are not tied down to a single leaf or to leaves in the immediate vicinity, with the otherwise normal habits of the miner.

Case-bearing miners are so far only known among Lepidoptera; case-bearing beetle larvae (e.g. *Cryptocephalus*) have not hitherto been found to feed by mining. Among Lepidoptera, species of this type are found in the genera *Apterona* (Psychidae) and *Coleophora* (Tin. Coleophoridae), the latter genus having a very large number of species. Larvae of the genus *Apterona* can easily be recognised by the tight coils of the case, which is covered with sand and earth; these cases are somewhat similar to a snail-shell (Fig. 71); in *Coleophora* they are not so tightly coiled, although indications of a spiral coiling of the case can be detected in some species. It must be mentioned here that not all species of *Coleophora* are miners; a number of species have quite different larval habits such as living in the stem, as fruit-eaters, or, among the species which live on the substance of leaves, as larvae which consume the whole flesh of the leaf or perforate it by window feeding or skeletonization. Anyone wishing to identify a *Coleophora* mine found on a plant, must in most instances look for the case which is often attached in quite a different place; special care is required since both leaf-miners and fruit-eaters may occur from this genus on the same plant; this applies particularly to many

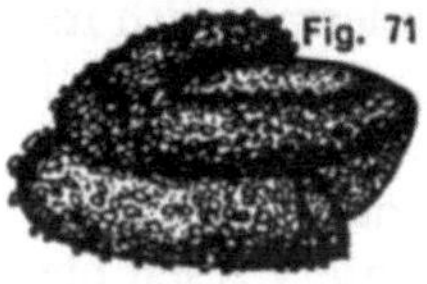

Fig. 71

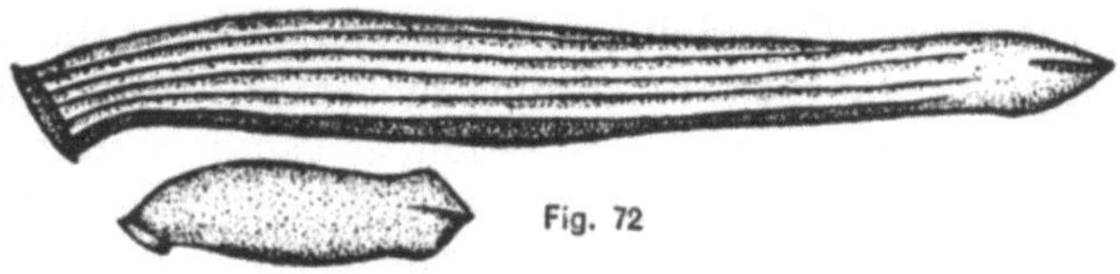

Fig. 72

Fig. 71. Salvia: Case of *Apterona crenulella* Brd. resembling a snail-shell.

Fig. 72. Camphorosma: Above: Tubular case of *Coleophora macrobiella* Cst. (miner). Below: *C. camphorosmella* Cst. (seed-eater), same enlargement.

plants of the family Compositae (Fig. 72). In such instances it will be necessary to investigate whether a species found living in a case is really producing mines or is leading a different kind of existence. If the larva has not yet webbed down its case for pupation, it will be advisable to ascertain its methods of feeding by putting it under observation for a day, with an adequate supply of leaves in a tightly closed breeding-cage.

After some practice the distinguishing features of a *Coleophora* mine can soon be recognised without fail, even when the larval case is no longer to be found attached in the vicinity. The whole mesophyll between the upper and lower epidermes is always eaten away and the resulting blotch mine never has any traces of frass, while at the same time there is a circular opening, the point of entry of the larva, in one or other epidermis. Normally this round hole is made in the lower epidermis. The combination of these two characteristics is found only very rarely in other miners: *Bucculatrix noltei* Petry produces coleophorid-like mines in the leaves of Artemisia vulgaris L., but not from within a case (Fig. 73). This species can be recognised if one recalls that in its infancy the *Bucculatrix* larva makes a linear mine distinguishable by a thin central line of black frass which resembles the channels of species of *Nepticula*, which do not, however, live on this plant. The opening to the mine is a slit on the edge of the leaf, a characteristic of the *Bucculatrix* genus, and is not found in *Coleophora*. The round opening of *Coleophora* is always well removed from the edge of the leaf in the surface of one or other epidermis.

The larvae of some species make in addition to the opening in the one epidermis, another exactly opposite in the other epidermis, so that at this point the mine completely perforates the leaf. This is an indication that a *Coleophora* larva is involved, which has been enlarging its case by the addition of transverse pieces of leaf; examples are the polyphagous *C. paripennella* Z. and *ahenella* Hein which are so common in autumn.

Feeding areas similar to *Coleophora* mines are formed by some weevils in the adult stage and these areas must not be confused with true mines. The adult beetle feeds by piercing the epidermis with its rostrum and then, by turning movements, eats away the mesophyll around the point of penetration, as can be seen in *Ceuthorrhynchus contractus* Mrsh. on Cruciferae and in *Rhynchaenus rusci* Herbst on Betula. Even though the resultant feeding patterns closely resemble the blotch mines of *Coleophora*, they cannot be considered as mines, if one accepts our earlier definition of a mine as both feeding a n d living quarters, for there is no question of the adult beetle using the interior of the leaf as its living quarters — it merely inserts the fore part of its head into the leaf tissue. Such feeding areas made by Curculionidae can be easily distinguished from genuine *Coleophora* mines by the smaller area which has been mined around the opening, and also by the hole in the epidermis always being smaller than one produced by a *Coleophora* larva. When in the early instars the *Coleophora* larva does sometimes produce very similar blotch mines; however, at this stage it does not move far, so that the case is invariably to be found on the mined leaf.

Separation of *Coleophora* mines from those of *Apterona* is not difficult. In a *Coleophora* mine, the point of entry of the larva is always

Fig. 73. Artemisia vulgaris L.: Coleophoroid blotch-mines of *Bucculatrix noltei* Petry, right the Nepticuloid early mines of the same species.

Fig. 74. Malus: Left: *Coleophora hemerobiella* Sc. ("mouth 5"), right *C. serratella* L. ("mouth 3") on the petiole. Above them in the base of the leaf, cut out areas with the early two-lobed cases remaining attached.

an exactly circular hole. On the other hand in *Apterona* the edges of the hole are always more or less irregular, never exactly circular and sometimes whole strips of epidermis are detached, so that the differences between mining and window feeding tend to disappear. This fact demands particular attention in the identification of frass-less blotch mines from the Mediterranean area, since the species of *Apterona* found more frequently in Southern Europe are extremely polyphagous and will undoubtedly be found on many more plants than have so far been reported. In addition, *Apterona* larvae more often mine leaves lying directly on the ground, but this preference is also shown by a number of *Coleophora*.

The life cycle of case-bearing miners differs in many ways from that of mining species which make no larval case. With the exception of *Apterona*, in which the female lays its eggs inside its former larval case, the ova are inserted into the tissue of the leaf, where the larvae later hatch. Subsequent habits are largely determined by the manner in which the later larval case is made. In many species of *Coleophora* (although this has been described in detail in very few), the first instar larva makes a blotch mine, which in contrast to the later stages contains grains of frass in the mine and is therefore (and also on account of the absence of the circular opening in the epidermis) not at this stage recognisable as a *Coleophora* mine. At the end of this instar however, the larva usually cuts out an elliptical piece of the leaf with which to construct its first case, and with which it then proceeds to wander around on the leaf, seeking out a spot at which to feed. From this stage onwards the typical *Coleophora* mine is made. When the larva has attained a certain size it is compelled to construct a new case, since its first case has become too small. It therefore makes a mine which is often more elongated than is usual and frequently runs along the edge of the leaf and out of this it cuts a larger case. Since the two longitudinal edges of the case must be closed by spinning, it saves itself having to spin down the one edge by constructing its mine along the leaf-edge. The larva always carries this case in such a way that the former leaf-edge is uppermost, with the result that any existing serration of the leaf stands out as a saw-like edge on the back of the new larval case. However, this is a characteristic of little value in the determination of larval cases, since the extent of leaf serration varies considerably, being less marked at the base, more marked at the tip of the leaf; on the other hand in the course of time this leaf serration becomes more and more worn down. The first larval case is usually cut out of the centre of the leaf and while this may also occur with subsequent cases, it is not so frequent.

The shape of the cut-out leaf portion is of particular importance, since in most instances this is sufficient to enable determination, even when the actual larval case cannot be found. On the one hand this cut-out accurately indicates the contours of the larval case which was made from it, and on the other hand in very many instances the first larval case is fastened to the edge of the area cut out and this frequently facilitates deductions as to the species responsible (Fig. 74). When collecting *Coleophora* mines, one should therefore never neglect to gather nearby leaves with cut-outs, and thus miss no opportunity of establishing the identity of the species concerned.

In certain circumstances more than two larval cases may be constructed by species which normally only make two. For example, *C. fuscedinella* Z. (on Corylaceae) constructs in late summer its first irregularly curving case; after hibernation this is exchanged for the final case, after the larva has first mined for a short time from within the previous year's case. However, when this species feeds on B e- t u l a n a n a L., it often finds the leaves immature to begin with and too small to permit the case to be made as large as is desired; when this happens it frequently first makes a short second case, which is never found when the species feeds on other plants, and only later when the leaves of B. nana have grown larger does a further change of case take place, resulting in the construction of the final case.

The nature of the case is of extreme importance for identifying species of this genus. The moths which in superficial appearance are very similar and can only be distinguished by examination of their genitalia often have widely different larval cases which provides a means of identification, especially if the food-plant is also known. Special attention must be paid to the material from which the case is constructed, to the structure of the rear end of the case and to the mouth of the case from which the fore-body of the larva protrudes during feeding and walking.

The material used in constructing the case varies with the mode of existence of the larva. In the species with habits as just described the case consists of parts of the leaf lined with silk. In all such instances the outer wall of the case largely conforms to the nature of the leaves; it is very hairy if the leaf has coarse hairs (*C. onosmella* Brahm on Boraginaceae) and the individual leaf veins can often still be recognised. All species do not cut their case out of the leaf in the way described; some construct it largely or wholly from silk and to this group belong in particular the species which make the so-called "pistol cases". These species make the case entirely or almost entirely of silk ranging in colour from blackish to snow white, curving more or less downwards at the rear end to resemble the handle of a pistol. Situated laterally on these cases there is a shield-like structure, the lateral shield ("pallium"), which usually appears to be composed of individual scales (Fig. 75). There are considerable differences in the construction of the case and of the lateral shield in closely related species. In this respect the variation shown by pistol-case bearing species living in the Palaearctic region on oak is interesting.

In many species a successive enlargement of the case occurs throughout the period of feeding. Normally, fresh material usually consisting of parts of the leaf or of silk, is added from time to time to the fore-end of the case. At the same time an increase of thickness of the case also often occurs, as the larva slits open the case along its length (always on the ventral side) and interposes a new layer of silk into this longitudinal slit. Some species fix grains of earth or sand or pellets of frass to the case and these foreign bodies are usually arranged on the case in longitudinal rows, due to their being woven into the case while the thickening process just described is in progress. Other species fix small bits of leaf in rows running either lengthways or crossways and link them up with silk. Inside this structure there is always a tube-like cavity, wherein the larva spends its periods of rest. The completely irregular bits of leaf, such as are cut of the leaves of many trees by *Coleophora siccifolia* Stt. produce a very striking appearance. These cases in which the tube containing the larva is to be seen only on the one side, resemble shrivelled up leaves and are therefore very difficult to detect. Only a very experienced eye will be able to discover them attached to the twigs after the leaves have fallen in autumn. Many *Coleophora* cases closely resemble withered bits of leaves and for this reason alone are often overlooked. Clear proof of mimicry is given by the behaviour of the larvae which, on completion of feeding choose a spot for pupation where the case is not readily visible and tones either with the surface of the twig or with the bud scales.

The rear end of the larval case may have two or three flaps. The first of these is the primary flap, formed when the larva cut the case out of the leaf. The two flaps or valves at the end of the case correspond to the two epidermes of the leaf. The construction of a case with three flaps involves special activity by the larva which by cutting and spinning forms three flaps out of the former two. Why it should do this is by no means clear. A number of species first construct a case with two flaps and only in the final instar are three flaps made; many others are satisfied with two flaps throughout their larval life. Since the presence of three flaps may be of special importance for identification, any larvae found feeding should be kept several days in a breeding-cage to discover whether a third valve is subsequently added to any case with only two flaps. The end of the case is either straight or curved downwards. In the latter instance the portion bent down is frequently the initial case of the first instar larva, and was

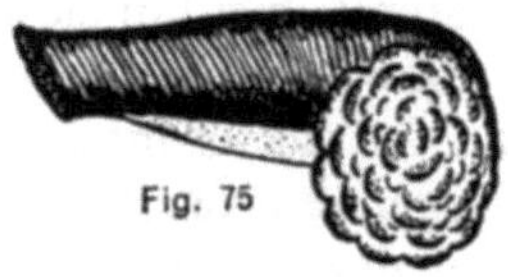

Fig. 75. Tilia: Pistol case of *Coleophora anatipennella* Z., with lateral shield ("pallium") at the read end (miner).

Fig. 76. Quercus: Pistol case of *Coleophora palliatella* Z. with enlarged lateral shields ("pallium") (not a mining species).

made in the preceding year and is distinguishable from the other parts of the case by its darker colouring. Sometimes a hypertrophic lateral shield may enshroud the whole case and even conceal the end of the case from a superficial examination (Fig. 76).

The position of the "mouth opening" is also very characteristic and almost always constant, so that it may be used for identifying a species whose case construction shows little variation, as for instance many species living in a tubular case on Rosaceae. This circular mouth opening, often having winged edges may be found with a great degree of variation in the angle it forms with the longitudinal axis of the case. Agreement was reached earlier on certain numbers to represent various positions of the mouth opening; Herrich-Schäffer proposed that "mouth 1" should represent the fact that the plane of the mouth opening lies parallel with the longitudinal axis of the case. In this instance the case lies directly adjacent to the leaf on which the larva is feeding. "Mouth 5" refers to the plane of the mouth opening forming a right-angle with the longitudinal axis of the case, with the larva thus feeding from a case standing vertically in relation to the leaf, as may often be seen in *C. hemerobiella* Sc. on fruit trees (Fig. 74). If the case forms an angle of 45° with the mouth opening (and also with the surface to which it is attached), it is referred to as "mouth 3"; 2 and 4 are the intermediate positions.

Numerous species also have longitudinal keels, which are usually of no great importance when dorsad, as they are then largely dependent on chance peculiarities of individual leaves. However, the ventral keel (Fig. 79) is important for identification and its degree of sharpness and length are characteristic of individual species; it is constructed by the caterpillar when the case is split open as already described in order to enlarge it.

The change of food-plant in some species is also worthy of note.

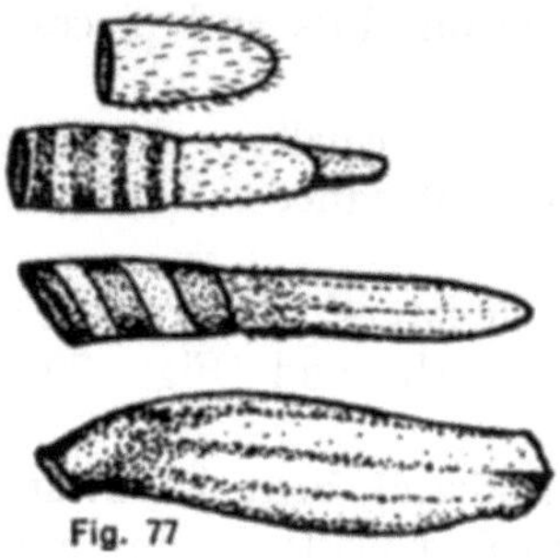

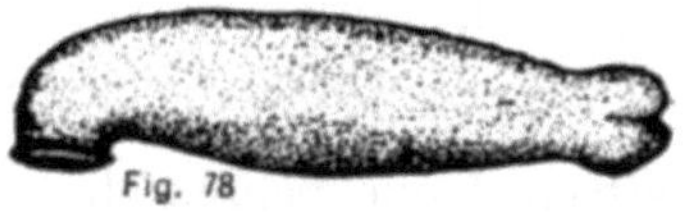

Fig. 77. Cerastium: *Coleophora prinziella* Krone, different stages of the case, first with two flaps ("mouth 5"), then with three flaps ("mouth 3"). (After Klimesch).

Fig. 78. Pyrus: *Coleophora trigeminella* Fuchs ("mouth 1").

Fig. 79. Artemisia: *Coleophora ditella* Z., case with well developed ventral keel.

For instance the young larvae of some species live in the calyx or seeds of Labiatae in autumn, while after hibernation they construct a case from grasses, in which they now mine until pupation and leave the Labiate calyx, which they used as their first case.

The following key sets out the terminology used in lepidopterology and minology for the various types of cases in the genus *Coleophora:*

Identification Table of the types of larval case in *Coleophora*

1. Anal end of the case with three flaps 2

- Anal end of the case with two flaps 3

2. Case cylindrical and straight, somewhat thickened only in the centre, constructed either of bits of leaf or silk (Figs. 72, 74) *Tubular case*

- Case irregular, constructed of fruit capsules or the involucrate heads of Compositae, with a short projecting end spun of silk and with three flaps .... *Seed case*

3. Case made up of various bits and pieces 5

- Case completely uniform, usually consisting of a homogeneous often parchment-like substance 4

4. Case with considerable lateral compression and possibly bent slightly downwards at the end (Figs. 79, 80) .... *Sheath case*

- Case round and cylindrical, bent downwards at the end like the handle of a pistol; usually having a blister-like lateral shield ("pallium") on each side and generally with a marked wrinkled appearance .... *Pistol case*

5. Case consisting of separate parts attached one behind the other longitudinally, or of a single leaf or pine needle (in this instance resembling tubular cases, but with two flaps) (Fig. 81) .... *Leaf case*

- The individual parts of the leaf are arranged crossways to the longitudinal axis of the case....................... 6

6. The bits of leaf are attached to each other crossways to the longitudinal axis of the case and at least in front stand out conspicuously from the case. The thinner end of the case may have more or less snail-like coils (Fig. 82) .... *Lobe case*

- The bits of leaf adhere closely to the case and are arranged obliquely, slanting from the upperside to the rear (like the wing sheaths of a moth pupa), the cases usually have considerably lateral compression (Fig. 83) .... *Pupa case*

When the larva is full-grown it frequently moves away from the area where it has been feeding to a protected spot, where it webs down its case and pupates inside. Frequently it attaches the case to the stem or a twig of the food-plant, where it is difficult to detect; on other occasions the case may be spun onto quite different plants, so that it is not always easy to link it up to the food-plant it has been mining. After spinning up, the *Coleophora* larva turns round in its case and the moth emerges from the case through the anal end. In *Apterona* and *Coleophora* the pupal shell remains in the case; in *Apterona* also so does the ovipositing female.

All case-bearing miners in our geographical region have only a single generation, and some species even spend two winters in hibernation. Parthenogenetic development has been observed in the female abdomen of some species of *Coleophora* (larviparous reproduction),

Fig. 80

Fig. 81

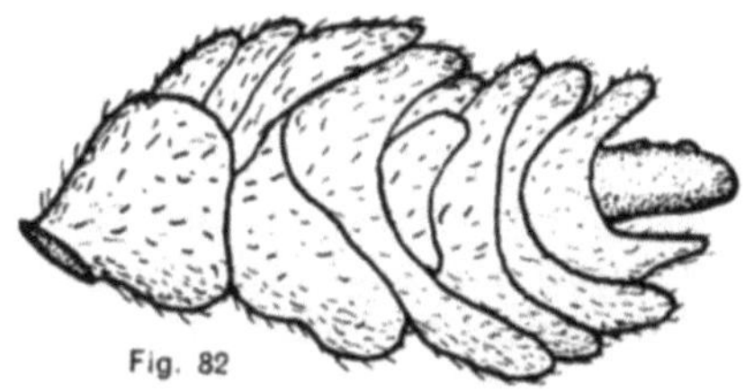

Fig. 82

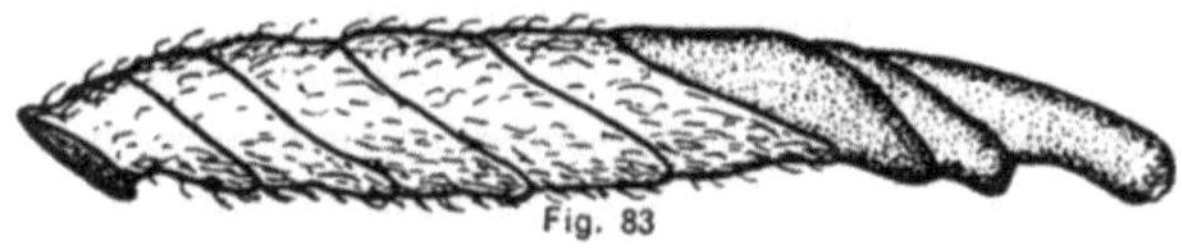

Fig. 83

Fig. 80. Ulmus: *Coleophora badiipennella* Dp., sheath case with "mouth 2".

Fig. 81. Satureja: *Coleophora bifrondella* Wlsgh. Leaf case, of two leaves one behind the other.

Fig. 82. Sarothamnus: *Coleophora saturatella* Stt., Lobe case, with the early case at the rear end.

Fig. 83. Dorycnium: *Coleophora perserenella* Rbl., Pupa case.

while in *Apterona* in the northern areas of the region this is the normal rule. *Coleophora* hibernate successfully in captivity only with difficulty and it is essential that they have conditions as closely as possible approaching the wild state. It is advisable to collect the cases of species which continue feeding in spring only shortly before pupation takes place.

Oligophagy and xenophagy occur particularly frequently in case-bearing miners on account of their greater freedom of movement. *Apterona* is undiscriminatingly polyphagous; while oligophagy is common in many species of *Coleophora*, closely related plants are usually found to be attacked. A clear example of xenophagy (Voigt, 1932) has been confirmed in *Coleophora fuscedinella* Z., which normally lives on Betulaceae but has been found on Ribes and even on the perianth of Caltha.

11.

CONNECTION BETWEEN MINING HABITS AND LARVAL MORPHOLOGY

The degree to which the highly specialised habits of mining insects are reflected in modifications in the body of the larvae in relation to their free-living relatives offers a most attractive field of investigation. Consistent investigations of this type have not yet been undertaken with equal thoroughness in respect to every part of the body and for all Orders of mining insects; results achieved so far, however, show that an immense amount of research remains to be done which will reveal a great many surprising facts. These modifications in the larva's body are of general biological interest; they show the extent of the influence of environment in the development of the species. In some cases the modification of the larva goes so far that even greater morphological differences of every type can be observed in different stages of the larva of the same species (with varying types of mine) than between species of different families which construct similar mines. Irrespective of whether we investigate the general shape of the body, the legs, the antennae, the eyes or the mouth parts, in each case strange modifications can be detected, which appear to have been determined by the habit of mining. It is a reasonable assumption that internal, histological differences have developed parallel to these morphological differences, although no investigations have been made into this aspect. It will scarcely be considered possible that the two larvae shown in Fig. 85 belong to the same species, so greatly do they vary in every respect. However, if one is aware of the variations in the larva's habits during the course of its life and if one takes into account the changed environment in which the differing types of larva develop, these differences become more readily understandable.

It is of particular interest that these morphological differences which appear to us as adaptations to their life as miners are developed to a varying degree in the different insect Orders, in other words each Order is not affected in the same way. The most marked differences are found in larvae of Lepidoptera; then come Coleoptera, followed at some distance by Hymenoptera and very last the Diptera. It would seem obvious to seek the cause of such a varying differentiation primarily in the systematic position of the species concerned. We know that among the mining Lepidoptera a very large number of different families have developed specialised habits of this type. During the course of time it has become ever more apparent that the "Tineidae", which were formerly considered as a single unit, must be split up into a number of clearly distinguishable families. Current taxonomy still fails to pay sufficient attention to this fact; the investigation of the larvae, however, will serve to stress its importance. Among the Diptera, where we believe there is the least evidence of special adaptation, only a few families contain miners and these largely belong to the family group of the Acalyptrata. The division into families is still of a rather tentative nature, since the external family characteristics have frequently been found to be indistinct and replaced by intermediate forms. The same applies to Hymenoptera, where only a few groups of the Tenthredinidae are represented by miners and these are very closely related to each other. It thus appears that the very large degree of variation among mining Lepidoptera should be ascribed less to their specialised habits than to their widely varying phylogenetic origin.

These striking morphological differences occur, however, not only in different species of the same genus but also in the different instars of the same species, which clearly suggests that in the latter case there is a direct connection with the variation in the type of existence. It seems reasonable to assume that the larvae of Lepidoptera possess great flexibility of reaction to environmental influences, which finds its expression in morphological modifications. Furthermore, Lepidoptera show the widest variation in types of mine. The mines of Tenthredinidae, whose larvae in other respects resemble most closely those of Lepidoptera, are considerably more uniform; they never produce epidermal mines and are seldom limited to the upper surface of the leaf; they are normally found in full-depth blotch mines, in which the whole parenchyma between the upper and lower epidermis is consumed. The larvae of the mining Diptera of the family group Acalyptrata have in any case such considerably reduced legs, antennae and mouth parts that no possibility remains of further conspicuous modifications. However, careful, detailed investigation will probably reveal differences in relation to externally-feeding forms even in this case.

The changes in the larva's body which develop parallel with mining can be detected by a superficial examination in a difference in the external shape of the body. Most insect larvae possess a round, cylindrical body, which is equally wide both at the fore- and hind-ends, as long as they feed externally on parts of plants. They frequently bear a variety of different surface structures, such as warts, bristles, the horn-like extension of the hawk-moths, etc. The mining larva is restricted in its living space; the leaf tissues which confine the mine cavity exert a certain pressure on it and parallel with this a flattening of the body has developed, which is more pronounced in some species than in others. The flatter the mine in the leaf, the more striking is the flattening of the larva. We can observe, for instance, in many mining lepidopterous larvae a dorsoventral flattening, which appears to be dependent on the depth of the mine. This is developed most conspicuously in epidermal miners, the sap-feeders, but is also clearly detectable in parenchymal miners, in cases where the mine is limited to one or the other side of the leaf and is only of slight depth. The effect of such differences is shown by a comparison of larvae from the genus of Coleoptera, *Rhynchaenus*. *Rh. fagi* L., *quercus* L., and others, which live in full-depth mines, are generally still quite round and not so flattened as *Rh. populi* Fbr. which only feeds in upper surface mines. Tenthredinidae, which almost without exception live in full-depth mines of more considerable depth, as yet show no evidence of the tendency towards flattening. However, this modification is to be observed not only in flat leaf mines, it occurs frequently in a more pronounced form in stem- or cortex-mining species. In such cases the mine cavity is particularly low; the outward growth of the stem narrows still further the width of the mine and the larva within is subject to greater pressure than in leaf mines, where the epidermis is less rigid and can be stretched by the mining larva. We thus find striking similarities in the shape of the larva between, for example, epidermal miners among the Lepidoptera and the stem-mining Buprestidae *(Aphanisticus)*. These similarities not only apply to the dorso-ventral flattening but also to the thickening of the fore-end, which is equally striking. In cases where increased resistance causes the larva difficulties in penetrating into the leaf, its fore-end appears thickened in relation to the rest of its body, in fact wedge-shaped (Fig. 84a). This thickening applies especially to the first thoracic segment, which in its width and thickness exceeds all the other segments of the body. In the dorsal and ventral areas there are also in many cases more strongly chitinised plates, which serve to increase the larva's pressure when it penetrates into the plant tissues and at the same time to protect the more sensitive internal organs lying below.

The d o r s a l p l a t e of the prothorax is of particular taxonomic importance and frequently serves to identify the larva in the mine, especially in the genus *Elachista*. The identification of *Elachista* mines is frequently a matter of some difficulty, since the narrowness of the Gramineae leaf militates against the development of any characteristic mine form. Characteristics of the larva must be looked

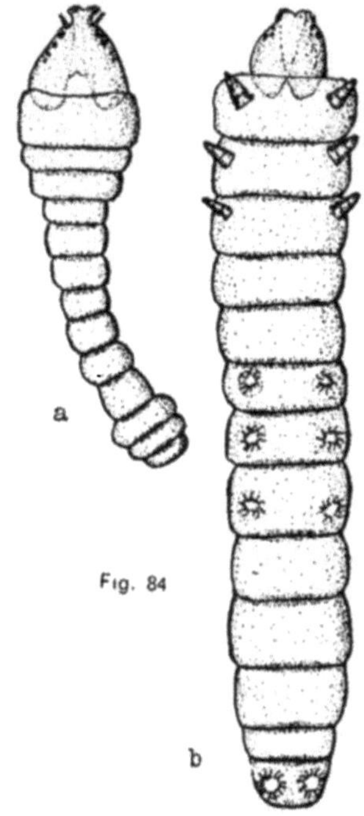

Fig. 84. Lithocolletis corylifoliella Hb. Larva as a sap-feeder, b. tissue-feeder (ventral view).

for to distinguish the different species feeding in the same food-plant. In this genus substantial differences occur in the shape and colour of the dorsal-plate, which can be successfully utilised as a means of distinguishing the spciees. This can also be made use of to differentiate the larva of the genera of Gracilariidae, *Lithocolletis* and *Parornix*, which in their early stages both live in very similar tenti-forms mines and are then difficult to differentiate, until one later sees that the species of *Parornix* leave their mine and, no longer mining, live under a folded-over leaf-edge. It is known, however, that as an exception even *Parornix* caterpillars may live as miners throughout their life and in such cases, if doubts arise, it suffices merely to examine the differences in the dorsal plates: in *Lithocolletis* there is only a single plate on the prothorax, which may be divided by a less chitinised longitudinal line (Fig. 85a), while in *Parornix* this plate is broken up into four small shields, so that four black points can be observed on the prothorax (Fig. 85b). In many cases such dorsal plates are not only found on the prothorax but also on one or several of the following segments. For instance, most species of *Coleophora* have also a smaller plate on the mesothorax (Fig. 86).

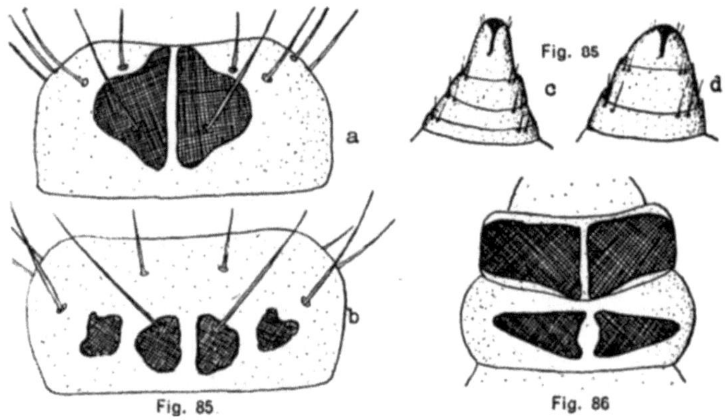

Fig. 85. *Lithocolletis ulmifoliella* Hb. (a and d) and *Parornix betulae* Stt. (b and c), Prothoracic plate and leg. In *Parornix betulae* Stt. the prothoracic plate is split up into 4 small plates, the thoracic legs have one segment more than in *Lithocolletis*.

Fig. 86. The first two thoracic segments of *Coleophora milvipennis* Z. with prothoracic and mesothoracic plates.

This does not result from the leaf substance, on which they live, being particularly rigid and firm; for species of *Coleophora* are found on the most varied plants. It must be assumed rather, that the necessity for the development of a second dorsal plate is connected with the especially frequent changes of mine in this species. The larva is compelled, as it constructs each new mine, to bore a hole in the epidermis time after time, before it reaches the nourishing tissue of the leaf, the parenchyma; and we know that the epidermis in its capacity of protective tissue of the leaf is tougher than the parenchymatous tissue, particularly when it is in addition covered by a cuticle. The necessity for an increase in the protecting dorsal plates can be clearly linked up with this frequent transfer from one mine to another.

The dorso-ventral flattening of the body is entirely absent in all dipterous larvae, even in the stem-miners and at first it seems strange that this tendency has not developed in this insect Order at least in individual cases. The reason for there being absolutely no cases of this flattening is to be found in the special way in which dipterous larvae feed inside their mine. The larvae of the other three Orders always feed lying either on their back or on their belly; in contrast to this the dipterous larva feeds lying on its side. It therefore has to be continually rolling over from one side to the other and for this type of movement, a cylindrical shape is the most suitable. A fly-larva flattened dorso-ventrally could not undertake the necessary changes in position as conveniently as a round, cylindrical larva. The larvae of Tendipedidae, which live inside mines in aquatic plants wash their food consisting of plankton organisms into their mine channels by twisting movements of their body and it is obvious that for a twisting movement of this type the round, cylindrical-shaped body is the only one suitable. When the larva has finished its feeding and pupates (in the higher Diptera inside the last larval skin, the puparium), the need for a cylindrical-shaped body disappears; accordingly, a dorso-ventral flattening is frequently found in the puparia which remain in the mine. This is especially striking in species of *Ophiomyia* and *Melanagromyza* and occurs most conspicuously in species whose puparium remains in the stem cortex.

The detailed modifications of the larva's body, which are the result of its mining habits, ensue on the one hand from the changed conditions of movement inside the mine, on the other hand from the different type of food and methods of consuming it, in relation to free-living species. The former consist primarily of changes in the extremities. We must recall how movement is achieved in the majority of insect larvae which have legs: the larva raises the centre of its body from the surface where it is resting, shortens and thickens itself by contraction, so that the segments bearing the thoracic legs can reach further forward and at the same time the last pair of legs on the terminal segment assist this movement with an additional push. The larva thereupon stretches out again, simultaneously raising its thoracic legs which in consequence move forward. This normal type of movement is made difficult within the mine; the thickening of the body attendant on the first contraction, as well as raising itself, become distinctly difficult in a flat mine cavity. For this reason we find that the legs are more and more reduced in mining larvae, according to the degree of specialisation and the flatness of the mine. This process begins in Lepidoptera and Tenthredinidae with the loss of the abdominal feet. In the Tenthredinidae the abdominal feet have no chitinous hooks, as is also the case with some lepidopterous larvae. The loss of these hooks on the proleg pads of the caterpillars is the first stage in the process of reduction to be observed as a result of the mining habit. The majority of mining larvae possess pedes coronati, in which the hooks are arranged in one or more circles (these species belong to the Stemmatoncopoda, the more scientific term for "micros"). Only a few species have the hooks arranged in a semi-circle on the outer margin of the foot — pedes semicoronati (these species are Harmonocopoda or scientifically speaking, macrolepidoptera). Of the latter we may perhaps mention *Cidaria incultaria* HS. whose caterpillar lives in leaves of Primula and morphologically does not essentially differ from the other "loopers". Among the mining larvae having hooks on their feet, the hooks are found in every stage of atrophy.

The process begins by the pad on each foot retaining merely one anterior and one posterior transverse row of hooks, while in other forms the actual number of hooks is also reduced; in *Bucculatrix* only a single pair of hooks remains at the front and the back and finally these also disappear. In species of *Nepticula* the hooks are already entirely lacking; the prolegs appear merely as fleshy outpushings of the body; no essential difference can now be detected between the prolegs of such lepidopterous larvae and the larvae of Tenthredinidae. In the last stage of atrophy, these fleshy projections are also absent and then, as in coleopterous larvae, there is no further trace of prolegs. A similar process of atrophy affects the thoracic feet. These are four-jointed in Lepidoptera, and correspond to the coxa, the trochanter and femur, the tibia and tarsus and the pretarsus (clawing joint) of the imago. The process of atrophy of the thoracic legs usually commences with a reduction of the basal segments; similar species can often be distinguished by this means. The genera *Lithocolletis* and *Parornix*, which are often difficult to distinguish, also differ in their thoracic legs: *Lithocolletis*, which passes its whole larval life in its mine, only has three joints to its foot (Fig. 85d), while *Parornix*, which on leaving its mine continues to feed externally under a leaf-fold and also changes this habitation several times, still possesses 4 joints on its thoracic legs (Fig. 85c), thus complying with the caterpillar's greater demand for movement. Further reduction processes also affect the claw on the terminal joint, which ultimately disappears; in *Tischeria* only a single-jointed, strongly chitinised, conical projection remains in place of each thoracic leg, while *Nepticula* retains only fleshy, expandable swellings similar to its prolegs, which can scarcely be considered as legs. Exactly similar processes occur in the larva of Coleoptera and Tenthredinidae, which also in the course of development lead to completely legless forms.

When the legs vanish in this way as organs of movement, other modifications ensue, enabling the larva to effect a certain degree of movement in the mine. The more pronounced constrictions between the individual segments of the body in many miners are explained in this way; the sides of the segments can be utilised as projections. They are also often covered with small hairs or bristles, frequently having a fine thorny consistancy and this, especially in linear mines, provides the larva with any motility which may still be required.

It is remarkable that the body bristles do not also disappear in the course of these reduction processes affecting the larva's body. In all mining caterpillars bristles are still clearly present, although sometimes shortened, even in species which live in the narrowest of linear

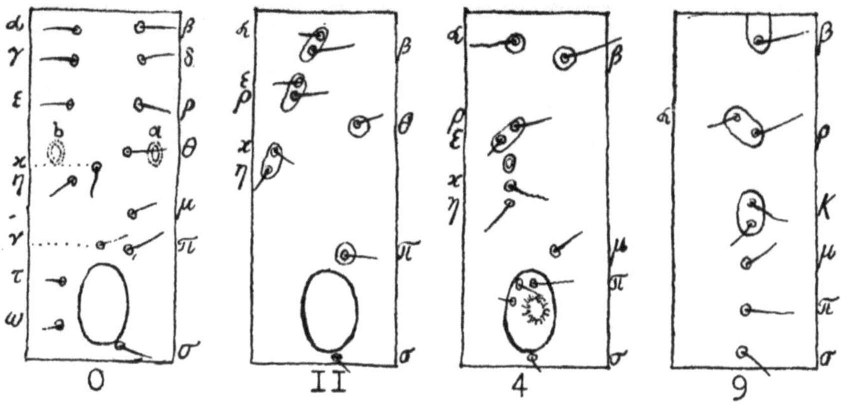

Fig. 87

Fig. 87. Diagrammatic arrangement of bristles in larvae (after Fracker, 1915). A rectangle represents the whole left side of a segment, from the dorso-median to the ventro-median line; the ovals above the lower edge represent the legs. O = setal pattern of a hypothetical segment with all setae which can occur; II = 2nd thoracic, 4 and 9 the corresponding abdominal segments of a Tortrix, *(Carpocapsa pomonella* L.*)* which show the divergences from the normal in the arrangement and the number of setae.

mines. The arrangement of bristles on the larva's body is of particular systematic and taxonomic importance and deserves more attention than it has hitherto received in descriptions of caterpillars. In the first instar, in the caterpillar which has just hatched from the egg, the arrangement and number of bristles is largely similar, in all later instars, however, there are basic differences which make it possible to distinguish families and even genera by this means. Consideration of the arrangement of body bristles is especially important in mining larvae, which are without the striking colours and patterns of their free-living relatives. Fig. 87 therefore shows the arrangement of bristles on caterpillars, in which 0 represents a hypothetical segment, with all possible bristles included, (a. shows the position of the prothoracal spiracle, b. the position on the abdominal segments). The different bristles are indicated by letters of the Greek alphabet after the method of Fracker (1915), and I I, 4 and 9 illustrate how homologous bristles may very in number and arrangement from the hypothetical primitive arrangement. It would be most useful if all future descriptions of mining larvae were to utilise this scheme of Frackers when referring to the arrangement of bristles; previously published details are quite inadequate.

In place of the designation of the bristles proposed by Fracker using Greek letters, for a considerable time now Roman numerals have also been employed. These were first introduced by Dyar (1895), were somewhat amended by Forbes (1910, 1916) and are today frequently used in the modern form employed by Gerasimov (1935). They correspond to FRACKER'S designation for the various parts of the body according to the following table:

	Prothorax	Meso- & Metathorax	Abdomen
I	β	α	α
II	δ	β	β
III	ϱ	ϱ	ϱ
IIIa	ε	ε	ε
IV	ϰ	ϰ	ϰ
IVa	-	-	-
V	η	η	η
VI	θ	θ	μ
VIa	-	-	λ
VIIa	π	π	π
VIIb	ν	ν	ν
VIIc	τ	τ	τ
VIId	ω	ω	ω
VIIe	φ	φ	-
VIII	σ	σ	σ
IXa, b	γ	γ	γ
Xa, b	α	-	-

One important conclusion can be drawn from this fact. As far as is known, there are no mining saw-fly larvae, which possess genuine body bristles like the Lepidoptera, that is to say with a skin structure having a basal, cuticular, ring-shaped callosity (except at their head). The presence of such genuine bristles therefore provides the possibility of distinguishing mining larvae of Lepidoptera from those of Hymenoptera. We previously noted that the more numerous prolegs (in *Nepticula* also), as well as the existence of only a single eyespot (in *Nepticula* and *Eriocrania* also) by which means saw-fly larvae can easily be recognised, do not apply to miners, especially when all the legs are in any case completely atrophied! In all cases of doubt therefore, the presence of genuine body bristles will indicate a lepidopterous larva, their absence will indicate the larva of a Tenthredinid.

Among Diptera two recognisable pairs of feet are still found only in the Tendipedidae (Chironomidae) (Fig. 32). In this respect mining species do not differ from others. Gall midges (Cecidomyidae), which are sometimes also found in mines (it has not been

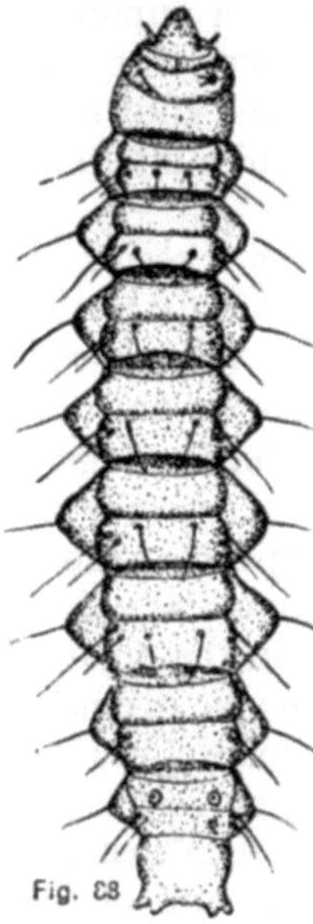

Fig. 88

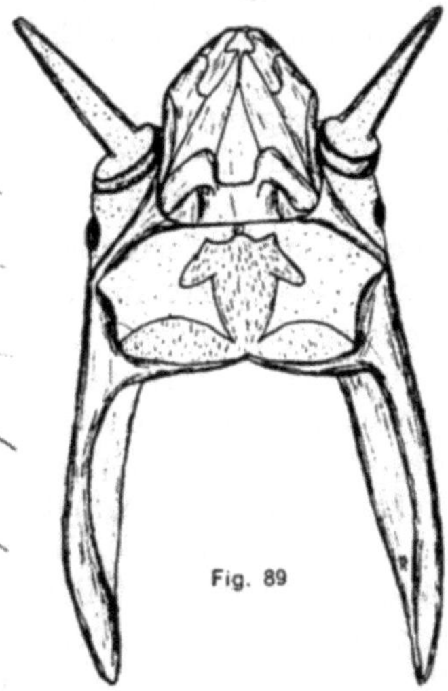

Fig. 89

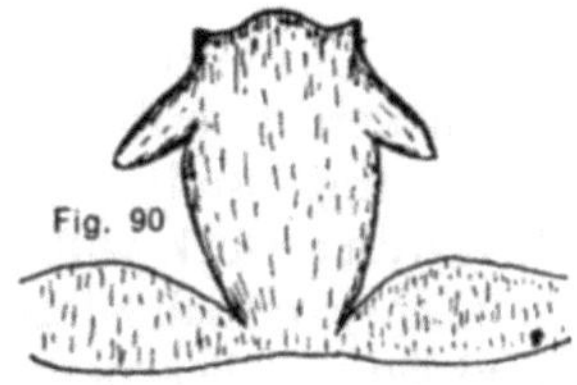

Fig. 90

Fig. 88. General appearance of a gall midge larva found in mines *(Coprodiplosis syringopais* Hering, a parasite on *Phyllocnistis nepenthae* Her.*)*.

Fig. 89. View of head with sternal plate of *Coprodiplosis syringopais* Hering.

Fig. 90. The sternal plate typical of Cecidomyidae, an organ of movement.

proven whether these are genuine miners or parasites on miners), no longer have any feet but possess a well-developed arrangement of bristles, so that they could easily be confused with highly atrophied lepidopterous larvae. They have, however, one characteristic organ, which places them inmistakably among the Cecidomyidae. Fig. 88 illustrates the general appearance of a larva of a parasitic gall-midge found in mines; the view of the head shown in Fig. 89 illustrates the position of the sternal plate, which in Fig. 90 is shown enlarged. According to the most recent investigations this is an organ of movement. Not even rudiments of legs can still be recognised in the majority of mining Diptera, belonging to the alcalyptrate Muscidae. Movement is achieved by parts of the segments furnished with bands of minute backwards directed hooks, which often vary considerably and may be of taxonomic importance; the larvae, however, do not essentially differ from related forms with different habits. Many of them have a sometimes considerable capacity for jumping, especially species of the genus *Liriomyza;* this, however, is not made use of inside the mine but only later when seeking out a pupation site. This jumping capacity is also found in mining as well as other types of feeders (for example flower-feeders); it is also well known in the larva of the cheese fly *(Piophila casei* L.*)* with its quite different habits.

The changes occurring in the region of the head as a result of mining habits are much more varied; they do not apply only to the actual mouth parts but also to the formation of the whole head capsule, the antennae and the eye spots. Particular attention must be paid to the difference between sap-feeders and tissue-feeders, since the changes which have occurred can be most easily understood in relation to the differences in their type of existence. The primary cause is again in this case the depression of the body we have already described. This leads to a series of fundamental changes in the organs, which were described and related to the mining habit in an extremely penetrating study by T r ä g å r d h (1913). The main lines of Trägårdh's work will be adhered to in the following survey.

If an externally-feeding caterpillar is observed, it is seen that the head capsule lies practically at right-angles to the long axis of the body. This is easily explained by the fact that the larva must eat something which is on practically the same plane as the ends of its feet, i.e. which is beneath it. The angle of this plane with the head results in the upper margin of the head capsule being considerably longer than its lower margin. A caterpillar which feeds in a spacious leaf-mine lives under conditions which do not greatly differ from those of external feeders; it also frequently eats matter lying beneath its body. At the same time, however, it consumes matter lying to its sides and in front. The feeding action of the mining larva, therefore, demands a h o r i z o n t a l setting of the head as opposed to the vertical setting of the external feeders. The flatter is the mine cavity, the more frequently does this horizontal head position become necessary. This is attained by the longer dorsal portion of the head being withdrawn into the first thoracic segment; the central portion of the capsule, which is much shorter, does not need to be withdrawn into the thorax in this way.

The recession of parts of the head capsule into the first thoracic segment naturally results in this appearing thicker and becoming somewhat arched at this spot. No difficulties ensue from this for larvae living in a sizeable mine, such as *Lithocolletis*, which artificially expand the cavity by their spinning. The situation is very different, however, for epidermal miners, the sap-feeders. They live in a single, thin layer of cells; they are not enclosed from the outside world by a whole row of cells, but only by a thin skin, the outer wall of the epidermis cells. There is thus a constant danger of the movements of the larva rupturing this skin and opening up the mine to the outside. This would inevitably result in the larva perishing; epidermal miners, whose mine has been opened, are not able to continue normal development nor are they able to burrow into the leaf at another place. For them, however, the horizontal setting of the head is an inescapable necessity, since they never eat food lying beneath their body but only eat what they find in front or beside them. Any considerable thickening of the first thoracic segment caused by the recession of the head, might easily lead to the rupture of the delicate membrane, which encloses the mine. In this case therefore another principle finds application; in the sap-feeders the head has become completely flat and the horizontal setting is achieved by the lower margin becoming longer, and thus becoming less short in relation to the dorsal margin than in the tissue-feeders. Their head is in fact no longer withdrawn into the thorax and the most that occurs is that the rear edges are covered by the front edge of the first thoracic segment. Simultaneously, the whole outline of the head capsule changes. In tissue-feeders it is still oval and egg-shaped, while in the sap-feeders the sides become straight, the whole head capsule now appears distinctly wedge-shaped, and this makes it much easier for the larva to penetrate into such tissue as the epidermis cells. These differences are manifested not only in the phylogenetic but also in the ontogenetic development: this differing head structure is also found in different instars of the same species according to its feeding habits. Many Gracilariidae live first as sap-feeders, later as tissue-feeders and they have accordingly first the wedge-shaped, dorsoventrally depressed head and later, after a change of skin, get the more normal head of the tissue-feeders, which is no longer strikingly different from that of external-feeders (Fig. 84). Owing to the sides of the head in sap-feeders becoming extended and almost straight, all organs on this part of the head are subjected to an increased strain, especially where the larva lives in linear mines; the greater friction between the sides of the head and the tissue of the leaf has resulted in a process of atrophy developing in the organ affected, which will be discussed in more detail later. It may be assumed that differences even among epidermal miners will appear in this respect; some of them live throughout the larval stages as linear miners *(Phyllocnistis)*, while others which commence with a linear mine, extend this in the following instar to a sizeable blotch, in which the strain on the sides of the head is no longer so marked. Presumably therefore, differences will occur between the first and second instars of the larva in the Gracilariidae *(Lithocolletis)*, which belong to the second group.

Fig. 91 may be of assistance in elucidating the position of the

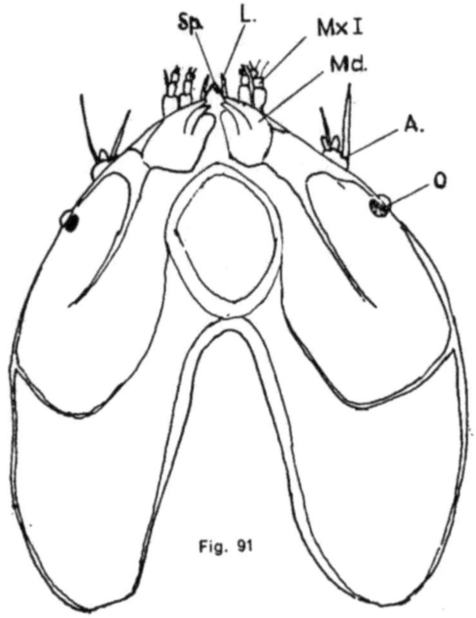

Fig. 91. Larval head of *Nepticula subbimaculella* Haw. (A - antenna, L- labial palpi; Md - mandible, MxI - 1st maxillae, O - ocellus, Sp - spinneret. The labrum which covers the mandibles has been left out. (Dorsal view).

different head parts which undergo modification as a result of the mining habit. The dorsal view of the head of a caterpillar shows in the centre the varyingly constructed apodemata, chitinous plates which extend into the centre of the head capsule and serve as points of attachment for the muscles. Right in front are the actual mouth parts, of which the most conspicuous are the powerful and very dark, chitinised mandibles (Md.). They are covered by a labrum, which has been left out of this drawing and will be discussed in detail later. Further ventrad are the 1st maxillae (Mx 1) and still further the 2nd maxillae, modified to a labium. This bears the labial palpi (L) and the spinneret (Sp). The actual mouth cavity, which is usually membranous and through which food enters the body, is visible in its dorsal portion as the epipharynx, and in its ventral portion usually as the hypopharynx. Behind the mouth parts are the antennae (A), which are not more than three-segmented, and one to six eye spots or ocelli (O), more correctly termed stemmata, since they are not homologous with the ocelli of the imago. When developed normally they are easily recognisable by their black pigmentation.

The majority of mining larvae have the dorsal apodemata very differently shaped from external feeders. The Y-line which normally stands out on every larval head and defines the position of the endoskeletal ridges, which unite to the rear in a single line, is no longer developed; on the contrary the two apodemata tend to diverge to the rear and a tendency develops for these two branches to be united by a chitinous bridge (Fig. 92). The more these apodemata tend to become parallel, the more highly developed does this bridge become. At the same time it can be seen that the apodemata are no longer so knife-like as in free-living larvae but they become broader; however, they do not extend so far inside the head capsule. This fact can be explained by the general dorso-ventral flattening of the head; the apodemata are, so to speak, compressed. This is a development which can already be detected in tissue-feeders but which is found even more clearly in sap-feeders. They develop particularly strong lateral arms (Figs. 93 and 94), which are suitable for supporting the dorsal angles of the head capsule and a further branch may then arise from them, directed towards the rear angle. The genus *Phyllocnistis* (Fig. 94) again in this respect shows itself as the most specialised. In some species one also finds that the sides of the head capsule are

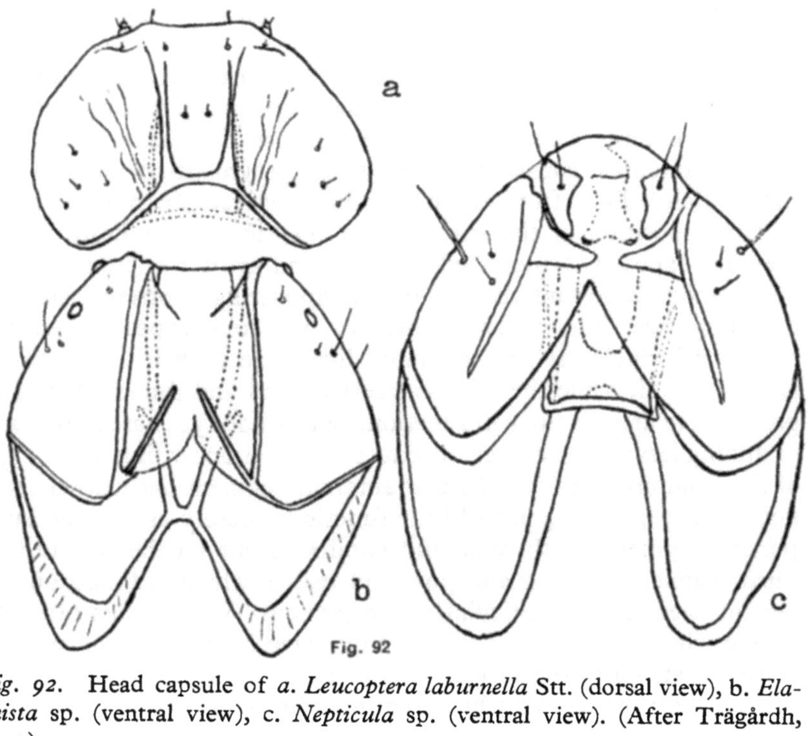

Fig. 92. Head capsule of *a. Leucoptera laburnella* Stt. (dorsal view), b. *Elachista* sp. (ventral view), c. *Nepticula* sp. (ventral view). (After Trägårdh, 1913).

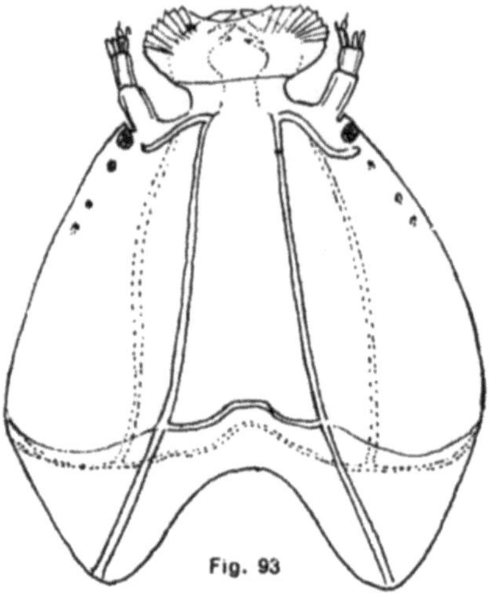

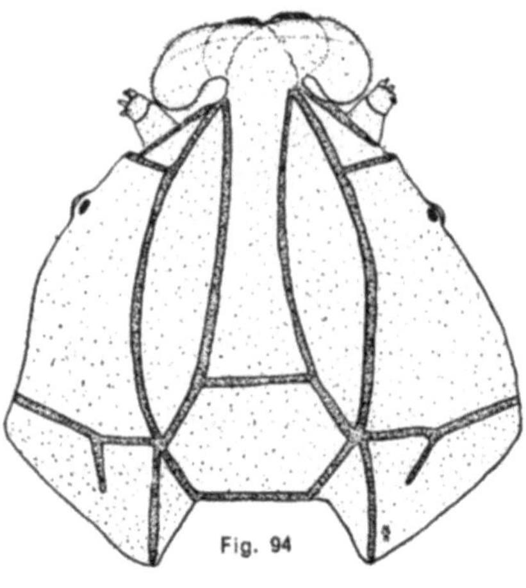

Fig. 93. Head of *Lithocolletis ulmifoliella* Hb. (young).
Fig. 94. Head of *Phyllocnistis nepenthae* Hering (dorsal view).

more strongly chitinised; this results from the greater strain on the sides of the head capsule when the head burrows into the plant tissue.

Reduction processes affecting the antennae are particularly extensive in mining larvae. Even external feeders have considerably atrophied antennae, by comparison with those of the imagines. They all have antennae consisting of only three segments, which do not substantially differ from those of mining larvae. In miners, however, the position of the antennae on the side of the head, where they are exposed to increased pressure and greater friction, results in the most important sensory-organs appearing shortened and reduced in size. The antennae usually bear two types of sensory organs, of which the first are more like setae or hairs and convey tactile stimuli, while the others appear conical and thin-skinned and presumably are olfactory in function. Both these types are also found in mining larvae; the degree of reduction varies according to the type of their mining activity. The first antennal segment has neither setae nor sensory papillae; probably it functions merely as a basal, supporting segment. Both setae and papillae occur on the second and third segments. A long and powerful bristle which is found on the second segment is particularly conspicuous. When attempting to decide which segments of the antenna have been subjected to processes of atrophy, the task is facilitated by looking for this strong bristle, the presence of which signifies the retention of the second segment.

Tissue-feeders and sap-feeders vary in the degree of atrophy. At first sight it seems strange but the reduction processes have been carried furthest in tissue-feeders and are less marked in sap-feeders. This is, however, easily explained by the difference in structure of the head-capsule. In sap-feeders, the region bearing the actual mouth-parts projects somewhat like a lobe (Fig. 93) and behind, there is a depression in which the antennae are implanted. In this depression this organ is less exposed to pressure from the surrounding leaf-tissue than in tissue-feeders, whose mouth-parts do not lie on a similar projection. For this reason their antennae are not situated in a protective depression and are much more exposed to pressure from the leaf tissue and are, so to speak, ground down. This explains the at first sight strange fact, that sap-feeders, in whom the process of reduction has affected the structure of their body more extensively in every other way, nevertheless have less atrophied antennae.

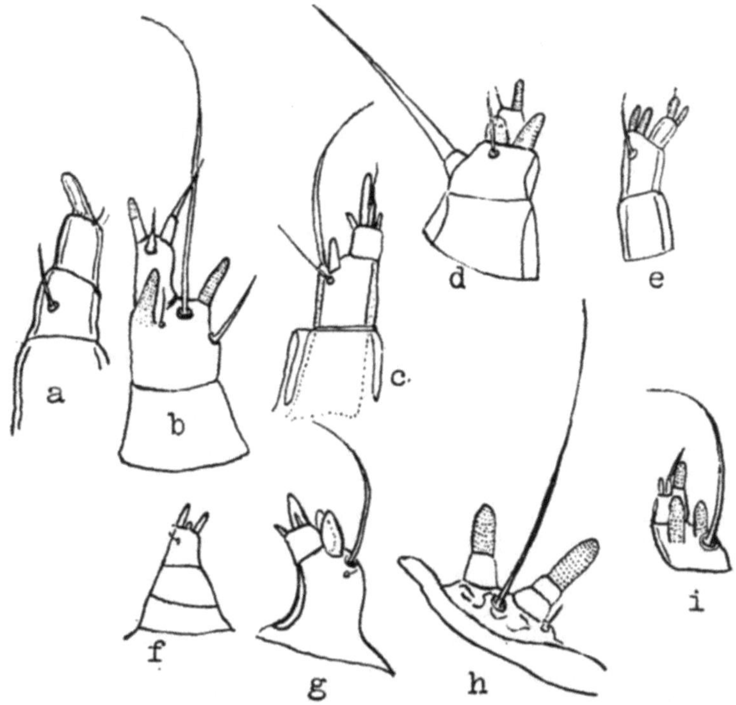

Fig. 95

Fig. 95. Antennae of mining larvae: a) *Metallus pumilus* Klg., b) *Gracilaria syringella* F. (full-grown). c) *Tischeria ekebladella* Bjerk., d) *Eriocrania* sp., e) *Lithocolletis stettinensis* Z., f) *Phyllocnistis suffusella* Z., g) *Gracilaria syringella* F. (infant larva), h) *Nepticula* sp., i) *Leucoptera* sp. (b—d and f—i after Trägårdh, 1913).

Modifications apparently develop at both ends of the antennae, in the basal segment as well as in the distal segment. The second segment is retained longest and can always be recognised by the strong bristle. The most important parts of the antenna, the sensillae, containing the olfactory organs are retained longest, even when the antennal segments on which they occur are already largely atrophied.

In tissue-feeders the tactile bristle on the outer edge of the second antennal segment bends forwards and, at its tip, inwards. In this case it clearly possesses a secondary function and protects the more sensitive sensillae from pressure from the plant tissue beside it. In sap-feeders there is no such obvious necessity for this and consequently we find that they have this essential bristle more or less atrophied and frequently only retained as a small spicule. Fig. 95 shows the various degrees of modification in mining larvae; for comparative purposes the antenna of a mining Tenthredinid is also illustrated (Fig. 95a), on which the strong bristle on the second antennal segment is also clearly visible; its sensillae, however, are not as highly developed as in the caterpillars. At (b) the antenna of a full-grown, non-mining caterpillar of *Gracilaria syringella* Fbr. is illustrated: all three segments are retained and the third segment is forked at its tip, bearing on one side a sensilla covered with sense-organs (dotted) and on the other a small, tactile seta. The second segment bears in addition to the large setae illustrated also some small setae and two sensillae. The mining *Tischeria ekebladella* Bjerk. has an antenna which is only slightly atrophied and with a simple terminal segment. *Lithocolletis* (e) still has all three segments but the long bristle on the second segment is considerably shortened. In *Phyllocnistis suffusella* Z. (f) we find the first case of atrophy of a segment; the terminal segment is merely a small invagination, smaller than the sensillae of the second segment; the base is strengthened by a swelling of the body. In the infant larva of *Gracilaria syringella* F., as well as in *Leucoptera* (g, i) the process of atrophy has commenced at the opposite end of the antenna. In this case the basal segment has disappeared, while segments 2 and 3 are still present with approximately normal development. The greatest degree of atrophy is found in *Nepticula* (h); the basal as well as the terminal segments have become atrophied. The sole remaining second segment bears its distinguishing strong bristle and well-formed sensillae; the whole antenna projects only slightly above the surface of the head. Various additional differences also occur in individual species; for instance in a larva of *Phyllocnistis nepenthae* Hering from Sumatra a more developed third segment can still be found (Fig. 96). In the last instar of *Phyllocnistis*, in which no more food is consumed, the antennae are completely atrophied; there remains only an oval area, situated ven-

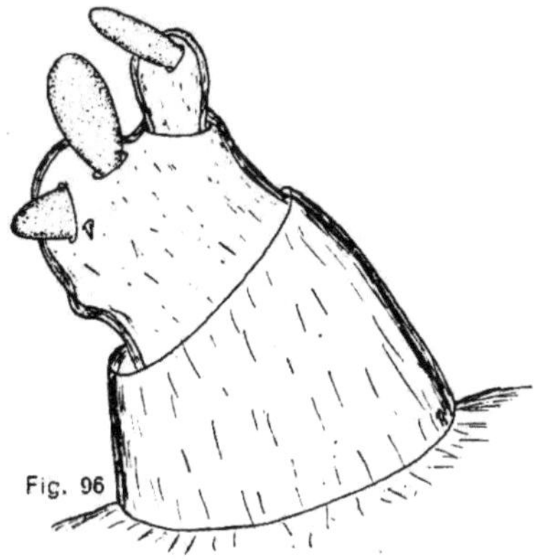

Fig. 96. Larval antenna of *Phyllocnistis nepenthae* Hering.

trally, having two sensillae and one hair, representing the vestige of the second antennal segment.

The larva's eyes, the stemmata (ocelli), are also subject to extensive modification as a result of the mining habit. In all external feeders the stemmata, invariably six in number, lie directly on the surface of the head behind the mandibles, in an irregular semicircle, open below. As has been shown, the most important modification occurring in mining larvae is the marked dorso-ventral flattening of the head. If the arrangement of the stemmata did not also change in such a flattened head, with its narrow knife-like sides, they would be split up into a dorsal and ventral group, both directed towards tissue of no importance for the life of the caterpillar, so that they would be unable to fulfil their function. The modifications arising from the mining habit therefore affect both the arrangement and number of the stemmata. In the first case the tendency is for the stemmata to become arranged in a line and, as regards their number, the direction of modification is to retain only one stemma and to enlarge this at the cost of the others, which atrophy. Fig. 97 illustrates the normal arrangement of stemmata in an external feeder, *Callisto denticulella* Thbg. (a) which is no longer mining. In *Mompha* (Fig. 97 b) the tendency can already be observed towards aligning the upper five stemmata in a straight line, while one of them comes to assume a ventral position. The same direction of development can already be found in *Leucoptera* (c) in which, however, all six stemmata have moved dorsad, while in *Elachista* (d) only three of them lie on the dorsal side and two have moved ventrad. The flattening of the head has become most conspicuous in species of *Tischeria;* the sides have become completely thin and knife-like and a correspondingly extreme

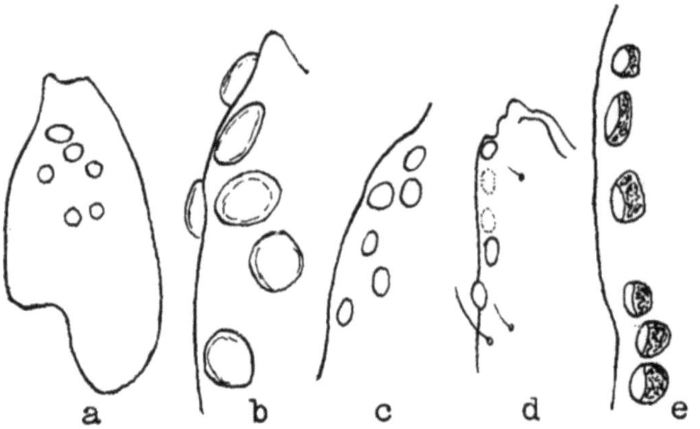

Fig. 97

Fig. 98

Fig. 97. Arrangement of stemmata in a) *Callisto denticulella* Thbg. (full-grown caterpillar), b) *Mompha* sp., c) *Leucoptera laburnella* Stt., d) *Elachista* sp., e) *Tischeria ekebladella* Bjerk. (After Trägårdh, 1913).

Fig. 98. Arrangement of stemmata in *Lyonetia clerkella* L. (after Trägårdh, 1913).

arrangement of the stemmata is found: all six lie dorsally in a line and all are developed about equally; but even here one can detect an arrangement into an anterior and posterior group, each with three stemmata (Fig. 97e) This grouping on the other hand is far more distinct in *Lyonetia* (Fig. 98); in this case the two groups are widely separated and the stemmata themselves have become much closer together; in each group two lie dorsally and one ventrally. This arrangement shows that, during its life in the mine, the caterpillar needs to see in two directions, both forwards and sideways.

The number, as well as the arrangement, of stemmata in the mining larva changes. Of the six eye spots of the normal caterpillar only three or four are present in *Lithocolletis* (Fig 93) and one can see that the rearmost of these are in process of atrophy; the foremost has become larger and the posterior ones are smaller, with less pigment and a simplified lens-structure, so that it is clear that only the anterior one has remained fully functional. We only find this development, however, as long as the caterpillar remains a sap-feeder. In the later instars as tissue-feeder, of the four stemmata three are large and lie in close juxtaposition well to the front; they are now no longer in a dorsal position but have moved ventrad. This is easily explained by the fact that the *Lithocolletis* caterpillar as a parenchymal miner, no longer eats what is in front but what is beneath it. When the caterpillar lives as a tissue-feeder in a vertically spacious tentiform mine, the space conditions are not so very different from those of an external feeder; for this reason there is such a great difference in the stemmata arrangement between two stages of caterpillars of the same species.

The genera *Nepticula* and *Eriocrania* demand separate consideration. Both have only a single stemma on each side. In *Nepticula* (Fig. 91) it lies directly on the side of the head-capsule; the caterpillar thus sees sideways and this is no doubt connected with the fact that *Nepticula* larvae practically all live, or at least commence their lives, in linear mines. In *Eriocrania* there is also only a single stemma but this lies on the ventral side of the head-capsule. This may well be connected with the fact that the species of *Eriocrania* are primarily producers of blotch mines.

The simplified number of stemmata in *Nepticula* and *Eriocrania* suggests that the two are closely related, more so than could be deduced from the morphology of the imagines. It is not, however, clear whether the simple stemma represents a primitive condition; this view might be supported by the fact that the larva of the genus *Micropteryx*, which is the most primitive in every other respect, also has only one stemma on each side. It would then have to be assumed that the six stemmata in all other caterpillars had only developed from the division of the original simple stemma. On the other hand the simple stemma may have evolved as the result of the reduction of the others, a process which would be confirmed from the process of atrophy found in *Lithocolletis* while still an epidermal miner (Fig. 93). This direction of development, carried still further in *Phyllocnistis*, which remains throughout its life a sap-feeder, then leads to the secondary development of only a single stemma (Fig. 94) and in this case even this disappears in the final larval instar, the larva finally living without any sort of optical organ.

All these processes of modification of the stemmata suggest that the caterpillar's eyes have a definite function in the life of the mining larva. Even though their whole structure shows that the optical impressions they convey can only be very weak and indistinct, the stemmata nevertheless are indispensible and, like other parts of the body, they are also adapted to the requirements of living conditions inside the mine.

We have throughout considered the mine from the dual angle of it being a cavity inside the plant tissue, serving as both living and feeding quarters for the mining larva. The changes in the larva's body we have discussed so far arose practically exclusively from the fact that the mine represents the living quarter of the mining insect. The resulting modifications of the body, and especially of the head, affected the organs lying externally on the head — the antennae and the stemmata. The modifications we will consider now which have occurred to the mouth parts under the influence of the mining habit are also connected with the characteristic of the mine as living quarter, but they are equally linked up with the feeding activity of the larva. As a result of these two influences we may therefore expect that the modifications which have occurred to the mouth parts will be particularly extensive. The part covering the actual mouth-parts dorsally is termed the labrum. The labrum, or upper lip, is developed in the form of a transverse plate in free-living larvae; its anterior portion has a median fold, which thus divides it into two lateral lobes. Dorsally it never bear more than six pairs of hairs and ventrally it has three pairs of short setae. The majority of externally-feeding larvae first attack a leaf from its edge; the median incision facilitates this feeding, since they are unable to bite off a larger piece of leaf. Furthermore, the labrum makes it possible for the larva to retain a firm grip on the portion of leaf it is attacking; in certain species the front edges are also serrated, which facilitates a secure grip.

The modifications of the larva's labrum as a result of the mining habit are extremely complex and may be partially explained by the widely varying phylogenetic origin of the genera concerned. In this brief description it will suffice to mention only the most striking of these modifications. They show different directions of development.

1. In mining larvae there is further development of the median fold. In *Lyonetia* a deepening of this incision can be noted; in *Leucoptera scitella* Z. the incision is also deepened but has in addition at its base a circular, hole-like extension (Fig. 99a). Presumably this arrangement makes it easier for the larva to imbibe the plant sap which has collected above its mouth-parts. In other genera secondary incisions develop at the base of the main incision; they are first flat, as in *Nepticula* (Fig. 99b), but in *Parornix* (Fig. 99c) have become so deep that a third, median lobe is formed.

2. Antero-lateral tufts of hairs develop and also hairs and thread-like appendages on the underside of the labrum which can be observed in many mining larvae. These hairy outgrowths function together with others on the mandibles, the palpiger or other mouth-parts. They make it easier for the caterpillar to convey the particles of food into the actual buccal cavity. Some of them are intended for the conveyance of liquid food; together they function rather like small strands of blotting-paper, others constructed more robustly are used for transporting more solid matter.

3. In other cases distinct teeth are developed on the labrum (Fig. 100a - c), which the larva quite clearly finds of use in cutting up the food-tissue. The labrum is brought into action by means of movements of the whole head. In some cases, especially in *Lithocolletis*, the hypopharynx projects over the anterior end of the labium; the larva is thus not able to use this organ for breaking up the layers of tissue in front of it; the labrum probably functions in support of the mandibles. After the flat mandibles have cut into the vertical walls of the

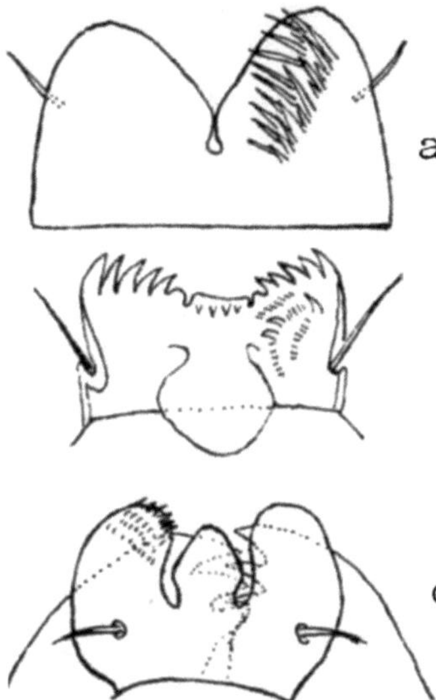

Fig. 99

Fig. 99. Labrum of a. *Leucoptera scitella* Z. (ventral), b. *Nepticula* sp., c. *Parornix* sp. (dorsal), (after Trägårdh, 1913).

cells, the labrum cuts them down to the level of the horizontal wall; that this is its function is supported by the fact that the effective surface of these teeth lies dorsally.

4. In *Phyllocnistis* an entirely different structure is found (Figs. 94 and 100d). Here the labrum is enormously enlarged and bears on its surface numerous small, vertical hairs. This genus represents the only miners which throughout their life live as sap-feeders solely in the relatively flat cells of the epidermis. It is obvious that with this type of existence the adult larva runs the danger of cutting through the epidermis with the sharp edges of its mandibles; in other words, as a result of the considerable pressure to which it is subject, there is a possibility of the larva breaking through the outer walls of the epidermis cells, thus opening up the mine to the outside; this would lead to the larva's death. There is little doubt that it is correct to assume that a labrum of this type serves as a protective cushion and keeps the edges of the mandibles away from the outer wall of the epidermis cells. In this case the modification of the labrum has not resulted from the feeding activity but has developed from the peculiarities of the too-confined living quarters. It becomes very clear that this explanation is correct, if one considers the labrum in relation to the mandibles (Fig. 101). One sees that the extension on both sides prevents any possible contact between the outer edges of the mandibles and the portions of tissue lying to the sides of the labrum.

Modifications of this type, which appear to have been induced by the specific type of mining, can also be detected in other insect orders, as T r ä g å r d h (1910) showed in his investigation of the

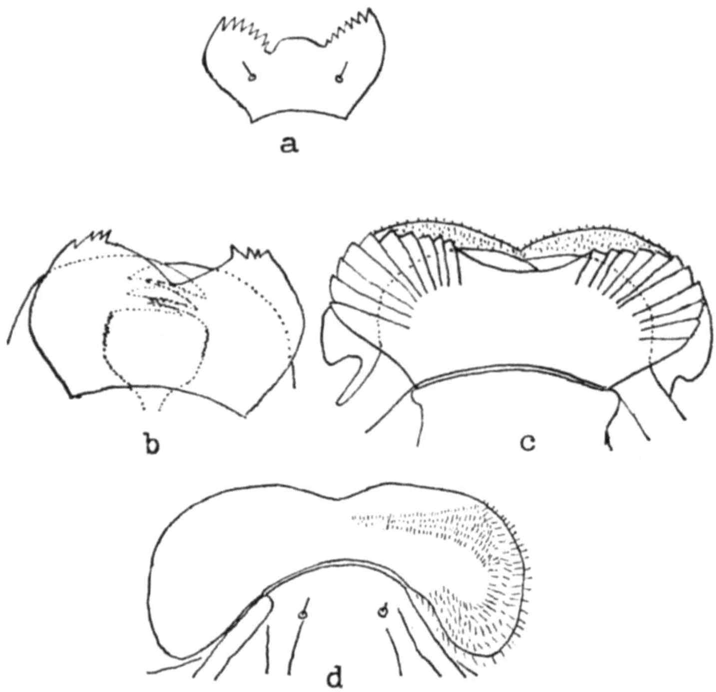

Fig. 100

Fig. 100. Labrum of: a. *Callisto denticulella* Thbg. (dorsal) b. *Gracilaria syringella* F., c. *Lithocolletis* Sp., d. *Phyllocnistis suffusella* Z. (after Trägårdh, 1913).

genus *Rhynchaenus*. In this genus the setal pattern of the upperside of the labrum varies with the habits of the larva (Fig. 102). In *Rhynchaenus fagi* L. and *Rh. quercus* L. the typical setae of the upperside of the labrum are only visible in very reduced form as a result of the fact that both species first live in a narrow linear mine; their whole head, therefore, has to be forced like a wedge into the plant tissue. In this type of feeding the setae run the danger of being damaged or of suffering extensive wear; they thus become atrophied. In *Rh. populi* Fbr., which at no stage produces a linear mine but merely consumes the surrounding tissue from within a central blotch, there is no development of a wedge-shaped head and the dorsal setae grow to their full size.

The mandibles, the actual biting organs of the larva, appear to be particularly influenced by the mining habit. In external feeders the mandible represents a shovel working in a vertical plane, convex on the outside, concave on the inside, on which the five teeth are arranged

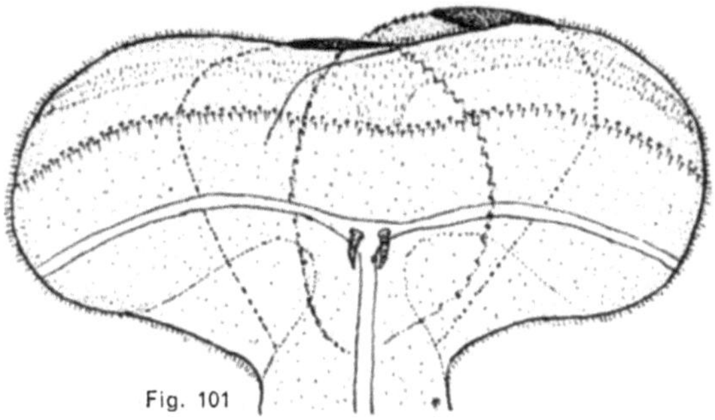

Fig. 101. Labrum and beneath it the mandibles of *Phyllocnistis nepenthae* Hering.

one below the other. The mandibles may be compared to cupped hands, in which the fingers correspond to the teeth; when the mandibles are closed this corresponds to clasped hands and the teeth on one side fit into the gaps between those on the other side and are then interlaced in a similar way to fingers on clasped hands, when these are held thumbs uppermost. There are four teeth on the edge of the underside of the mandible and a fifth lies ventrally on the inside (Fig. 103). As a result of the dorso-ventral flattening a change in the position of the mandibles occurs, in the same way, to retain our simile, as if the thumbs of the hands were turned outwards and the little-fingers inwards. In this way a horizontal arrangement develops from the former vertical arrangement. Just as there could no longer be contact between all the fingers, if the hands were turned away from each other, so the teeth of the mandibles could no longer work jointly and would lose their function. Since the central teeth of the mandible are the longest they would bore into the tissue of the leaf but would no longer cut the cell, owing to the absence of pressure from the opposite teeth. Special modifications to the mandible are therefore necessary to enable them to fulfil their function. These changes occur from two different angles, of which the one applies particularly to parenchymal, the other to epidermal miners. It is understandable that in both cases the direction of development is towards a reduction in the number of teeth, since it is impossible to continue joint action with the corresponding teeth of the opposing mandible, as occurred when the mandible was arranged vertically. In both tissue-feeders and sap-feeders the horizontal arrangement of the mandibles is the result of the horizontal position and the dorso-ventral flattening of the head. The manner in which the resulting difficulties are overcome varies considerably.

In parenchymal miners the tendency develops towards bringing

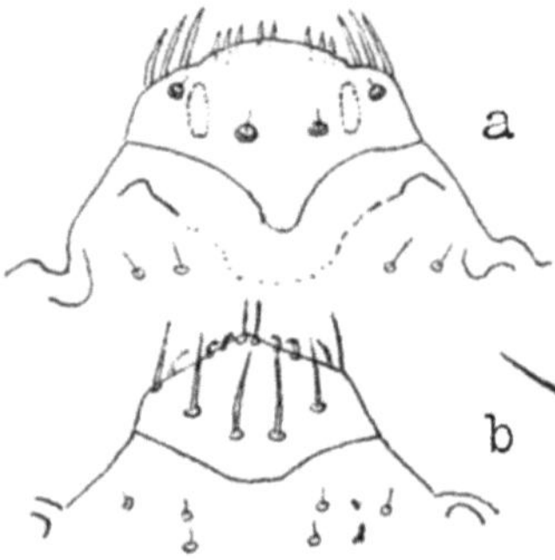

Fig. 102

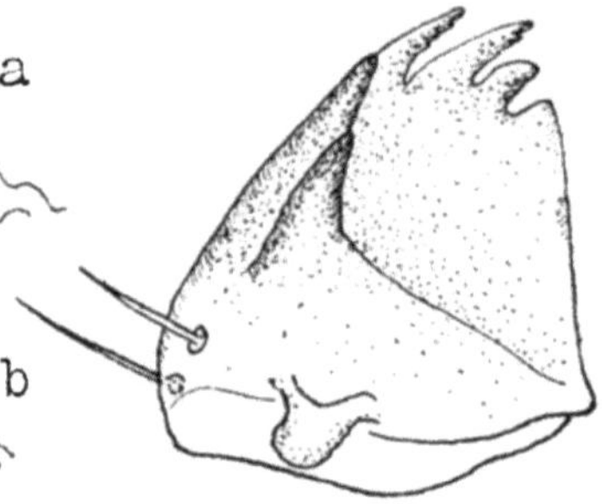

Fig. 103

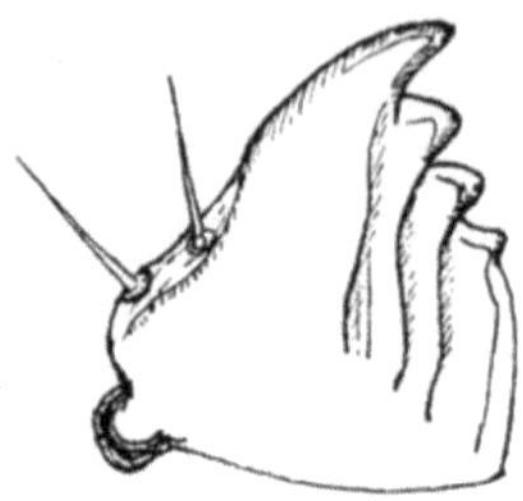

Fig. 104

Fig. 102. Labrum of *Rhynchaenus:* a. *Rh. fagi* L., b. *Rh. populi* F. (after Trägårdh, 1910) (dorsal view).

Fig. 103. Mandible of *Gracilaria syringella* F., full-grown. (after Trägårdh 1913) (ventral view).

Fig. 104. Mandible of *Leucoptera* sp. (after Trägårdh, 1913).

the tips of the teeth more and more into a straight line with the inner margin. This occurs at the expense of the development of the median area of the mandible and the number of teeth. In *Eriocrania* the number of teeth has already diminished to three, and the same process of modification is found in *Tischeria;* in both genera the median edge bears certain setae, so that when the mandibles are closed a sort of tube is formed together with the setal structures on the ventral side of the labrum, through which the nutriment is directed into the mouth-orifice. It should not be forgotten that the caterpillars in their narrow habitations can carry out only very limited movement while eating by comparison with those of an external-feeder. There is thus a danger of some of the severed bits of tissue by-passing the sides of the larva's head and not reaching its mouth. Many species in linear mines never return to spots they have left behind and any food that has by-passed their mouth-orifice is lost. This explains the formation of such organs to assist in the consumption of food. In species of *Nepticula* the tips of the teeth lie in an almost completely straight line with the median part of the mandible, which is much reduced in size, and the ventral tooth has become vestigial; in *Leucoptera* (Fig. 104) the aim is practically reached: the ventral tooth has entirely disappeared but the tips of the others lie in an approximately straight line and are able to function jointly with those of the opposite mandible. A reduction of the median area develops in consequence.

In sap-feeders the modification processes assume a quite different form. As they live exclusively in the flat epidermis cells the head has undergone a particularly well-marked flattening and consequently the mandibles have become quite thin, lamina-like blades. A change of function ensues, in that the former teeth become more and more atrophied, while the median portion of the mandible is greatly enlarged. This develops a fine serration, so that these median parts work like a circular-saw and merely serve to cut up the cells (Fig. 105). In *Phyllocnistis* this extremely highly developed median portion partly covers the basal part of the single remaining tooth which becomes functionless; presumably it will be possible to discover among them species of permanent epidermal miners which have even lost this last tooth. In this case reduction takes place in the opposite side of the mandible, which we have compared with the side of the hand bearing the thumb, while the other side which was atrophied in tissue-feeders, undergoes an especially striking development in sap-feeders.

Among the Tenthredinidae there are in general not such striking modifications to the mandibles; this may perhaps be explained on the one hand by the fact that very closely related species are involved, and on the other hand by the fact that mines of saw-flies show very little differentiation and are all developed uniformly as full-depth blotch-mines. It is all the more surprising to find in one genus a quite different shape of mandible in the final instar (Fig. 106). This occurs in the genus *Phyllotoma* but is possibly to be met with in other genera and is merely more difficult to detect than in the species concerned, which pupates in the leaf. While the mandible is developed in all stages except the last in a similar way to the majority of external-feeders and merely appears to be arranged somewhat more horizontally, in the last instar it has become rather like a dagger and

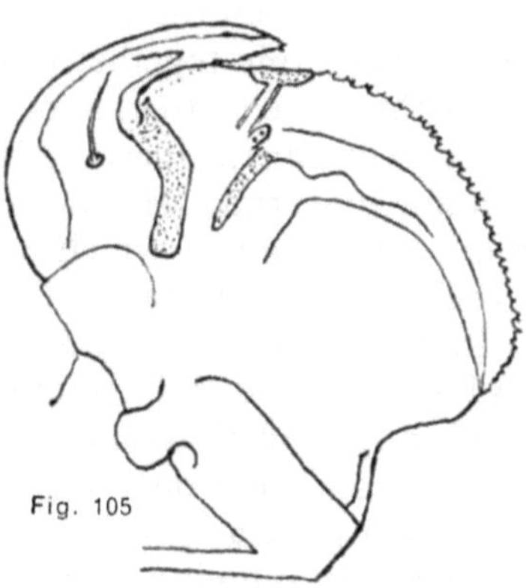

Fig. 105

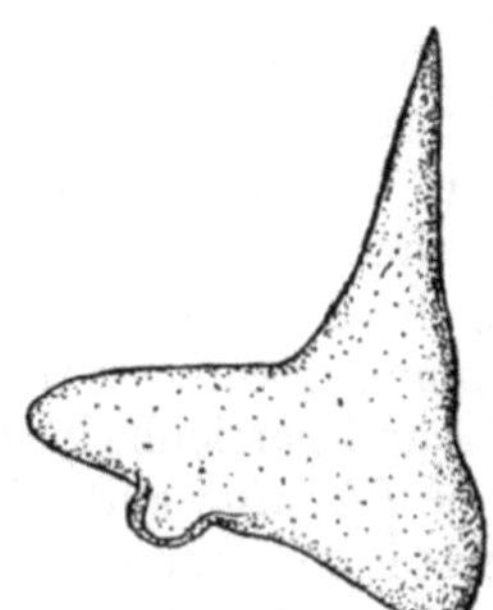

Fig. 106

Fig. 105. Mandible of *Phyllocnistis suffusella* Z. (after Trägårdh, 1913).
Fig. 106. Mandible of *Phyllotoma vagans* Fall., last instar.

possesses only a single tooth in the median area projecting forwards. H. Weber (1940) who first discovered this structure in *Phyllotoma aceris* McLchl., which cuts a circular piece out of the upper epidermis of the leaf and when pupating uses this for enclosing one side of its lenticular cocoon, assumed that the peculiar modification was somehow connected with cutting out this piece of leaf. This, however, cannot be so, since the same modification is found in all representatives of the genus, even in those which construct their cocoon inside the leaf and do not cut the epidermis at all. The large degree and strange type of modification of the mandible suggests some connection with the construction of the cocoon, although I am unable to form any picture of how such a modified mandible might be of use in this process.

The processes of modification also affect the most ventral area of the mouth-parts, the first maxillae and the second maxillae merging into the labium. Maxillae and labium together form a functional unit and must be considered as such here. They appear to have a special importance in miners for conveying the pieces of bitten-off tissue into the actual buccal cavity. In this way the lower lip (labium) and the maxillae which flank it, function as a shovel with which the bits of tissue are shovelled up from the floor of the mine and conveyed into the mouth. A second very important organ, however, adjoins the labium, namely the spinneret. Spinning is of great importance for the caterpillar almost throughout its entire life. The secretion it produces is not only required for the construction of the pupation and any possible moulting cocoon, but silk threads are constantly being produced to assist its movement: the majority of caterpillars, especially when they are on smooth surfaces, spin a zig-zag line of silk on which they can gain a more secure footing. In other cases they lower themselves from the tree on a silk thread; the majority of Tortrices and Pyrales spin together leaves as their larval habitation. The spinneret is therefore of great importance for them. This also applies to the majority of miners, inasfar as members of the families under discussion are concerned. However, even among the Tineoid families there are many genera which employ their spinning secretion throughout their life, as for instance *Atemelia*, *Bedellia* and particularly *Tischeria*. In *Nepticula*, *Leucoptera*, *Lyonetia*, some *Elachista* and others, no silk is produced while mining; the spinneret is correspondingly atrophied and any clash with the shovelling function of the lower lip is obviated. They only commence spinning at the time of pupation. In such genera with this limited spinning activity, the lower lip, where the spinning apparatus is found becomes smaller, the spinneret and also the labial palpi degenerate and recede from the forward edge to the rear. The shovelling

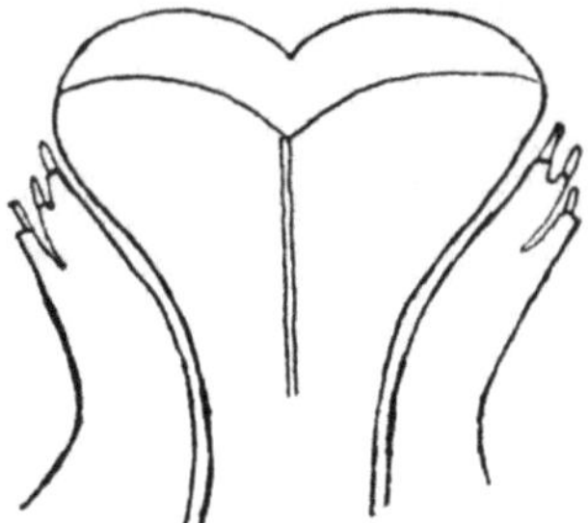

Fig. 107

Fig. 107. *Gracilaria syringella* F. first instar, top of labium and palpi (after Trägårdh, 1913).

function of the lower lip is assumed by the enlarged hypopharynx, the ventral portion of the anterior mouth-cavity, and it is assisted by the basal parts of the maxillae, especially by the palpiger, the basal joint bearing the maxillary palpi. In *Tischeria* a different method of modification is adopted. The spinneret is used throughout the larva's life; however, it is not "in the way", since it is concealed beneath a horizontal fold in the mentum, while simultaneously the palpigers of the maxillae are greatly enlarged and lamina-like and thus together form an almost uniform plate, which in function assists the lower lip.

In sap-feeders we can observe the same tendencies of development in these structures, i.a., degeneration of the lower lip, assumption of its shovelling function by the hypopharynx and modification of the basal parts of the maxillae; the latter, however, undergo quite excessive development in two different directions, so that it is impossible to establish any uniform process for this development. In *Gracilaria syringella* Fbr. we find in the first instar a considerably modified labium (Fig. 107), which is flanked on both sides by the likewise largely atrophied maxillae. The labium has become a thin lamina with an incision in its anterior margin. Its anterior portion is covered with extremely fine hairs and in fact is the hypopharynx, which has merged with the labium, (the hypopharynx is always characterised by these fine hairs); the posterior portion is the mentum. Spinneret and labial palpi have become completely lost. In the most extreme cases in epidermal miners the maxillary palpi also completely disappear and only remain as small setae near the more fully developed spinneret, as for instance in *Phyllocnistis*.

On examining the later stages in sap-feeders considerable differences can be observed within individual genera, which for their part are presumably connected with morphological differences. *Callisto* and *Parornix* after leaving their epidermal mine live under a folded-over leaf-edge or in a leaf-roll or leaf-cone, the latter often being cut out of the leaf with great skill (Fig. 22). They now feed in the same way as other free-living micro-lepidopterous larvae. The genus *Lithocolletis* does not in fact leave the mine and with few exceptions also pupates in the mine but the type of mining is quite different; the caterpillar no longer lives in the epidermis cells but now feeds as a tissue-feeder. Since the mine-cavity is made quite spacious by the formation of folds in the epidermis as a result of its spinning, its type of existence does not really greatly vary from that of its relatives *Callisto* and *Parornix*, which have given up mining altogether. However, its head becomes more pointed and the hypopharynx is strongly developed and it shares these characteristics with other tissue-feeders.

Phyllocnistis remains a sap-feeder throughout its life but in the last instar consumes no further food. During this period it is scarcely recognisable as a caterpillar; it no longer has any stemmata, the antennae which have become ventral are as much atrophied as in *Nepticula;* at the end of the mentum there is a raised plate, adjoined by the greatly enlarged spinneret and on each side of this is an area covered by setae, representing the remnants of the maxillae; the labial palpi have completely disappeared. From these modifications one can see that in *Phyllocnistis* this final instar is entirely devoted to spinning its cocoon.

The formation of the different parts of the head in all species which live first as sap-feeders and later as tissue-feeders in a certain sense contradicts the biogenetic principle that individual development broadly reflects phylogenetic development and also another law, that organs that have once atrophied do not re-appear. In their early instars we find distinct processes of atrophy of widely varying parts of the body, while later instars show a more normal development. According to the biogenetic principle the reverse should be the case. For example, atrophied stemmata in the sap-feeder instar re-appear during life as a tissue-feeder. One sees from this that all these modifications are closely connected with the type of feeding and should be considered as mere adaptations, of value for phyletic speculations only in very few cases. This also explains why the modifications described do not follow on logically step by step from one related genus to the next but occur at widely varying places in the taxonomic system. Subsequent investigations will have to separate these modifications into those inherited phylogenetically and those which are mere adaptations to the mining habit. It can already be accepted that in many cases numerous modifications can be ascribed to the type of plant tissue in which the larva is living.

The colouration of the larva in relation to its mining activity has so far been very little investigated. In general, mining larvae are practically never so brightly coloured as many of their free-living relatives. Endophagous larvae, i. e. those which live inside plants are invariably paler or almost colourless. This results from the fact that light is necessary for the development of pigment and is almost entirely lacking for species feeding, for example, in wood. Some light can penetrate into a mine, even though in greatly reduced strength. Mining larvae are accordingly pale, whitish or a bright greenish to yellowish. Only a few areas of the larva's body, such as head, pronotum and anal shield are usually of a darker colour, not as a result of greater pigmentation, but owing to the chitin being thicker in these places. The intestines are usually darker as a result of the food pulp they contain, as also are the testes and the Malpighian tubes. Their pigmentation which is often quite striking is difficult to explain. In many species of *Nepticula* the ganglia are dark and are clearly visible through the skin as dark patches. One has to take care not to confuse them with genuine dark markings on the larva's body. It is not correct that the degree of dark colouration of the ganglia depends on which side the larva turns to the light; it is therefore of no importance for the development of the larva, whether it mines with its back or its belly uppermost. A green or yellow colouring of the caterpillar is an important characteristic for identifying species, especially with many species of *Nepticula*.

In many larvae the green colouration changes in the final instar into a shiny red, although they also spend this instar in the mine. This applies especially to many *Gracilariidae*, for instance to *Leucospilapteryx omissella* Stt. (in Artemisia) and to *Cosmopteryx eximia* Stt. in Humulus. With the former and many of its relatives similar transformations from green to red occur in the colour of the leaf; these phenomena will therefore be discussed later in the chapter on the colour and discolouration of the mine.

12.

THE MINERS' CHOICE OF FOOD

The various aspects of mining which have been discussed in the previous chapters have shown clearly that the mining habit represents a high degree of specialisation in the insect larva. The construction of a mine means that the larva creates for itself special living quarters, of a kind rarely encountered in other types of existence, but bearing some resemblance to feeding channels in stems or roots. This endophagous method of feeding, however, also represents specialisation in the type of food. In contrast to external feeders, mining larvae only consume certain plant substances; the larva has become more selective than free-living forms and no longer indiscriminately eats the whole tissue of the plant but has specialised in certain substances, on which it relies for its nourishment. In contrast to the external feeder, it avoids eating certain plant tissues which it finds of little value, such as the epidermis which is of relatively low nutritive content; furthermore, the epidermis is frequently covered with a protective skin, the cuticle, which can only be digested with difficulty or not at all and which is of course eaten by the external feeder and merely represents superfluous ballast in the digestive system. Although the mining habit in itself represents a specialisation in the choice of food, we nevertheless find that miners carry this specialisation very much further. This specialisation should be considered from three different angles: first, which t i s s u e of the plant is eaten, secondly, which o r g a n s of the plant are attacked and, finally, it is of interest to know which s p e c i e s o f p l a n t are preferred. Miners, or at least the great majority of permanent miners, unlike external feeders, affect only a single species of plant on which the egg has originally been deposited; in general this food plant cannot be changed and it is particularly interesting to investigate which plants are favoured by each mining species and which, on the other hand, are rejected. The problem of the selection of the food plant devolves in most species on the ovipositing imago; this, however, in no way detracts from our observations. It will be seen that in the majority of cases the selection of the food plant is not haphazard, but is subject to certain regular patterns which often clearly illustrate the ability of the species concerned and give a highly interesting insight into the inter-relations between plants and insects.

a) Selection of tissue by mining insects.

In many cases but not always, the manner in which a larva feeds on a plant is dependent to some extent on the place of oviposition and in such cases the selection of the food rests with the ovipositing female. The genuine, mining larva — apart from sap-feeders, the epidermal miners — feeds on the green tissue of plants which contain chlorophyll. It is not yet known in detail which substances from the plants' cells containing chlorophyll are necessary to the larva for its nourishment and growth. Investigation of its frass has shown that a considerable part of the substances contained in the cells are defecated without having been digested. The larva does not digest carbohydrates — starch and sugar — which are subsequently voided in the frass. The green granules of chlorophyll become darker than in the living plant cells after passing through the digestive tract. This, however, is not because the chlorophyll has played any part in the nourishment of the larva; presumably, under the influence of certain digestive ferments, it undergoes a secondary change from chlorophyll to the darker chlorophyllan, without the larva deriving any benefit from the process. Similarly, other characteristic mineral products of the cell, especially the crystals or crystalloid structures consisting of calcium oxalate and calcium carbonate, are voided undigested and usually unchanged. Naturally, products of this type may be indirectly used by the larva in the course of certain adaptations, as occurs with the trichoblasts in the tissue of Nymphaeaceae; these plants are used by the mining larva of *Hydromyza livens* Fall. in the construction of its puparium, but this is a rare exception. (This case is discussed in greater detail in the chapter on miners in aquatic plants). The ability to digest cellulose, either with or without symbiotic organisms which are found in many wood-eaters, is lacking in most miners; Haberlandt (1918) discovered, however, that *Leucoptera* is able to digest cellulose to a limited degree. Thus, of the products contained in the plant cell, protein products are alone of value to the larva as food and it is these proteins which are affected by the specialisation of the mining larva. We know that the number of different proteins is extremely large and it is on these differences of the proteins that sero-diagnostic investigations of plants (and animals) are based. Plant protein is found especially in the cytoplasm and the nucleus of the cell and, to a lesser extent, also in the granules of chlorophyll. Selection according to the type of protein in the plant cell thus represents a specialisation of the mining larva and we will see later that this selection is developed to a quite surprising degree.

It would be extremely important for our subsequent investigations to know which protein of the cell is of the greatest value for the mining larva — the protein of the cell cytoplasm or of the cell nucleus. It is usually accepted that the nucleus of the cell is the determinant of heredity and also of relationship between species but in some quarters it is claimed that heredity is conveyed by the cytoplasm. Unfortunately it has not yet been determined whether the mining larva selects primarily the nucleus or the cytoplasm. The question must therefore, for the time being, remain open. The remaining products contained by the cell, such as fatty and aetherical oils, glucosides, alkaloids, etc., are clearly of no importance for the larva, even though their presence or absence is frequently decisive when the larva is selecting its food. They play no part in the larva's feeding but it is just possible that they have some influence as a means of enticing the ovipositing female. All experience, however, shows that they play a very subsidiary role and that their importance in the selection of the food has been much exaggerated.

There is little doubt that different types of protein occur in the different types of cell in a leaf, even although present chemical research has not been able to determine such differences. When a tissue such as the palisade parenchyma serves solely for assimilation and the spongy parenchyma primarily for ventilation or secondary assimilation, it can be assumed with some degree of certainty that the differing function corresponds to a differing structure of the protein contained in the cells, even though this has not yet been proved. The protein in the almost liquid content of the epidermis cells will likewise be different from that of the parenchyma. Although our chemists have not yet been able to prove these protein differences, there is little doubt that they exist and their presence is suggested by the differentiation in the taste of the mining larvae, of which some prefer one, others a different tissue. Such a difference is also indicated by the differing serum reactions when used to prove plant relationships. It has been found that serums from fruits had a quite different range of reactions to that from various leaf parts. In serum diagnosis use is made of the differences between specific types of protein which occur in leaves and the so-called reserve proteins in the fruits of the same plant. These represent the extreme differences in the protein of the same plant and there is little doubt that future research will be able to show that the proteins in the different tissues of the same plant organ also differ in certain respects, and this differentiation is already indicated by the specialisation of mining larvae.

It can be accepted that the greatest difference is found in the protein occurring in the epidermis cells, and that the proteins of the green plant tissue differ from each other only slightly, according to whether the palisade or the spongy parenchyma is concerned. Thus the miners of epidermis cells represent a distinct group, in as far as they adhere to this type of food throughout their lives. Included in this group are primarily representatives of the genus *Phyllocnistis* Z. which mine exclusively epidermis cells throughout their lives. Members of this genus are represented throughout the world and show a preference for no particular family of plants but at all times are only found in epidermis cells and appear to be completely dependent on epidermal protein. However, there are in addition numerous species which mine the epidermis in their infant stage, as for instance many Gracilariidae and also some Agromyzidae. They give up this type of existence after a moult (usually the first) and then become parenchymal feeders. It has already been shown how far-reaching modifications to the mouth-parts develop parallel with the change-over from being a sap-feeder to a tissue-feeder. One can be certain that such basic changes do not take place merely externally but that, after such a vital moult, the digestive tract also undergoes a far-reaching modification. The fore and hind-intestine at least, which are renewed during the moult, probably undergo changes which would not be too difficult to detect, while changes in the mid-intestine, where the actual absorption of food takes place, might not be so easy to confirm but doubtless occur. Thus, with the change-over of the larva from food with epidermal protein to food of parenchymal origin, modifications probably occur in the intestinal organs which are as extensive although perhaps not so striking as those which occur in the mouth-parts. This opens up a wide and fruitful field for future histological research, which will probably only be made possible by the employment of subtle chemical reactions.

Tracing this differentiation still further, one must also assume that the epidermal protein of the upper surface of the leaf differs from that of the lower surface. This idea is not immediately obvious, since in both cases the function of the tissue is the same, on both upper and lower surface the epidermis serves as a protection. However, we know that this protective skin is not completely identical on the upper and the lower surface; the epidermis on the underside differs in certain respects from that of the upperside, especially for instance, by the existence of stomata, organs for excretion of water (hydathodes) and sometimes in the number of cell layers. It can therefore probably be assumed that, at least in certain cases, the cells differ in their type of protein, according to whether they are on the upper surface or lower surface of the leaf. This is also confirmed by the behaviour and differentiation in taste of mining larvae. *Phyllocnistis suffusella* Z. lives on Populus nigra L. and in this case there is no preference for one or the other epidermis; the species is found as frequently on the upperside as on the underside. *Ph. labyrin-*

thella Bjerk. lives likewise in the epidermis on Populus tremula L. However, it occurs predominantly in the underside of the leaf and only rarely in the upperside. One might assume that the two species of Populus differ in that in Populus nigra L. the cell content of the upper epidermis differs less from that of the lower surface than in Populus tremula L. It will be shown later that a similar differentiation affecting the parenchyma can be suspected in these two species. These examples show, however, the degree to which the specialisation in taste has developed in mining larvae, as expressed by the selective evaluation of the upper and lower epidermis of the same leaf. In certain species a larva hatching from an egg laid by mistake on the wrong side of the leaf is unable to live in these cells, showing that the larva itself is adapted to a specific type of epidermis cells; this fact indicates that the preference of the ovipositing female for one or other epidermis does not rest on perhaps purely mechanical differences in the leaf cells. For instance, one finds not infrequently that eggs of *Lithocolletis emberizaepennella* Bché. are deposited on the upper surface of leaves of Lonicera and Symphoricarpos. An unnatural-looking mine is then formed in which the larva sooner or later dies, before being able to construct its subsequent parenchymal mine. Other species, on the other hand, seem to be less specialised in their tastes; certain species can develop normally in either the upper or the lower surface, of which we may perhaps quote as the best-known example *Callisto denticulella* Thbg. which is common on Malus and which in some years develops also in considerable numbers in lower surface mines. Normal development from upper surface mines may also occur in *Lithocolletis rajella* L. on Alnus. Normally, however, the genus *Lithocolletis* shows a marked consistency in the side of leaf in which its epidermal mine is constructed and this is a good guide for distinguishing individual species.

The change-over from epidermal to parenchymal mining seems to be quite obligatory in all species, presumably always after a specific moult. This change of skin is reflected in a change in the selection of essential food and it will doubtless in due course be possible to show parallel changes in the intestinal tract, similar to those already described affecting the mouth-parts. A change from epidermal to parenchymal feeding is the normal rule, especially among most species of Gracilariidae, but is also found elsewhere, as in some Agromyzidae. A strange intermediate stage is interpolated in *Lithocolletis corylifoliella* Hb. which mines the upper surface of Betula leaves. This represents the only so far proven case of a "double mine," a mine in a mine. Investigation of its upper surface tentiform mine which is usually formed above the mid-rib or a large lateral vein, shows a differently coloured, usually lighter, patch in the centre

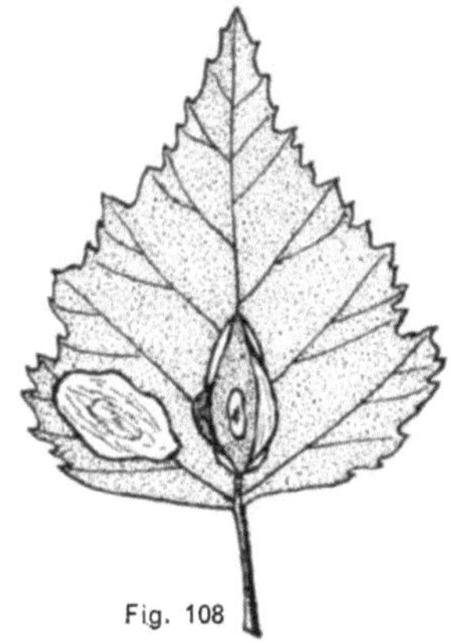

Fig. 108

Fig. 108. Betula: *Lithocolletis corylifoliella* Hb. The mine opened up on the midrib shows a mine within the mine.

of the mine. If the outer wall of the mine, consisting of the outer walls of the eaten-out epidermis cells, is opened up (Fig. 108) another small mine is found lying in the parenchyma. A microscopical cross-section of this "double mine" (Fig. 109) shows for the primary mine the normal picture of epidermal mines (M 1). Beneath the remaining inner walls of the epidermis cells, however, there is a second cavity (M 2), the inner mine. It seems that in this case the caterpillar has merely removed the epidermis walls and has opened the upperside of the parenchyma cells in order to consume their sap, in other words, has continued to live in the internal mine actually as a sap-feeder. The parenchyma cells are completely eaten up only later and the caterpillar commences feeding in the manner typical of the genus. This behaviour is confirmed by the fact that the larva's characteristic sap-feeder head is retained in the second instar and the previously described transformation of the mouth-parts and head-structure do not take place till the third instar. Perhaps more thorough investigation will also reveal the construction of a double mine of this type in other species of *Lithocolletis.*

A considerable differentiation of taste can also be shown in many parenchymal miners. Frequently the position of the mine is constant for the species concerned, either on the upperside or the underside of the leaf. We previously referred to the fact that starch and sugar are formed in the assimilation parenchyma and that these substances are then transferred in another, easily moveable form to the conducting parenchyma. The carbohydrate content is not therefore identical in the parenchyma of the upper and of the lower surface of the leaf. Carbohydrates are utilised as food by the mining larva

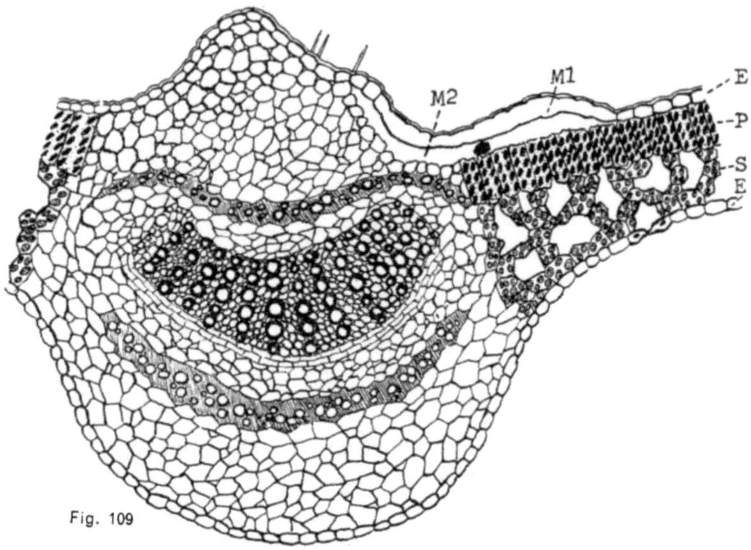

Fig. 109. Betula: Double mine of *Lithocolletis corylifoliella* Hb. near the mid-rib in a cross-section of the leaf: M1 - primary (epidermal) mine, M2 - secondary mine lying below the epidermis (E - epidermis, P - palisade parenchyma, S - spongy parenchyma).

either not at all or only to an insignificant extent; in this respect, therefore, there is scarcely any essential difference for the larva between the two types of parenchyma. Meanwhile, it must be assumed that the protein of the palisade parenchyma is differently constructed, in accordance with its differing function, to that of the spongy parenchyma; this difference in the proteins certainly accounts for the fact that certain species prefer one or the other parenchyma.

The simplest situation is found in those insect larvae which construct full-depth mines, where therefore the whole mesophyll between upper and lower epidermis is consumed. In such species no distinction is made in the type of protein food; they obviously find one as good as the other and in these species no differentiation of taste has as yet developed. It is thus to be expected that the larvae of Tenthredinidae which are primitive in many other respects will always construct full-depth mines (apart from species of *Fenusa*, in which the early part of the blotch-mine is always somewhat shallower than the subsequent mine). The Eriocraniidae are likewise primitive and, similarly to saw-fly larvae, live predominently in full-depth blotch mines.

A large proportion of miners, however, construct their mines consistently in one or the other side of the leaf, and this is found especially in the Agromyzidae, which are so highly specialised in other ways and in the Gracilariidae, which have such differentiated habits. If the mine is restricted to a single layer of parenchyma, in most cases the species of insect concerned likewise constructs either upper-surface or lower-surface mines. Exceptions are encountered only in very few species. The linear mine of *Nepticula trimaculella* Hw., for instance, in the leaves of Populus nigra L. is encountered equally often on the upperside as on the underside (this applies also to *N. populetorum* Frey & Boll found in the United States on the same plant), and *Phytagromyza populi* Kltb. also mines either side of the leaf indiscriminately. We must point out, however, the peculiar conditions encountered by epidermal miners in this species of Populus, which we have mentioned previously. According to A. Braun the lower surface of the leaves of this species also contains assimilation parenchyma, so that species frequenting these leaves find similar conditions on either leaf surface.

It is worthy of note that in primitive forms such as the Tenthredinidae and Eriocraniidae the mine is always upper-surface, apart from cases where the whole mesophyll between upper and lower epidermis is removed. If a thin layer of parenchyma remains, it is always the spongy parenchyma of the underside of the leaf. One might deduce from this that the assimilation parenchyma is in fact more palatable for the mining larva. This is actually easy to understand since presumably, if carbohydrates are absorbed at all, those originating in the assimilation parenchyma can be consumed more easily than those already prepared for removal in the conducting parenchyma.

The differentiation of taste may go even further. The assimilation parenchyma frequently consists of several layers and theoretically it may be assumed that the protein also differs according to which layer of the parenchyma it is in. Such differences are suggested by the fact that there are species of miners which do not feed indiscriminately in the assimilation parenchyma but confine themselves to a specific layer of it. A number of species, for instance, feed on the lowest parenchyma layer but do not touch those above. This is found particularly frequently in *Phytomyza* species. *Ph. affinis* Fall. for instance, lives in the lower layer of the assimilation parenchyma of Carduus and Cirsium; the upper layer remains untouched and merely appears paler (Fig. 5). It has not yet been investigated whether this is due to a process of degeneration in the upper layer or whether the larva perhaps without eating this layer nevertheless imbibes sap from it. Feeding of this type can always be recognised by the mine assuming a peculiar yellowish-green colouration. In other cases the upper layer of the assimilation parenchyma is apparently not affected in any way and the larva living in this way and also its mine is extremely difficult to detect, especially for instance, the early mine of *Phytomyza heringiana* Hend. in the leaves of Malus. In many cases the larva lives primarily in the lower layer of the assimilation parenchyma but eats small holes into the upper layer which reach the epidermis. This makes the presence of the mine easy to detect, such as, for instance, many Umbelliferae mines of the type of *Phytomyza heracleana* Hering on Heracleum, while, for example, *Ph. pauli-löwi* Hend. on Peucedanum oreoselinum L. does not form similar holes. Such interparenchymal, ("binnenläufige") mines always occur only in the assimilation parenchyma and never in the spongy parenchyma, which has no division into separate layers.

A large number of miners change the side of the leaf they mine during the course of their larval development. There are species which first mine on the lower surface, then later on the upper surface and vice versa. It also occurs that a species commences on the underside, continues feeding on the upperside and finally returns to the underside of the leaf. In many cases this represents an ill-developed differentiation of taste: the larva thrives equally well on spongy parenchyma as on assimilation parenchyma. In other cases this change is obligatory, possibly linked up with moults. Very closely related species behave differently in this respect. For instance, a pair of species live in linear mines in Clematis vitalba L., of which the mine of *Phytomyza vitalbae* Kalt. runs solely on the upper surface; a very closely related species, so far only recognised as a separate species from the morphology of the larva, commences with a short stretch of mine on the lower surface which coils around the point of oviposition, and only then does the larva visit the upper surface. The mine of *Phytomyza rectae* Hend. also follows an obligatory course, first as a short spiral coil on the underside of the leaf, and then immediately leads on to the upperside, so that in this limited area the mine appears quite transparent. In *Ph. kaltenbachi* Hend. on the same plant the lower surface channel is very shallow and extremely long, before transferring to the upper surface. Consistent differences of this type are very valuable for distinguishing otherwise similar mines of different species on the same plant. In a whole series of Compositae miners related to *Phytomyza sonchi* R.-D. there is first of all a considerable length of channel on the underside of the leaf and finally, a portion of varying length on the upperside. There are also species such as the polyphagous *Ph. atricornis* Mg., which apparently finds the side of the leaf quite immaterial; it occurs both on the upper and lower surfaces and no rules for its behaviour can be detected. However, the final stretch of mine which ends up as the pupal-blister is constructed almost exclusively on the underside of the leaf.

That it is in general of importance for the larva which parenchymal tissue it eats, is clearly shown by the attention paid to this by the female. Oviposition, both on the upperside and underside, does not take place, as might be assumed, under the influence of light stimuli. Smulyan (1915) showed by experiments involving the reduction of light that the ovipositing female is not influenced in its choice of leaf surface by variations of light intensity.

Changes also occur in the depth of mine produced by the larva. Mines commencing as upper or lower surface mines later become full-depth. It need not be assumed in such cases that an originally existing differentiation of taste is lost. A mine which starts by being limited to one surface and subsequently becomes full-depth usually results from the single layer of parenchyma no longer providing the larva with sufficient food for it to complete its development. If it is not able to transfer to another leaf, it is then forced to eat tissue, which it originally found unsuitable. Such cases are of very frequent occurrence. If, however, the contrary occurs and the larva which first produces full-depth mines, later develops a preference for one or other surface of the leaf, it can be assumed that a differentiation of taste has evolved. For instance, the larvae of *Philophylla heraclei* L. on Umbelliferae, the various *Pegomyia* species on Polygonaceae and Chenopodiaceae and *Microsetia hermannella* Fbr. on Chenopodiaceae, which commence by forming full-depth mines, later continue feeding only in the upper surface. Such a change of taste is probably linked up with a moult.

Finally, in connection with the choice of leaf tissue to be eaten, one further peculiarity must be mentioned; certain species of miner make a temporary distinction between the "living" and "feeding" quarters of their mine. A mining larva will sometimes mine a considerable distance in the tissue of the leaf without consuming any of the tissue it removes. In this way substantial lengths of the mine channel are found in which the severed cells lie on both sides of the channel but have not passed through the larva's body; accordingly, such portions of the channel are without any traces of frass. *Dizygomyza luctuosa* Mg. produces long, upper surface mines in the leaves of Carex, which do not differ from those of related species. Finally, however, it constructs a rather long length of channel in which it no longer feeds and which is therefore free from frass; this part of the channel is made only in the leaf sheath, where pupation takes place. In a similar way the mine of *Scaptomyzella incana* Mg. in small-leaved species of Caryophyllaceae frequently passes from one leaf through the stem into another and during its passage through the stem the larva takes no food, clearly because the parenchyma of the stem of, for instance, Stellaria, is not as palatable as that of the leaves. In contrast to this, one finds in *Nepticula headleyella* Stt. on Prunella, where several leaves are mined, that a continuous line of frass also occurs throughout the stem and that therefore the larva has also eaten the stem parenchyma on its way. The construction of lengthy mine channels in which the larva consumes no food occurs frequently in aquatic plants; in the existing channels containing air, divided up by only a few partitions, or in the more compact air tissue ("aerenchyma"), the larva has no great difficulty in covering considerable distances, without having to make much use of its mouth parts.

The vascular bundles of the plant running through the leaf and stem tissue and which are visible externally, particularly in the leaf veins, are of considerable importance in the life of the miner. Many miners avoid the vascular bundles, others favour them. There are one or more vascular bundles in each leaf vein, petiole or stem, and each vascular bundle is of varying composition. They contain cells, which are woody and very much thickened and promise only a small yield of protein for the mining larva. The veins include especially the thick fibres of sclerenchyma which function as strengthening elements, but also the organs of the water-conducting system, which consist of thick tubes possibly strengthened by rings or spirals and which have no or very limited protein inclusions. Besides these there are other elements, suited to the transport of carbohydrates and protein which have greater nutritive value for the caterpillar; these are the sieve-tubes. In the water-conducting parts of the vascular bundle (xylem, hadrom), as well as in the parts conducting plastic substances (phloem, leptom), there are companion cells, which provide the larva with even better nourishment. Finally, the vascular bundle also contains the formative tissue or meristem and, as a covering for the whole bundle, a parenchymatous tissue which differs least of all from the remaining parenchyma of the leaf. The mining larva is thus able to find sufficient food even in the vein. One finds, however, that most species of miner have a tendency to avoid leaf veins when constructing their mine. In many cases this results in the characteristic shpae of the mine channel, which avoids crossing leaf veins where they are strongly developed and, instead, runs along beside them; only when the vascular bundle has branched many times and is consequently not so strong, is it crossed. This, the serpentine or intestinally-coiled mine (the heliconome and visceronome) can be ascribed to the desire of the larva to avoid the vascular bundles. This explains also why linear mines towards the centre of the leaf often wind about in a striking way but that, when the egg has been deposited on the margin of the leaf, where the vascular bundles are quite tender, the same species produces a completely straight channel. This also explains the course of the mine along the serrated edge of the leaf, which is particularly marked in *Nepticula marginicolella* Stt. In many cases the young larva avoids crossing the leaf vein but does so later without difficulty when the more adult larva has developed stronger mouth-parts. Variations according to the species can be noted, however, even in these cases. While *Nepticula* larvae invariably avoid crossing the mid-rib (they do so only as the most rare exception), in *Lyonetia clerkella* L. one frequently finds that the mine channel crosses the mid-rib even in its earliest instar.

By contrast, there are species which particularly favour the leaf veins and in which the greater part of the mine runs through the mid-rib or one of the stronger lateral veins, with only small offshoots to the sides. In almost all these species the larva returns to the channel in the veins during pauses in feeding and this would appear to be influenced by the need for protection. The mine constructed in the vein is far less visible and this type of existence thus offers the larva protection against certain enemies. Furthermore, the covering above the mine is much stronger than the simple epidermis on the other parts of the leaf. Mines in the mid-rib are encountered very extensively, for instance in the dipterous larvae of *Trypeta zoe* Mg. on Compositae and *Phytomyza ramosa* Hend. on Dipsacaceae, in *Phytomyza alpigenae* Hend. on Lonicera, in *Agromyza apfelbecki* Str. on Compositae and in many others. In the first-mentioned, the lateral branches are very long, in the latter they are very short, while the others occupy an intermediate position. If these offshoots are shorter they are more numerous, while the longer ones are less frequent. The greater protection found here presumably explains the fact that pupation frequently takes place in the mine in the vein, but this is not always the case; for instance, it does not apply to the commonest and most polyphagous species of this group, the well-known *Liriomyza strigata* Mg. This habit is also found not infrequently among the Lepidoptera, the bestknown example being *Gnorimoschema acuminatella* Sirc. on Carduus and Cirsium. The presence of certain of *Nepticula* related to *N. argyropeza* Z. must be considered from a special point of view; the caterpillar which hatches in the parenchyma of the petiole proceeds by the shortest route into the petiole near a vascular bundle; the special influence which this feeding near the vascular bundle exerts on the formation of "green islands" will be described in detail in a subsequent chapter.

In general, one may say that living conditions in the leaf vein are very similar to those the miner finds in the stem, particularly in relation to the numerous, hard, strengthening elements of the plant. We need not, therefore, be surprised that, for example, most species of the mining genera *Melanagromyza* and *Ophiomyia* have adapted themselves to the stem or leaf veins.

b) Selection of the plant organ by the mining larva

The same species of miner will invariably select the same specific plant organs, in which to construct its mine. Our definition of a mine has already shown that only those parts of plants can contain mines in which there is chlorophyll-containing parenchymal tissue. Such parenchymatous tissues occur not only in leaves and stems but also in unripe fruits, in parts of the flower and occasionally in the root or stock. The occurrence of mines in other parts of the plant than in leaves is determined in most cases by the point of oviposition by the female. Only a few species such as those of *Coleophora* are able to move over to another plant organ during the course of their life as a larva. In Chapter 3 on "Leaf mines and mines in other parts of the plant" an outline was given of the possible presence of miners in the various organs of the plant. The majority of all miners live in leaves and, within the leaf, primarily in the leaf-blade, less frequently in the petiole. A miner rarely occurs in the petiole purely by chance but this is invariably part of the regular life cycle of the species. Only in cases when a single leaf does not provide sufficient food for the larva, does it proceed through the petiole into another leaf. However, one finds quite frequently that a larva which normally only mines a leaf, may occur as a stem-miner. H. Buhr, for instance, observed that when the mines of *Liriomyza strigata* Mg. were formed in the small, uppermost leaves of the plant they frequently continue on into the parenchyma of the stem, so that finally the main part of the mine was developed in the stem. We previously pointed out that living conditions do not greatly vary for the miners in the mid-rib and those in the stem. But Buhr also observed similar habits in other species, for instance in *Phytomyza matricariae* Hend. There is thus no well-defined dividing line between leaf and stem miners. Some species spend their infancy in the leaf and only later become stem miners, such as *Ophiomyia proboscidea* Strbl. on Hieracium and Aster. *Lixus* species appear to be even less restricted in the choice of plant organ; they frequently commence with a channel in an Umbelliferae leaf and immediately proceed to the stem, where they no longer mine but feed in the pith and finally end up in the root. Sometimes a change of plant organ is only apparent: when we find that the minute mines of *Ptochomyza asparagi* Hering occur in the branches of the stem as well as in the needle-shaped leaves, this does not represent a change of organ, for these needle-shaped leaves of asparagus, in correct botanical terminology are not leaves but phylloclades, in other words modified shoots and therefore correspond to the branches. Mines formed in stems are found especially frequently when the parts of the stem are rich in parenchyma. This applies primarily to herbaceous plants but mines are also constructed in the cortex of shrubs and undershrubs when this is particularly rich in parenchyma; species of Genista (Leguminosae) may be quoted as an example, in which many miners live in the stem. A stem-mining habit is particularly suitable for species which live on plants whose leaves are too small for the construction of mines, such as *Lithocolletis ulicicolella* Stt. on Ulex. For the *Lithocolletis* species, which even form folds in the epidermis of the stem cortex, this must be particularly difficult on account of the greater tension in the stem cortex.

No essential differences exist between the type of food consumed by leaf miners on the one hand and miners living in the stems, flowers, fruits etc. on the other hand. Flowers possess scarcely any chlorophyl and accordingly miners are found in them only very infrequently and as an exception. However, exceptions do occur. T u r n e r (1907), for instance, observed that the larva of *Coleophora genistae* Stt. shows a striking preference for flowers. While it normally lives in leaves like other related species, it leaves them immediately the flowers appear and feeds on the flowers.

However, species which have already developed a considerable differentiation of taste do not find the transfer from one plant to another so easy. This makes all the more noteworthy those cases in which the mining larva lives in any other but the part of the plant it normally frequents. The term o r g a n o x e n y (Voigt, 1932) is used to describe the occurrence of phytophagous insects on other organs of the plant than those they normally frequent. Organoxenic mining larvae are encountered relatively seldom but most frequently in the genus *Coleophora*, whose larvae possess the greatest freedom of movement among the miners. We should mention here the occurrence of one species of *Coleophora*, *C. fuscedinella* Z. in the flowers of Caltha, in which it fed as a genuine miner, after it had presumably fallen from the Alnus trees above. *Coleophora serratella* L., *hemerobiella* Sc., and *anatipennella* Hb. have been observed in unripe fruits of Malus (Velbinger, 1939), while the first-mentioned species was found by Voigt (1932) also in the unripe fruits of Prunus avium, in which it was even eating the seeds. The same habits also apparently apply to *Coleophora ochrea* Hw., which, according to B e n a n d e r (1938) sometimes mines seed capsules of its food plant Helianthemum instead of the leaves. This again shows the close connection between mining and seed-eating species of *Coleophora*, which represent the two main ecological groups in this genus. When we recall that a large proportion of *Coleophora* species feed on seeds of the most widely varying plants, we may perhaps not be mistaken in assuming that this is the primary habit of the genus and that mining represents merely a secondary development; the case reported by Voigt would thus merely represent a return to the habits of the ancestors! *Phytomyza hellebori* Kalt. which normally mines in the leaves of species of Helleborus has also been fed up on the green perianth segments of Helleborus viridis L. Organoxeny is not involved in all those cases where mining in a different plant organ is a part of the normal life cycle. For instance, *Phyllocnistis saligna* Z. first mines a short length in a leaf, then constructs the major portion of the channel usually in the stem cortex, finally completing the mine in a second leaf. Since the caterpillar mines exclusively in epidermis cells, it will detect no great differences between its food in the epidermis cells of the leaf and that in the stem cortex; probably its differentiation of taste is not so highly developed as in the previously mentioned *Phyll. labyrinthella* Bjerk.

Organoxeny is involved in all cases where the mine occurs on a plant gall. Such cases have so far only been observed very infrequently. For instance, *Liriomyza strigata* Mg. was found on one occasion on a stem gall of *Gnorimoschema cauliginella* Schmid on Silene. Buhr (1941) on one occasion found the linear mine of *Ophiomyia alliariae* Hering, which is normally a stem mine, on the stem gall of *Ceuthorrhynchus chalybaeus* Germ. on Alliaria. The occurrence of a miner on galls is in some rare cases habitual and cannot then be considered as organoxeny; a case of this type was reported by Noble (1938) of a species of Lepidoptera from Australia, whose larva mines the galls on Acacia flower buds caused by the saw-fly *Tepperella trilineata* Cam. When the galls are of normal size, the larva mines only in the parenchyma of the gall and in no way damages the gall-producer; on the other hand if the gall is abnormally small, the larva eats into deeper layers of the gall and this leads to the death of the hymenopterous larva.

When *Phytomyza crassiseta* Zett. mines the calyx instead of the leaves, as observed by Buhr (1932), this organoxeny does not appear so surprising as the regular occurrence of certain Nepticulids of the genus *Zimmermannia* in the cortex of Fagus and Quercus. Here green parenchyma is very sparsely developed but nevertheless the larva is able to complete its mine normally. In such cases one may refer to organoxeny within the family Nepticulidae, applying to the genus *Zimmermannia* (and *Ectoedemia* in the United States with analogous habits). This is undoubtedly a particularly interesting problem, since the possibility of mining in another organ of the plant may provide indications that the two organs contain one or more specific substances in common and is therefore of importance for the botanist. All cases of organoxeny which are observed should therefore be carefully reported.

Although the cases described may have been more or less chance examples of organoxeny, we nevertheless know of certain cases which can be considered as habitual temporary organoxeny. Caterpillars of *Coleophora* of the *lixella* group live in their infancy always as miners either of the calyx or seeds of Labiatae. They give up this type of existence either in late autumn or in early spring and become genuine miners exclusively in Gramineae. This temporary organoxeny is coupled in this instance with disjunctive oligophagy, which will be discussed in the following section. Strange as a consistent change of host of this type may at first sight appear, it is nevertheless paralleled in plant parasites; certain fungi, which similarly in one generation live on quite different plant families from the subsequent one, for instance the rust, Gymnosporangium, on pear trees, which is limited to Rosaceae and the related Saxifragaceae, while in the following generation it affects exclusively Cupressaceae, thus changing even to conifers. It is difficult to find any sort of plausible explanation either for the temporary organoxeny of our *Coleophora* or for the change of hosts of so many rust fungi (heteroecious habits). This survey shows, however, how important for general biological knowledge the choice of food of mining insects can be.

c) Selection of host plant by the miner

Mining insect larvae are dependent for their development on one or more plants and a knowledge of these is an essential prerequisite for the study of mines. The expert on mines must therefore be sufficient of a botanist to know or to be able to identify the vascular plants in his collecting area. The plant species on which a mine is found, (the host plant or food plant) is of great assistance in identifying the mine-producer. The identification of mines generally means that from two known facts — name of the host plant and characteristics of the mine — the third unknown is discovered, namely the name of the mine-producer. In this respect the miner's food plant is one of the most important aids to identification. This applies especially to species of great similarity even in the imago, which can be accurately determined with difficulty and frequently only after examining the genitalia. We may mention as an example the two so similar species, *Lithocolletis cerasicolella* HS. on Prunus cerasus L. and avium L. and *L. spinicolella* Z. on Prunus spinosa L. Externally the moths are largely identical; but if they have been bred from their mines, no doubts will arise as to their identification, owing to the knowledge of the food-plant. However, not every mining species is limited to a single food-plant; there are very many which accept several definite species, genera or families of plants and are able to complete their development successfully. The surprising fact then transpires that species which live on several plants usually visit hosts, which are themselves related. This is a peculiarity not limited to miners; it is of general application to all phytophagous animals and is also often found in parasitic fungi. Miners, however, deserve special discussion in this respect; they are more suited to investigations of this type than externally-feeding phytophagous insects, since they are usually not able to change their host-plant, as frequently occurs in external feeders. They mostly spend their whole life in a single leaf of the plant and they have no possibility of visiting another plant. Errors will not therefore creep into investigations of this type so easily, as a result of the occurrence of larvae on plants on which they normally do not feed at all or only in an emergency. It in no way detracts from the value of comparisons of this type that in miners the choice of host devolves solely on the ovipositing female. Many examples show that definite parallels exist between miners feeding on more than one species of plant and the relationship of their hosts; frequently phytophagous insects suggest relationships between plants, which are only later recognised by botanists. For instance, there was a plant genus in South America Brunfelsia, which was included by botanists in the Scrophulariaceae. A species of caterpillar from the genus *Thyridia*, belonging to the Neotropidae, fed on this Brunfelsia, while all other species of the genus normally occur on Solanaceae; this larva thus represented a noteworthy exception. Later, however, it was discovered that the systematic classification of Brunfelsia was incorrect, that the genus likewise belonged to Solanaceae, as had been suggested by the presence of the *Thyridia* larva. This is the classic case reported by F. Müller of the choice of food by phytophagous insects indicating plant relationships. Such cases can now be frequently encountered in mining insect larvae — especially owing to their limited freedom of movement — and the choice of host plant can in this way provide valuable indications of plant relationships for the botanist. It should not, however, be forgotten that in such cases there is no p r o o f of plant relationships, the possibility of such a relationship is merely indicated. Interesting and important as the study of host plants of phytophagous insects undoubtedly is, especially as regards miners, its value in explaining phytophyletic relationships should not be overestimated. On the other hand the parallels between selection of food and plant relationships enable us, when identifying mines which have not yet been bred out and whose producer is not yet known, to predict merely from the food-plant which genus or species is likely to be the producer. For instance, if full-depth mines with a coleopterous larva are found in a genus of

Cruciferae, in which no mines had previously been reported, one can say with some certainty that *Phyllotreta nemorum* L. or *Ceuthorrhynchus contractus* Mrsh. are the species concerned. Lepidopterous mines on Gramineae can usually be ascribed to a species of *Elachista*, large blotch-mines on Umbelliferae will probably have been produced by *Philophylla* and in this way details of the producer can frequently be deduced from knowledge of the plant.

Below are listed the families of plants, in which miners have been discovered in Europe and the countries of the Mediterranean basin. Opposite the name of each plant family are indicated the Orders of the miners which have occurred on them. The abbreviations are as follows: C: Coleoptera, D: Diptera, H: Hymenoptera and L: Lepidoptera.

Ascomycetes

1.	Hypocreaceae	D.

Lichens

2.	Peltigeraceae	D.

Bryophyta

3.	Marchantiaceae	D.

Pteridophyta (Cryptogams)

4.	Polypodiaceae	D.L
5.	Osmundaceae	D.
6.	Equisetaceae	D.

Gymnospermae

7.	Taxaceae	L.
8.	Pinaceae	C.L.
9.	Cupressaceae	L.

Monocotyledons

10.	Typhaceae	D.L
11.	Sparganiaceae	D.L.
12.	Zosteraceae (Potamogetonaceae)	D.L.
13.	Alismataceae	C.D.
14.	Butomaceae	C.D.
15.	Hydrocharitaceae	D.L.
16.	Gramineae	C.D.L.
17.	Cyperaceae	D.L.
18.	Araceae	C.L.
19.	Lemnaceae	C.D.

20.	Juncaceae	C.D.H.L.
21.	Liliaceae	D.L.
22.	Dioscoreaceae	L.
23.	Iridaceae	D.L.
24.	Orchidaceae	C.D.L.

Dicotyledons

25.	Salicaceae	C.D.H.L.
26.	Myricaceae	C.L.
27.	Juglandaceae	L.
28.	Corylaceae (Betulaceae)	C.D.H.L.
29.	Fagaceae	C.H.L.
30.	Ulmaceae	C.D.H.L.
31.	Moraceae	D.L.
32.	Urticaceae	D.L.
33.	Santalaceae	D.L.
34.	Loranthaceae	L.
35.	Aristolochiaceae	L.
36.	Polygonaceae	C.D.L.
37.	Chenopodiaceae	C.D.L.
38.	Amaranthaceae	D.L.
39.	Portulacaceae	C.D.
40.	Caryophyllaceae	C D.L.
41.	Nymphaeaceae	D.L.
42.	Ranunculaceae	C.D.H.L.
43.	Lauraceae	L.
44.	Berberidaceae	L.
45.	Papaveraceae	D.L.
46.	Capparidaceae	C.D.L.
47.	Cruciferae	C.D.L.
48.	Resedaceae	C.D.
49.	Crassulaceae	C.D.L.
50.	Saxifragaceae	D.L.
51.	Platanaceae	L.
52.	Rosaceae	C.D.H.L.
53.	Leguminosae	C.D.L.
54.	Oxalidaceae	D.L.
55.	Tropaeolaceae	C.D.L.
56.	Geranicaeae	C.D.H.L.
57.	Linaceae	D.L.
58.	Zygophyllaceae	L.
59.	Rutaceae	D.
60.	Polygalaceae	C.D.
61.	Euphorbiaceae	D.L.

62.	Callitrichaceae	D.
63.	Buxaceae	D.
64.	Anacardiaceae	D.L.
65.	Aquifoliaceae	D.
66.	Celastraceae	L.
67.	Staphyleaceae	D.L.
68.	Aceraceae	D.H.L.
69.	Hippocastanaceae	L.
70.	Balsaminaceae	D.L.
71.	Rhamnaceae	L.
72.	Vitaceae	L.
73.	Tiliaceae	C.H.L.
74.	Malvaceae	C.D.L.
75.	Guttiferae	L.
76.	Frankeniaceae	L.
77.	Tamaricaceae	L.
78.	Cistaceae	L.
79.	Violaceae	D.L.
80.	Passifloraceae	L.
81.	Begoniaceae	D.
82.	Thymelaeaceae	D.L.
83.	Elaeagnaceae	L.
84.	Lythraceae	C.D.
85.	Trapaceae	D.
86.	Onagraceae	D.L.
87.	Araliaceae	L.
88.	Umbelliferae	D.L.
89.	Cornaceae	D.L.
90.	Pyrolaceae	L.
91.	Ericaceae	L.
92.	Primulaceae	C.D.L.
93.	Plumbaginaceae	D.L.
94.	Oleaceae	D.H.L.
95.	Loganiaceae	D.
96.	Gentianaceae	C.D.L.
97.	Apocynaceae	L.
98.	Asclepiadaceae	L.
99.	Convolvulaceae	D.L.
100.	Polemoniaceae	D.
101.	Hydrophyllaceae.	D.
102.	Boraginaceae	D.L.
103.	Verbenaceae	D.L.
104.	Labiatae	D.D.L.
105.	Solanaceae	D.L.

106.	Scrophulariaceae	C.D.L.
107.	Pedaliaceae	L.
108.	Orobanchaceae	D.
109.	Globulariaceae	D.L.
110.	Plantaginaceae	C.D.L.
111.	Rubiaceae	D.L.
112.	Caprifoliaceae	C.D.L.
113.	Adoxaceae	L.
114.	Valerianaceae	D.L.
115.	Dipsacaceae	C.D.L.
116.	Cucurbitaceae	D.L.
117.	Campanulaceae	C.D.L.
118.	Compositae	C.D.L.

Of these 118 European plant families attacked by any type of miner the majority, namely 95, are infested by Lepidoptera. They are followed by Diptera which are represented on 85 families, Coleoptera on 37 families, while the smallest number of families — only 11 — are selected by Hymenoptera.

These figures are not final but may vary with the discovery of additional food-plants for highly polyphagous species, such as *Cnephasia* among Lepidoptera, *Phytomyza atricornis* Mg. and *Liriomyza strigata* Mg. among Diptera and *Orthochaetes insignis* Aubé among Coleoptera. New host families cannot be considered likely for hymenopterous miners.

On the other hand the number of plant families in Europe and the Mediterranean basin on which mines have not so far been reported is extremely small. Among vascular plants this applies only to the following families:

1. Lycopodiaceae
2. Selaginellaceae
3. Isoëtaceae
4. Ophioglossaceae
5. Hymenophyllaceae
6. Marsileaceae
7. Salviniaceae
8. Ephedraceae
9. Najadaceae
10. Scheuchzeriaceae
11. Amaryllidaceae
12. Artocarpaceae
13. Rafflesiaceae
14. Thelygonaceae
15. Phytolaccaceae
16. Cactaceae
17. Aizoaceae
18. Ceratophyllaceae
19. Coriariaceae
20. Simarubaceae
21. Droseraceae
22. Empetraceae
23. Elatinaceae
24. Punicaceae
25. Myrtaceae
26. Halorrhagaceae
27. Hippuridaceae
28. Lentibulariaceae
29. Bignoniaceae
30. Acanthaceae

It is possible that mines may later be discovered in one or other of these families, possibly Amaryllidaceae and Empetraceae; this list will not, however, substantially alter. In the first seven families of this list, belonging to the Cryptograms, it is unlikely that miners will still be found and among the Phanerogams also the majority of families have been examined so frequently for mines without success, that one can no longer expect new discoveries in these families. In general they are also largely scorned by other insects. Accordingly, of the 148 families mentioned, only 30 are not infested by mining insects, that is approximately 20%. Among them, there are families which are subject to mining in other areas of the world, such as Cactaceae and to an especially large degree Myrtaceae. In tropical regions a far larger number of families support mining larvae; for instance, an extremely large number of mines are produced by *Hispinae*, a group of Coleoptera which is represented in Europe by only three species. *Gracilariidae* also produce mines in numerous families of plants outside Europe. On the other hand the number of mines of Tenthredinidae is far smaller and the number of species producing dipterous mines is also reduced in the tropics in relation to the abundance of species, which have developed in the Palaearctic Region.

Special terms have been coined to indicate the different types of feeding of phytophagous insects — from the point of view of the number of different plants attacked by each species of insect. Monophagy is used when only a single species of plant is eaten; oligophagy when several plants are found acceptable, and polyphagy when many different plants from widely varying families may serve as food for a single species of insect. All three types are represented among miners; oligophagous species, however, are the most frequently encountered. These three basic types of feeding have been divided up still further so that the following types of food selection can be distinguished (modified, after Voigt, 1932 and Gieysztor, 1938):

I. Euphagy. The species feeds on its normal food plant.

a. Monophagy. Feeding on one plant genus or species.

1. Monophagy (first degree) Restriction to one species of plant, termed specific or strict monophagy. Primarily applies to plant genera, containing only a single species.

2. Monophagy (second degree). The insect lives solely on one section of a specific plant genus but within this section on several species.

3. Monophagy (third degree), generic monophagy. The phytophagous species eats many or all species of a plant genus.

b. Oligophagy (Pleophagy). Several plant genera are attacked.

1. Systematic oligophagy. Feeding on several, related plant genera but only on these.

α. Systematic oligophagy (first degree); all infested genera belong to the same family.

β. Systematic oligophagy (second degree): the genera selected belong to different but closely related families of the same Order.

γ. Systematic oligophagy (third degree): The attacked plant genera belong to families of different but related Orders.

2. Combined oligophagy. The phytophagous insect lives on different genera of one family plus a single plant of another, non-related family.

3. Disjunctive oligophagy. Several plants are consistently eaten, which are not closely related but belong to different Orders.

4. Temporary oligophagy. In its different stages of development one species of insect consistently attacks different, unrelated plants.

c. Polyphagy. Members of different plant Orders are eaten indiscriminately.

1. Polyphagy (first degree). Plants are attacked belonging to different Orders but the same class.

2. Polyphagy (second degree). The phytophagous species occurs in different plants, belonging to several classes.

d. Pantophagy is a polyphagy extending to different phyla and sub-phyla of the plant kingdom.

II. Xenophagy. This applies not to normal, regular habits but to cases where an abnormal secondary food-plant is attacked out of necessity or for some other reason. Such cases may arise owing to the absence, scarcity or unsuitability of the normal plant or owing to a sheer accident.

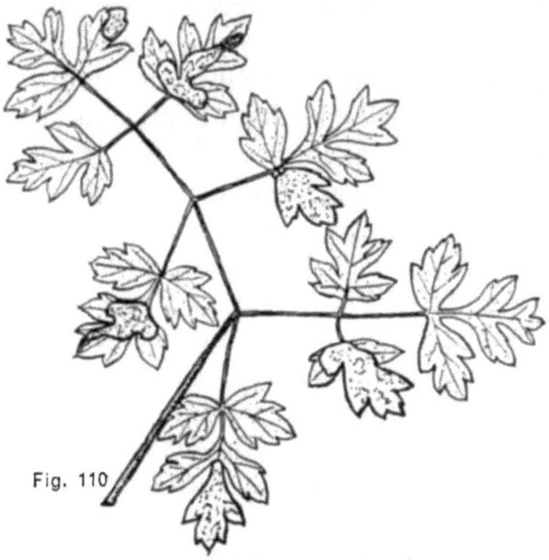

Fig. 110

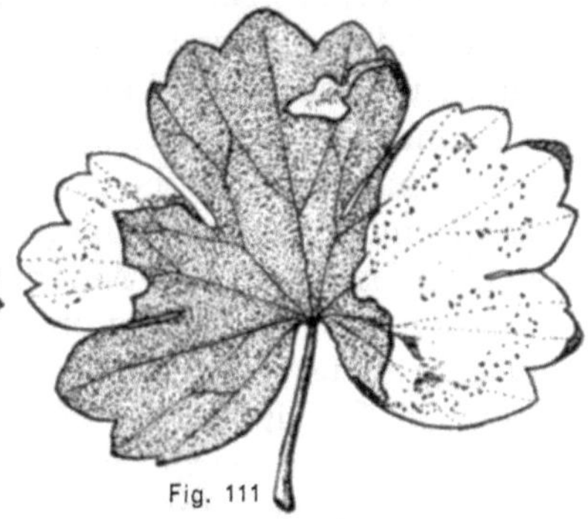

Fig. 111

Fig. 110. Peucedanum oreoselinum L.: *Phytomyza pauli-löwi* Hd., lives exclusively on this species. Monophagy (first degree).

Fig. 111. Anemone transsilvanica Heuff.: *Pseudodineura mentiens* Thoms. lives only on subgenus Hepatica, also on A. hepatica L.-Monophagy (second degree).

Phytophagous animals are also said to adopt generic or specific monophagy or oligophagy. This applies to cases where all members of an animal genus live on the same plant (generic monophagy) or where monophagy has only been shown in a single species of the genus (specific monophagy). This could also be sub-divided into several grades, since within the animal genera groupings occur in which all species feed on the same plant, etc. Similarly, in many cases family oligophagy can be observed. In this respect a more detailed classification is unnecessary for our purposes and will therefore not be discussed.

In order to illustrate the foregoing we will now quote examples of the various categories of food selection of mining insect larvae. It will be seen that for practically all types representatives of miners can be quoted, which clearly indicates the extraordinary variation in the specialisation in taste within the ecological group of mining insects.

I. a. 1. Monophagy (first degree). Specific or strict monophagy is naturally found in those species feeding on a plant genus, which is monotpyic, in other words possesses only a single species. *Liriomyza cannabis* Hend. which mines Cannabis sativa L. is monophagous in this way; since Cannabis includes no further species, a fly limited to this genus cannot occur on other species. Specific monophagy also occurs among miners in other forms, especially for instance, in plant families in which the "protein differentiation", which will be discussed later, has developed to a high degree, as in Ranunculaceae and Umbelliferae. Many examples of this strict monophagy can be quoted. Each species of the genus Peucedanum has its own mining fly of the genus *Phytomyza;* for instance, *Ph. marginella* Hend. on Peucedanum cervaria (L.) Koch, *Ph. pauli-löwi* Hend. on P. oreoselinum (L.) Mnch. (Fig. 110), *Ph. thysselini* Hend. and *Ph. thysselinivora* Hering on P. palustre (L.) Mnch. In the Ranunculaceae genus Clematis *Pseudodineura clematidis* Hering lives exclusively on Cl. alpina (L.) Mill., *Ps. clematidis-rectae* Her. and also *Argopus ahrensi* Germ. exclusively on C. recta L. *Endophytus anemones* Her. is only found on Anemone nemerosa L., *Pseudodineura heringi* Ens. only on A. silvestris L.

I. a. 2. M o n o p h a g y (s e c o n d d e g r e e) occurs when one section within a plant genus (a single group of species) is attacked, while the others remain immune. For example, *Phytomyza rectae pulsatillae* Hering feeds on Anemone but only on the subgenus Pulsatilla, parallel to the rust fungus Puccinea pulsatillae Klchb.; the same applies to *Pseudodineura parvula* Klg., while *Ps. mentiens* Thoms. and *Phytomyza abdominalis* Zett. live exclusively on the subgenus Hepatica of Anemone (A. hepatica L. and transsilvanica (Fuss) Heuff.) (Fig. 111). Such cases possibly occur more frequently than is at the moment realised and in all cases where a mining species is mentioned as occurring on a l l species of a plant genus, a careful check should be made to discover which species, if any, are in fact immune.

I. a. 3. M o n o p h a g y (t h i r d d e g r e e) is the commonest form of monophagy. In this case all species of a plant genus are attacked, without any preference being shown for a particular species. For instance, *Phytomyza hellebori* Kaltb. lives on all species of Helleborus, *Ph. ranunculi* Schrk. on all Ranunculi (Fig. 112), *Dizygomyza posticata* Mg. on all Solidago species, *Ph. erigerophila* Hering on all species of Erigeron, *Ph. digitalis* on all Digitalis species, etc.

M o n o p h a g y (t h i r d d e g r e e) often provides valuable indications of generic differences, which have not yet been recognised botanically. Earlier, Pyrus and Malus were united in one genus, incorrectly, as miners suggested. Numerous miners live either on the one genus or on the other, as in *Nepticula, Phytomyza, Gracilariidae* etc. The parasitic fungus Gymnosporangium sabinae (Dicks.) Wint. lives on Pyrus but not on Malus (Fischer & Gäumann, 1929). On the other hand the behaviour of identical species of miners on different plant genera may in some cases suggest a relationship

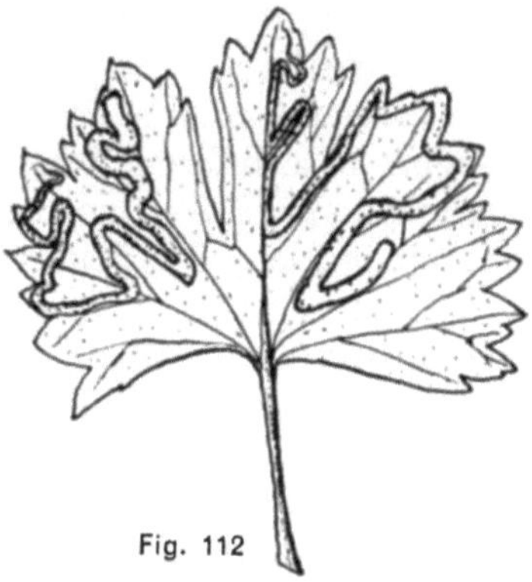

Fig. 112

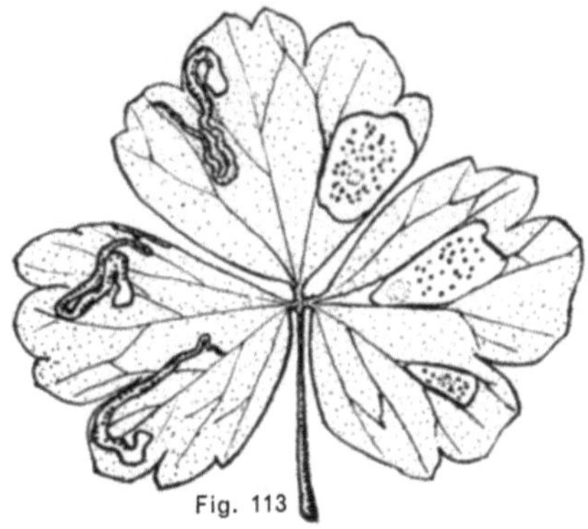

Fig. 113

Fig. 112. Ranunculus auricomus L.: *Phytomyza ranunculi* Schrk. lives on all species of Ranunculus, including Batrachium and Ficaria. Monophagy (third degree).

Fig. 113. Aquilegia: Left, linear mines of *Phytomyza minuscula* Gour., right, blotch-mines of *Ph. aquilegiae* Hardy, both occurring also on Thalictrum. Systematic oligophagy (first degree).

between the genera, for instance, the linking-up of Angelica and Archangelica owing to the identical miners on both is recognised as justified.

I. b. 1. *α*. Systematic oligophagy (first degree). The phytophagous insect species occurs on several plant genera, which all, however, belong to the same family. This is the commonest type of oligophagy. For example, *Nepticula promissa* Stgr. lives on Rhus, Cotinus and Pistacia, all genera of Anacardiaceae. *Dizygomyza morio* Zett. lives on the Rubiaceae Galium, Asperula and Sherardia. *Lithocolletis emberizaepennella* Bché., *Phytagromyza xylostei* R.-D. and *Phytomyza xylostei* Kltb. live on Lonicera and Symphoricarpos, *Trachys troglodytes* Gyllh. occurs on the Dipsacaceae Succisa, Knautia and Scabiosa. *Acidia cognata* Wied. mines Tussilago, Petasites, Adenostyles and other Compositae. *Phytomyza conyzae* Hend. lives in linear mines on Inula, Pulicaria, Buphthalmum and Telekia, thus in several closely related Compositae. *Liriomyza congesta* Beck. occurs on numerous genera of Leguminosae. *Phytomyza minuscula* Gour. and *Ph. aquilegiae* Hardy mine Aquilegia and Thalictrum (Fig. 113). *Phytomyza aconiti* Hend. and *Ph. aconitophila* Hend. form very similar blotch-mines only on Aconitum and Delphinium. *Philophylla heraclei* L. and also the North American *Ph. fratria* Lw. produce their blotch-mines on numerous genera of Umbelliferae. *Tischeria ekebladella* Bjerk., *dodonaea* Stt. and *decidua* Wck. live only on Quercus and Castanea. *Agromyza spiraeae* Klt.

constructs its common linear mines in a large number of Rosaceae genera. A very large number of *Elachista* species are found in the different genera of Gramineae. In all these cases the mining species in question is never found in plants belonging to a different family.

Similar habits can be observed in parasitic fungi. Peronospora parasitica (Pers.) Fr. lives on more than 100 species of Cruciferae. It can be divided into a large number of biological forms, according to its different food-plants, which can scarcely be distinguished externally but only by statistics of variation. This clearly indicates how species can evolve by specialisation on certain food-plants; it is probable that in the course of time these biological forms will develop into species which can also be distinguished morphologically.

I. b. 1. β. & γ. Systematic oligophagy (second and third degree) occurs when the species mines in different plant genera, which do not belong to the same but to another, closely related, family. Our definition already indicates that in this group the dividing-line between systematic and disjunctive oligophagy is no longer rigid. On the one hand the degree of relationship may vary; these are cases where the miners occur in closely related plant families, while the family relationship at other times may not be so well defined. Furthermore, the concept of relationship of plant families is still very much in dispute even among botanists. Some recognise a relationship, while others consider it to be non-existent.

Precisely these species, however, are of the greatest interest for our observations, since they show to what an extent the oligophagy of phytophagous insects such as our miners may give indications of the relationships of the food-plant. Views on the relationships of plant families have also changed in the course of time; many are now recognised, which not long ago would have been most strongly disputed.

An interesting example of this type is the occurrence of *Clidogastra* and *Chylizosoma* in leaf mines in Liliaceae and Orchidaceae and in this case an identical species *(Ch. vittatum* Mg.*)* is found in both plant families of different Orders. Although there is no doubt as to the relationship of these two, it has not yet been proved by serum diagnosis. Gramineae and Cyperaceae have for a long time been considered as closely related; this is supported by the occurrence on both of the genus *Elachista*, as also of certain species groups of *Dizygomyza*. However, the close relationship of both these families to the Juncaceae has not been proved by serum diagnosis but it can also not be doubted. The genus *Elachista* occurs predominantly on Gramineae and Cyperaceae; some species, however, also live on Luzula. *Phytomyza luzulae* Hering has its closest relative in the grass-mining *Ph. milii* Kaltb., thus indicating the connection between Luzula and the grasses; a single species, *Dizygomyza luctuosa* Mg. occurs exclusively

Fig. 114

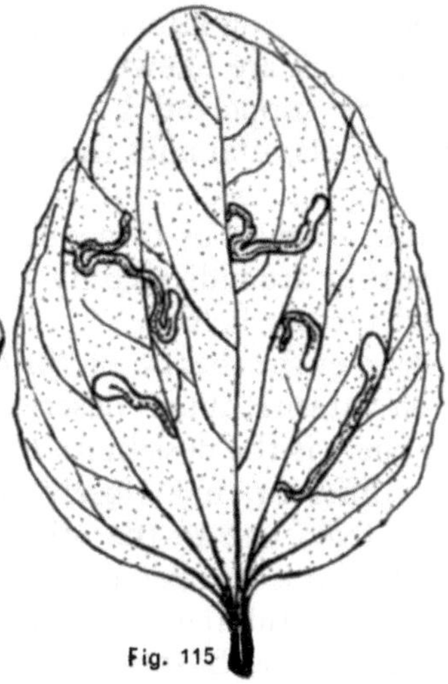

Fig. 115

Fig. 116

Fig. 114. Polygonum: Epidermal, lithocolletoid tentiform mines and leaf cone of *Euspilapteryx phasianipennella* Hb. (also on Chenopodium and Lysimachia). Systematic oligophagy (second degree).

Fig. 115. Plantago: Linear mines of *Apteropeda* (also on Labiatae and Scrophulariaceae). Systematic oligophagy (third degree).

Fig. 116. Tropaeolum: Bridging species of *Scaptomyzella incana* Meig. and *Sc. flava* Fall., for transference from Centrospermae and Cruciferae to Leguminosae.

in both Carex and Juncus, thus giving evidence of their relationship. The connection between the Caprifoliaceae and Oleaceae, (it is admittedly not very close) which appear on the same branch of the genealogical tree constructed on the basis of serum diagnosis, is emphasised by the occasional occurrence of the Oleaceae miner *Gracilaria syringella* Fbr. on Symphoricarpos. *Euspilapteryx phasianipennella* Hb. lives on Chenopodium and Atriplex and also on Polygonum (Fig. 114), showing the close relationship of the Chenopodiaceae and the Polygonaceae; it also occurs, however, on Lysimachia (Primulaceae), which in the former classification is far removed from these two families; according to Wettstein, and also as proved by serum diagnosis, the Primulaceae are closely related to the Polygonaceae—Chenopodiaceae. The close relationship of Myricaceae with other families of Amentaceae is supported by numerous examples, as for instance by the Myrica miners *Nepticula obliquella* Hein. (usually on Salix), *Rhynchaenus jota* L. (on Salix and Betula),

Caloptilia stigmatella F. (on Salix and Populus), *Rhamphus pulicarius* Hbst. (on Salix and Betula) and *Coleophora viminetella* Z. (on Salix). The *Apteropeda* species live on Labiatae, Scrophulariaceae and Plantaginaceae (Fig. 115), which are likewise closely connected. These examples could easily be supported by numerous similar cases. Mining species on different plant families thus faithfully reflect plant relationships. The so frequent occurrence of even identical species on related plant families permits us to deduce relationships, even though not yet proven, if they are indicated by identical mining species; this will be discussed further below.

In this connection the position of the theory on "bridging species" worked out during research on parasitic fungi, deserves some attention. According to this theory there is the possibility of two species of miner, which occur on two different, not closely related, family groups, being able to transfer from one to the other, via a certain plant genus or species which need not be related to either of the two families or family groups. In these bridging species there are certain substances which are essential for both species, so that the two species here find common ground. From this common denominator each mining species has the possibility of extending beyond its normal family group and transferring to those of the other species. As an example of such a bridging species used by miners we may mention the genus T r o p a e o l u m. Both European species of the genus *Scaptomyzella* Hend. are restricted to certain plant families, *Sc. incana* Mg. mining Chenopodiaceae and Caryophyllaceae (the two are grouped under Centrospermae owing to their close relationship), and also the unrelated Tropaeolum, Trigonella and Anthyllis. *Sc. flava* Fall. on the other hand is the species which mines on Cruciferae, Papaverceae and Resedaceae (all related) and also on Tropaeolum, Trigonella and Pisum, the three latter not being included within the relationship group of the first three. Both species therefore meet on the bridging species Tropaeolum (Fig. 116). Clearly, the food plant Pisum and therefore the Leguminosae must be considered as having some relationship to the Cruciferae. Initially, therefore, the two species have common ground on these bridging species and on the basis of the substances they contain, which are found palatable by both species, each of them, according to the theory, is able to penetrate the range of host plənts of the other, by gradually accustoming itself to the substances in the bridging species which are contained within the other family group. In this way the species living on the Centrospermae reached the Legume Anthyllis, on which it is now very common and it is to be expected that *Sc. flava* Fall. which is at the moment generally restricted to the Cruciferae group will in future succeed in colonizing also members of the Centrospermae.

A quite striking analogy is found in parasitic fungi. Puccinia subnitens (after Fischer & Gäumann, 1929) occurs at the present time on the related group Chenopodiaceae, Polygonaceae and Amaranthaceae on the one hand and on the closely related Cruciferae and Capparidaceae on the other hand. In this case also the bridging species for the two groups can be Tropaeolum. Mycologists can assume that the fungus has succeeded in transferring from the one group to the other via this genus. Tropaeolum is also collective host for other species of oligophagous parasitic fungi.

Tropaeolum appears to possess special characteristics as a bridging species; presumably other mining insects will also find their way by means of it to plants on which they have not so far been found. For instance, *Ceuthorrhynchus contractus* Mrsh. which mines Cruciferae or Resedaceae, *Plutella maculipennis* Curt. and *Liriomyza brassicae* Riley also live on Tropaeolum; there is some probability that from here they will succeed in making the jump to the Centrospermae. We know that Tropaeolum is distinguished by the myrosin cells it contains, similar to the families of the Cruciferae group, in which this constituent is an important characteristic. The absence of myrosin cells in Polygonaceae and Centrospermae leaves us in the dark as to the reasons why mining species of the Centrospermae group utilise this bridge. This intensely interesting problem of bridging species, which has not yet been investigated in relation to phytophagous insects deserves the special attention of entomologists. More detailed investigation will probably reveal forms of oligophagy, in which unrelated plant groups have been colonized by means of such bridges.

The possibilities of oligophagy are not so apparent when development is impeded owing to the unsatisfactory nature of the food. It is known that *Lyonetia clerkella* L. may mine on numerous genera of Rosaceae (and on Betula, Castanea and Salix). However, within the Rosaceae there are certain species on which a large number of mines fail to develop normally beyond their early stages. For instance, this markedly oligophagous species which always develops quite normally on Malus and Pyrus, as well as on Betula and Prunus cerasus L., is rarely found on Prunus serotina Ehrh., an American species, frequently planted in Europe. Even in years when it is abundant and when the species occurs on every possible type of Rosaceae and Betula in very large numbers, the mines on P. serotina are all extremely short and the larva invariably dies in an early instar (Fig. 117). Voigt (1932) also observed the same thing. A similar occurrence can be observed in *Gracilaria syringella* Fbr., which can develop on the most varied Oleaceae and even Caprifoliaceae; on the other hand the unsatisfactory food impedes its development on Chionanthes, as was observed likewise by Voigt (1932). In certain cases one can detect

Fig. 117

Fig. 118

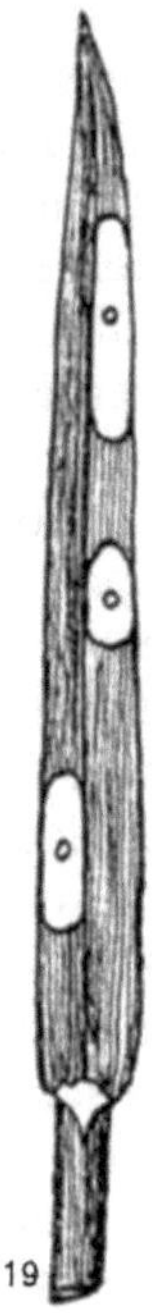

Fig. 119

Fig. 117. Prunus serotina Ehrh.: *Lyonetia clerkella* L. when abundant in 1946; all mines undeveloped. The light round patches are areas of oviposition without the development of larvae.

Fig. 118. Galeopsis: *Liriomyza eupatorii* Kltb. (also Eupatorium). Disjunctive oligophagy.

Fig. 119. Bromus: *Coleophora ornatipennella* Hb. Blotch-mines in early spring. In autumn its mines are formed in the receptacle of Salvia flowers — Temporary oligophagy.

a distinct *lethal* oligophagy, the causes of which are still not known.

Suggestions as to certain factors which might be responsible for lethal monophagy and oligophagy can, however, be made. Investigations carried out over many years in the Botanical Garden in Berlin showed that common miners by no means attack all species of a genus, even though they may not be limited to systematic sub-groups within the genus. The impression was gained that both in monophagy (third degree) and in systematic oligophagy differences occur according to the country of origin of the species concerned. If plants of various regions are grouped in a small area, as in a Botanical Garden, the greatest infestation of miners occur on plants which are found in the wild state within the normal range of the species. Plants from more distant localities are often attacked to a degree which diminishes as the distance from their natural habitats increases. For example, *Phytomyza actaeae* Hend. occurs very commonly on our local Actaeae and to the same extent also on the genus Cimicifuga; this might suggest that its ecological potentialities were sufficiently wide to enable it to attack at least Actaea species from other parts of the world. In fact, however, American species are virtually ignored by it, although in the Berlin Botanical Gardens they stand almost side by side with local species. The blister mines of *Pseudodineura enslini* Hering which are so common on Trollius europaeus L. are completely absent from the Siberian Trollius. North American species of Quercus are infested by a far smaller number of miners than those from Europe. The same thing applies also to plants growing outside the Botanical Gardens, which are virtually naturalized, such as certain species of Prunus; these are avoided by *Nepticula* and *Lithocolletis*, which occur in other European species of the same section. It would appear obvious to assume that a long period of isolation from Nearctic and Palaearctic species has led to the development of a marked differentiation between them; such a line of argument is not, however, valid. One can scarcely believe, to return to the above-mentioned example, that the differentiation between our native Prunus and P. serotina Ehrh. is greater than that between Prunus and Betula. For the time being there appears to be no explanation for the strange fact that in both monophagous and oligophagous insects the possibility of colonizing new hosts is in some way reduced by their removal from their natural habitat. There is no doubt that this represents an interesting problem.

Investigations of this type must be tempered by due caution. Consideration of ecological potentialities of miners has an importance which should not be underestimated. Certain species find the transfer to plants originating no matter how far away perfectly simple, while for others it is impossible. The American Quercus in the Botanical Gardens are accepted by some miners quite or almost as easily as the European species, such as for instance *Fenusella pygmaea* Klg., *Tischeria decidua* Wck. and *Acrocercops brongniardella* Fbr. This transfer seems impossible for certain *Lithocolletis* species *(L. lautella* Z.*)* and many *Rhynchaenus*. Clearly the first-mentioned have far wider ecological potentialities than others and this facilitates their transfer to other food-plants. It cannot be excluded that this potentiality increases during the course of time; we know from the vine pest *(Phylloxera)*, which originally avoided the American Vitis species, that by developing certain new forms it is now able successfully to transfer to it. This fact deserves special attention in relation to investigations into the indications provided by oligophagy of plant relationships. The absence of a species which occurs on many genera of a family but avoids others, can be used to deduce a lesser form of relationship, if plants are concerned whose original habitat was perhaps very remote from the present one. The presence of oligophagous species can be considered as indicating a plant relationship but their absence should never be explained by a lack of such relationship. Caution is always necessary when comparing the choice of food-plant of several mining species. The same rule applies as was discovered by Fischer and Gäumann (1929) with parasitic fungi: "we must consider the host group of a single parasite always as a separate entity"; only then can we make deductions concerning plant relationships from the occurrence of the parasite. Nevertheless deductions of this type will have all the greater validity, the larger is the number of cases in which oligophagous species of the most widely varying groups make the same choice of food.

It is conceivable that changes may occur in the choice of the food-plant within a measurable space of time. Species which originally favoured monophagy (first or second degree) may change over to monophagy (third degree). Bridging species may play a special role in facilitating this. Petersen, for instance, accepted the *Nepticula nanivora* Pet. which was restricted to Betula nana L. developed from *N. betulicola* Stt. mining B. pendula Roth. and B. pubescens Ehrh.; he supposed that this occurred via the hybrid Betula pubescens X nana which was quite frequent in the localities of *N. nanivora* Pet. This supposition seems quite valid and might well be applied to the occurrence of *Lithocolletis*, which are normally limited either to sallows or to willows, on species of the other group. Presumably a species might well originate in this way, by the transfer of strictly monophagous species to other food-plants via hybrids between the two. Ample opportunities exist here for conducting experimental minology. Similar experiments have frequently been undertaken with parasitic fungi, which showed that in certain cases the parasite became so accustomed to its new food-plant, that it could no longer thrive on its former food-plant.

In a similar way monophagy may be transformed into systematic oligophagy. To what extent this may occur under the influence of chimaeras has not yet been determined; it seems, however, theoretically possible. As is known, chimaeras consist of organisms containing various types of tissue, in which the one type is surrounded by the other. The genus Crataegomespilus is formed by external tissue of Mespilus enfolding internal layers of Crataegus and vice versa. Crataegomespilus was consequently studied with considerable care in the Berlin Botanical Gardens: so far, however, only species of Crataegus were found on it, such as *Nepticula oxyacanthella* Stt., *Paror-*

nix anglicella Stt. and others. It should, however, be emphasised that these chimaeras are very suitable for study, since in our latitudes Mespilus is infested by no monophagous species. Such plants, which are shown beyond doubt to be chimaeras, should be given particular attention by the minologist.

I. b. 2. Combined oligophagy occurs when a species of insect attacks all or practically all genera of a family but is found simultaneously on one other genus, which is in no way related. The term was coined by Heikertinger and as an example we may mention *Haltica oleracea* L. which infests practically all genera of Onagraceae but in addition occurs on Polygonum aviculare L. and only on this. This interesting case, for which we can give no explanation, has not before been observed in miners, but is known to mycologists, who speak in such cases of principal and subsidiary hosts. There is no strict dividing line between combined and disjunctive oligophagy. One cannot correctly consider this as an example of wider ecological potentialities, since the subsidiary host is only represented by a single species of a different genus and family.

I. b. 3. Disjunctive oligophagy. This applies to all cases in which a species of insect consistently occurs with the same degree of frequency on a number of plants (and only on these) belonging to different Orders of the plant kingdom, and which cannot thus be considered as related. The emphasis is on the consistency of the insect's appearance. Genuine disjunctive oligophagy is very infrequent in mining insects; the commonest example is the occurrence of *Liriomyza eupatorii* Kltb. (easily recognisable by its serpentine mine) on Eupatorium (Compositae) and Galeopsis (Labiatae), which are in no way related (Fig. 118); it possibly occurs also on Solidago (Linnaniemi, 1913) but this has only been reported once. One sees how the distinction between disjunctive and combined oligophagy begins to disappear. Among Lepidoptera *Lyonetia ledi* Wocke mines Ledum (and no other Ericaceae!) and Myrica but no other plants. This last case possibly indicates how disjunctive oligophagy arises: the two plants belong to the same plant community, are frequently mutually exclusive in their occurrence but at other times may appear together. (We will see later that genealogically there is some connection between Ericaceae and Amentaceae, which possibly facilitated the transfer from one to the other; the relationship has not, however, been proved by serum diagnosis). Disjunctive oligophagy may thus have developed on the basis of presence in the same community; the mining species then transferred to the same plant in other communities. *Liriomyza eupatorii* Kalt. now occurs on Galeopsis in places where Eupatorium could never grow. The rareness of this disjunctive oligophagy in mining insects justifies us to some extent in making deductions concerning plant relationships (which will be discussed later), which have not yet been proved botanically and are therefore inevitably rejected by the botanist. The numerous cases of systematic oligophagy are opposed by so few of disjunctive that the importance of the latter as a justification for discounting the significance of oligophagy for phytophyletic research can carry really very little weight. It should not, however, be forgotten that oligophagy merely provides indications but no proofs of plant relationships. At most they will be merely pointers.

I. b. 4. In temporary oligophagy a species possesses a number of hosts, which are not related but which are adopted consistently only in specific stages of its life. Here also the main emphasis must be again laid on the consistency of the insect's habits. The best known example among our miners are the species of the *lixella* group of *Coleophora*. In their first instar they live exclusively in mines in the calyx or seeds of Labiatae but in the late autumn or more frequently in early spring live in large mines in various Gramineae (Fig. 119). It is natural that they are unable to retain their initial type of existence; the long period they require for their development makes it impossible to find sufficient of the food to which they first became accustomed; in the early spring neither flowers nor seeds of these Labiates are available. It is difficult to explain what induced them to transfer to such a dissimilar family. The same problem applies here as in many parasitic fungi, where many rusts develop the two different types of spores within their life cycle on two quite different unrelated families. Of these heteroecious rust fungi, we already mentioned Gymnosporangium sabinae (Dicks.) Wint., which develops its aecidiospore generation on Pyrus (but not on Malus), while the teleutospores can only develop on Juniperus. Whether this represents an original polyphagy with subsequent specialisation on two specific hosts, or whether a former monophagy has been extended by transfer to secondary food-plants can no longer be determined. However, with the fungi, the first suggestion seems more plausible, while with our mining *Coleophora* species perhaps the second explanation is more probable.

I. c. 1. Polyphagy (first degree) is found in many species which feed indiscriminately on the most varying plant families, which are not related and even belong to different Orders. A preference may be shown for certain plants, while others are regularly avoided but the freedom of choice is so extensive that the miner seems to be in no way tied to any specific plants. One can never lay down on which plants they will not occur, since according to circumstance and opportunity one or the other plant which is normally avoided may be found acceptable. Among Diptera *Phyto-*

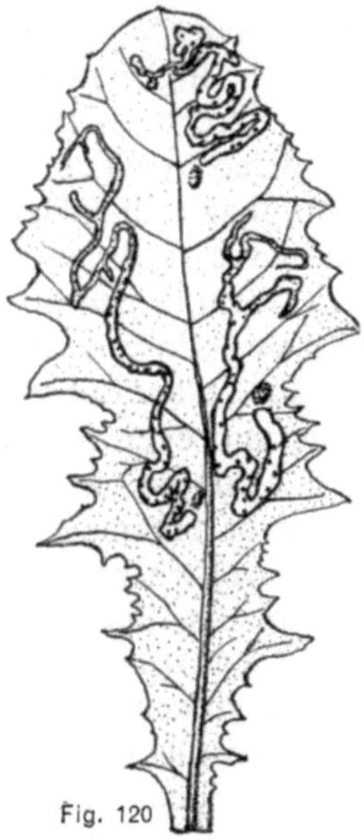

Fig. 120

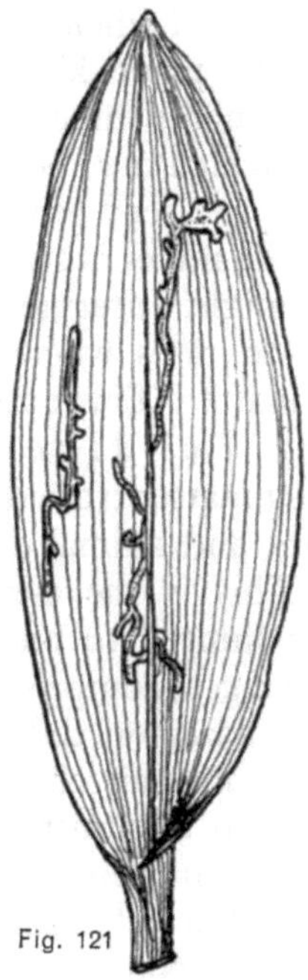

Fig. 121

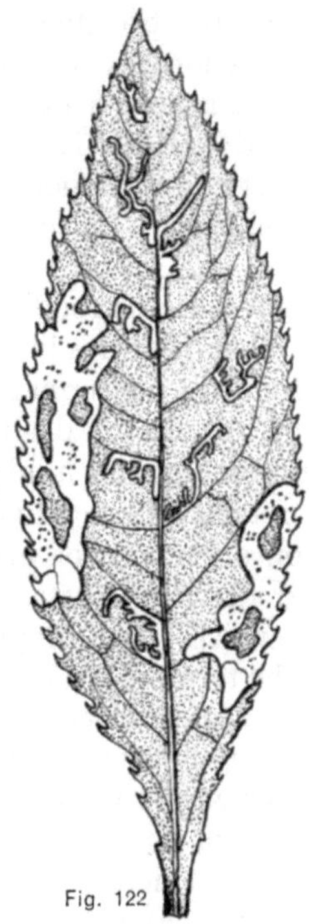

Fig. 122

Fig. 120. Taraxacum: *Phytomyza atricornis* Mg., the most polyphagous mining fly. Polyphagy (first degree).

Fig. 121. Himantoglossum: *Orthochaetes insignis* Aubé. It mines mono- and dicotyledons. Polyphagy (second degree).

Fig. 122. Senecio: Mines of *Lycoria* (also on Lichens and Marchantia). Pantophagy.

myza atricornis Mg. (Fig. 120) and *Liriomyza strigata* Mg. (Fig. 12) must be considered as definitely polyphagous species; they occur on a large number of herbaceous plants among the dicotyledons but despite their polyphagy they are absent from woody plants, unless they perhaps mine their cotyledons. The same applies to the mining species of the genus *Cnephasia*, of which three species especially are indefatigable in their selection of ever more host plants; they likewise prefer herbaceous plants.

I. c. 2. In polyphagy (second degree) this characteristic is even increased; the miners are thus found on plants belonging to different classes. The most striking example of this is *Orthochaetes insignis* Aubé, which, according to the large number found by Buhr and confirmed by breeding, constructs its mines indiscriminately in monocotyledons as well as in dicotyledons (Fig. 121).

I. d. Pantophagy is used when insects can feed on different phyla and sub-phyla of the plant kingdom (conifers, flowering plants, ferns and lichens); among miners this applies especially to species of *Lycoria (Sciara)*, which may occur on the highest Compositae (Fig. 122), as well as on the very primitive Marchantia. In such cases, however, the food of the miner is still provided by plant organisms and habits of this kind might be termed polyphagy (third degree) while limiting pantophagy to cases where the larva feeds not only on living but also on dead matter. This extreme form of pantophagy is found among the mining Tendipedidae (Chironomidae), which not only feed on living matter from the mined plant but also on plankton organisms of plant and animal origin, and even partly on decayed detritus matter. This is the clearest example of pantophagy, which is no longer limited to one type of organism.

II. The term Xenophagy comprises all cases where a species cannot be said to occur regularly on different plants. Such species are usually highly specialised in their choice of food; they are never polyphagous but under certain circumstances may occur on plants which cannot be included within their normal food-plants. For certain reasons which we are unable to fathom they suddenly and quite exceptionally adopt a different food-plant totally unrelated to their normal one. The majority of such cases are probably accounted for by an urge for oviposition in the absence of the food-plant, which compels the female to lay its eggs on an abnormal, secondary food-plant. We cannot say how often this happens but certainly much more frequently than is generally supposed; for only in a limited number of cases will the larvae be able to develop in the secondary food-plant and the larvae which have died before competing their development will usually escape the sharpest eye. Such exceptional host plants occur with all phytophagous insects. In 1932 G. Voigt published an interesting list of all such cases of xenophagy in mining insects. It is also noteworthy that numerous analogous cases of xenophagy are known on parasitic fungi; mycologists term such species xenoparasites (Fischer and Gäumann). However, the cause of xenophagy in these cases is more obvious. It was discovered that xenoparasitism was especially in evidence when the secondary, abnormal food-plant was injured in such a way as to cause a general weakening, thus giving the fungus the possibility of attacking successfully.

Xenophagy is found in mining insects expecially in the most mobile species of the genus *Coleophora*. For instance, *C. fuscedinella* Z., which clearly had fallen from an Alnus tree, was observed mining

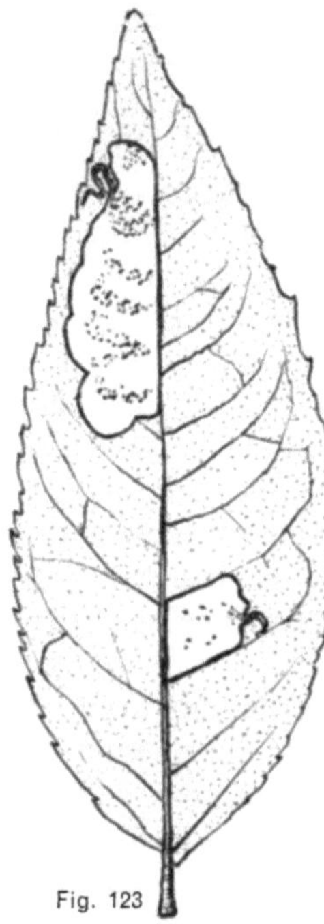

Fig. 123

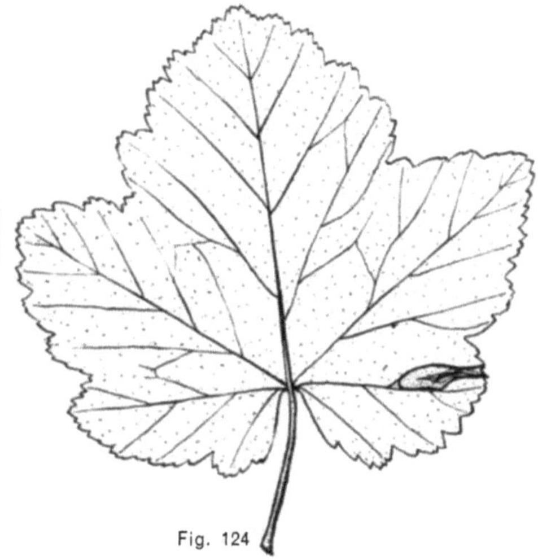

Fig. 124

Fig. 123. Buddleia (Loganiaceae): *Dizygomyza verbasci* Bché., normally on Scrophulariaceae.

Fig. 124. Ribes sanguineum Pursh.: *Lithocolletis ulmifoliella* Hb. Normally only on Betula. Xenophagy.

the flowers of Caltha. In this case xenophagy is combined with organoxeny. The same species was found beneath Betula mining in Lysimachia (B e n a n d e r, 1938), while the same author discovered *C. flavipennella* HS., which normally lives exclusively on Quercus, also on Helianthemum. It is interesting that in the latter case all the caterpillars were found to be parasitised; the weakness resulting from the parasites was undoubtedly the cause of the larvae being unable to maintain their hold on the Quercus leaves and falling. *Pegomyia bicolor* Wied., which normally mines on Polygonum was twice bred in some numbers on Begonia, a quite unrelated genus.

The conspicuous blotch-mines of *Dizygomyza verbasci* Bché., which mines only in Scrophulariaceae, were also found by V o i g t on the genus Buddleia, which belongs to the Loganiacede (Fig. 123). *Euspilapteryx auroguttella* Stph., usually living only on Hypericum, and *Ophiomyia proboscidea* Strbl., a Compositae stem-miner, were both found on Satureja, a Labiate. *Gracilaria syringella* Fbr., a miner of Oleaceae, was also observed by Voigt on the Saxifragaceae genus

Deutzia. De Meijere once discovered the normally monophagous *Phytomyza lappina* Gour., which lives on Arctium, also on Diervilla, a species of Caprifoliaceae. *Euspilapteryx phasianipennella* Hb. is restricted to Polygonaceae, Chenopodiaceae and Primulaceae. Buhr, however, also found the species on Lythrum; Lythraceae are in no way related to the normal food-plants of the species. Particularly striking is the occurrence of *Callisto denticulella* Thnbg., normally limited to Malus, on the genus Parthenocissus (Ampelopsis) belonging to the Vitaceae, as observed by Voigt. Very strange also is the appearance of *Lyonetia clerkella* L. on Humulus, while it is normally confined to Rosaceae and Amentaceae. *Lithocolletis ulmifoliella* Hbn. has twice been found on Ribes (Fig. 124). This cannot be considered as oligophagy based on the relationship of plants; the species is so strictly monophagous that it only occurs on Betula but not on the closely related Alnus or Corylus of the same family.

Although earlier suggesting that some compelling urge causing the female to deposit its eggs is responsible for the phenomenon of xenophagy, nevertheless in certain cases other factors play a part. In many cases it could be shown that the secondary food-plant selected was situated in the immediate vicinity of considerable quantities of the normal species. The ovipositing female was thus so affected by the scent of the normal species (and in addition perhaps experienced a sudden need to lay) that it lost its ability to discriminate between the two plants. The two species mentioned above, in which xenophagy was observed and which were feeding on Satureja, were on plants surrounded by numerous specimens of Hypericum and Hieracium. In one example of *Lithocolletis ulmifoliella* Hb. being found on Ribes, the collector specifically noticed that the Ribes bush was intergrown with a birch tree. This, however, still does not explain how the normal development of the larva is able to be completed on a secondary food-plant which is so very different. Nevertheless, some connection of the type indicated would seem to exist. The effective area of scent of the regular food-plant might also have led to the disjunctive oligophagy described in *Lyonetia ledi* Wocke on Ledum and Myrica.

Voigt, however, considers that the opposite may be possible. In the case referred to where species normally on Hypericum or Hieracium were found on Satureja, he considers that the scent area of the secondary food-plant may equally exert some influence. The great concentration of the Labiates's oil may so weaken the ovipositing female's powers of discrimination that it is no longer able to distinguish the primary from the secondary food-plant. Such an assumption appears highly plausible in this case, while the previous suggestion is perhaps more apposite in the previously mentioned case of disjunctive oligophagy on Ledum and Myrica. At least by human standards the scent of Ledum is stronger than that of Myrica, even though the latter is distinguished by a highly characteristic but weaker scent of the whole plant.

The influence of the extension of cultivation should not be underestimated as a factor inducing xenophagy. This has resulted in the

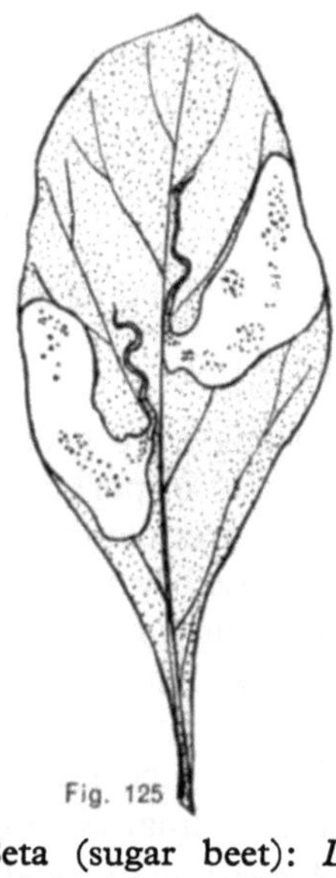

Fig. 125

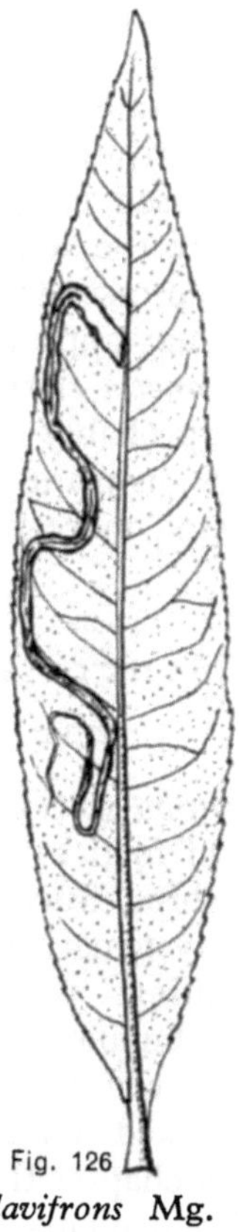

Fig. 126

Fig. 125. Beta (sugar beet): *Dizygomyza flavifrons* Mg. (normally on Caryophyllaceae). Systematic oligophagy (second degree).

Fig. 126. Salix: *Lyonetia clerkella* L. (also on Rosaceae, Betula, Castanea).

destruction of certain plants over wide areas, where now only relatively few wild plants are still found. Since mining larvae produce imagines with only very limited powers of flight, the necessity to lay is increased by the ovipositing female having in many cases to spend a long time searching for a suitable plant. Furthermore, cultivated plants are frequently cut or removed to a far greater extent than any comparable destruction resulting from natural conditions; when the female wishes to commence laying, it no longer finds the food-plant of the larva, from which it has developed and is compelled to divert its attention elsewhere. The change in agricultural crops is also not without its effect. Presumably the not infrequent occurrence of *Dizygomyza flavifrons* Mg., normally a parasite on Caryophyllaceae, on various cultivated types of Beta, especially on sugar-beet, can be explained in this way (Fig. 125). Clearly, a certain tendency towards oligophagy must have existed; for Caryophyllaceae were not lacking, as Buhr mentioned, at the spot where the beet was attacked. It is natural enough that such succulent leaves as have developed in our cultivated forms should prove a special attraction to miners. The raised temperature found in greenhouses and beside walls exposed to the sun etc., is also considered by Voigt to be a factor in some cases favouring xenophagy. Certain cases of xenophagy, which may appear strange, such as the occurrence of the oligophagous *Lyonetia clerkella* L., on Salix (Fig 126) (normally on Rosaceae) will probably in future be recognised as oligophagy, when different views on plant relationships have found acceptance. This will be discussed more fully in the subsequent section.

D. Leaf-Miners and the genealogical tree of the plant kingdom.

In the previous section several indications were given as to how the selection of food by miners may provide important pointers to plant relationships. Such connections have frequently been proved by botanical systematists only later or possibly they still remain largely suppositions. We will subsequently show how miners behave within various families of the plant kingdom and we shall consider the families according to the degree of their relationship on a genealogical tree worked out by botanists. It was mentioned at the beginning of this chapter that mining insects primarily affect the proteins of their food-plants. Views on plant relationships vary considerably, each being expressed by a different genealogical tree of the plant kingdom; we shall, however, work from an illustration for the greatest part based on protein similarities. Our views correspond most closely to the genealogical tree of the plant kingdom worked out by Mez (1924) on the basis of the results of serum diagnosis applied to plant relationships. There are two reasons for this: firstly, to some extent the genealogical tree established on the basis of serum diagnosis, in other words of the protein relationships, most closely reflects the natural relationships of plant families, since we assume that the bearers of heredity are certain protein bodies. It has not, however, been satisfactorily established whether these hereditary factors are primarily nuclear proteins or cytoplasmic proteins. While originally certain constituents of the cell nucleus were considered responsible for heredity, the view has also been expressed that cytoplasm proteins might contain the essential heredity factors. Secondly, information so obtained is large extent supported by miners, which are dependent on protein for their food.

It must not, of course, be overlooked that botanists have raised important and serious objections to the serodiagnostic method of establishing plant relationships. The method of Mez and his school reputedly obtained reactions indicating relationships, which under no circumstances could be considered valid. We may mention the similar reactions for Leguminosae and Gramineae, which could be supported by no botanist (but which remarkably enough find an analogy in the food selection of the *Agromyza* species of the *Domomyza* group, which mine predominantely in Gramineae and Leguminosae). Another factor, however, seems to be of greater significance. The investigations used primarily the fruits of the plants concerned; the fruits contain largely special reserve proteins, differing from the proteins of the green tissues. These reserve proteins cannot be considered to the same extent hereditary factors as the specific proteins occurring in the remaining parts of the plant. Such differences were proved in the case of Laburnum, in which no reaction could be found between fruit and cytoplasm serum corresponding to that produced by the leaves. Similarly, many unspecific reactions are obtained when using fruit protein of Cucurbita or Amygdalus with numerous plants with which there is not the slightest relationship. Furthermore, it is not possible completely to remove various associated substances, such as saponin: the result is often that the associated substances prevented the manifestation of specific reactions indicating relationship, while in other cases such reactions were artificially induced. This criticism of the plant genealogical tree worked out from serum diagnosis in no way invalidates the theory as such; it has, however, been shown that the methods so far employed have not been sufficiently subtle to exclude the possibility of error. One can be certain that future corrections undertaken to avoid errors will provide a more accurate picture of the real relationships in the plant kingdom. It should also not be forgotten that every possible method should be employed to establish these relationships and not only serum diagnosis; in the present situation this is only one of many methods and not even the most important. This is all of small account in our discussion, since we are not concerned with the more controversial families. We shall confine our attention

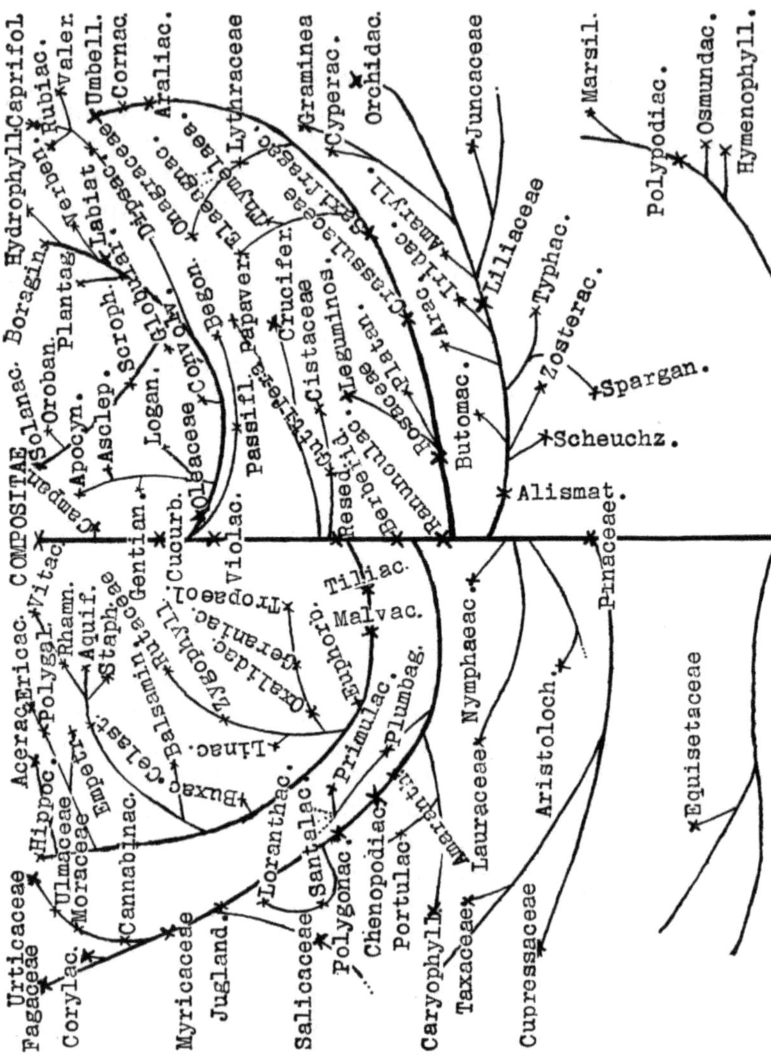

Fig. 127. Genealogical tree of plants worked out by serum diagnosis (Partially after Mez, 1924).

to the main families, which provide food-plants for miners. The genealogical tree as worked out by serum diagnosis agrees on most points with the relationships as determined by morphological and cytological investigations.

In the following survey of the parts of the plant kingdom where mining insects occur, we shall utilise the genealogical tree worked out from serum diagnosis; Fig. 127 shows an extract from the tree relating to the families on which miners are found in Europe. We shall discuss in turn the relations of the different plant families to mining insects and shall consider each family from three points of view, as follows:

1. To what extent is the position on the genealogical tree and the relationship with other families confirmed by the miners' choice of food?

2. To what extent does the miners' choice of food-plant conflict with the position of the family concerned, as established by serum diagnosis?

3. To what extent do monophagy, oligophagy and polyphagy of mining insects within a plant family provide an indication of the degree of protein differentiation which has developed within the family?

We shall begin our discussion with the vascular cryptogams, of which we see two branches at the foot of the tree represented in Fig. 127. They are little suited to consideration here, since the species of miner found on them were originally on the phanerogams and only transferred to the cryptogams as food-plants recently. They are thus highly specialised forms which have no real connection with the primitive position of the cryptogams. A limited systematic oligophagy can be observed in these species. For instance, *Phytomyza scolopendri* R.-D. and also *Teichobia verhuellella* Stt. live on several genera of ferns and primarily on Polypodiaceae. This is therefore essentially oligophagy (first degree). However, *Dizygomyza hilarella* Zett. which is usually limited to *Polypodiaceae* is also found as a complete exception on Osmunda, which belongs to the next family, Osmundaceae. The oligophagy is, however, so limited, or the protein differentiation in the families of ferns has developed to such an extent that no other families are found acceptable. With some families of water ferns this may be a result of the differing living conditions in water; however, the adjacent Hymenophyllaceae which inhabit the land are also avoided.

The results are of even less consequence if we investigate the first branch on the left of our genealogical tree, representing the Equisetaceae. Two species of *Liriomyza* are found on the genus Equisetum; these are both definitely monophagous but no investigations have been undertaken into the exact type of monophagy. The even more primitive families of plants have not been considered here. For instance, an Anthomyid which constructs mine-like channels in the fungus Epichloë (parasitic on grasses) and the Lycoriidae on Lichens and Marchantia have only adopted these abnormal food-plants in very recent times; pantophagous species like *Lycoria* are valueless for our considerations, while a one-sided specialisation, such as that of the *Paregle* species on Epichloë can in no way support our thesis.

Consideration of the Gymnospermae branch found somewhat higher on the tree adds little to our discussion. The species occurring on Coniferae have adopted these particular food-plants as a secondary measure and this is limited to certain miners in which adaptation to mining habits is not highly developed. There is also little to be said on the type of food selection found within the Gymnospermae; some cases represent clear examples of monophagy, such as the *Argyresthia* species on Juniperus (Cupressaceae), while others can be considered as oligophagy, extending from the Pinaceae at the base of this branch to the Cupressaceae at its end. *Ocnerostoma piniariella* Z. is a good example of this. It should, however, be pointed out that protein differentiation in the whole Gymnospermae branch is only slight; this explains the occasional occurrence of Pinaceae miners on Cupressaceae and even on Taxaceae. Miners in these families are almost exclusively lepidopterous larvae of a highly specialised type, which have especially adapted themselves to their host but whose nearest relatives live on flowering plants. Extensive specialisation of recent development explains, for example, the strict monophagy of *Argyresthia* on Juniperus.

Higher up our genealogical tree we find that the next branch on the right has many sub-divisions, representing families of monocotyledons. Their place in the plant system has been disputed on many occasions; they have not always been included as a primitive group, as has been done in the classification used here. The mining species discovered on this group of families are themselves not all primitive. The majority of them clearly went over to these plants in recent times and evolved from forms which lived on dicotyledons. The facts are more difficult to determine, owing to many of these families comprising aquatic plants; these are only colonized by miners which have specially adapted themselves to aquatic life, such as Ephydridae, of which the genus *Hydrellia* is almost exclusively limited to monocotyledons (one species lives on Nasturtium, a Crucifer). Only one species has been frequently found on terrestrial grasses, but this is a secondary transfer, which probably took place from aquatic grasses. Life on aquatic plants demands special adaptations even in miners and for these reasons these forms will be discussed separately in the following chapter. The aquatic life of these species is the primary factor determining their choice of food and it should not, therefore, cause surprise that one species is found far removed on the genealogical tree among the dicotyledons (on Callitriche). For the reason mentioned these forms provide little material indicating the relationships or the degree of protein differentiation among the monocotyledons. However, one very primitive mining family, the Scatophagidae, occurs exclusively on monocotyledons. One species, *Hydromyza livens* Fall. lives exclusively on Nymphaeaceae, which is very close to the base of the monocotyledon branch, while the other genera, *Clidogastra* and *Chylizosoma*, are restricted to Liliaceae and Orchidaceae and avoid all other families of this group, including the terrestrial ones. One species, *Chyl. vittatum* Mg. lives on both families and thus indicates a very close relationship between them, which has not even been proved by serum diagnosis. The most highly developed families among the monocotyledons are Gramineae and Cyperaceae, which are grouped together as Glumiflorae. In accordance with their high degree of development they also contain numerous species of miners. Oligophagy (generic or specific) of mining insects provides numerous indications of the close relationship of the two families. However, the Juncaceae are obviously closer to them both than would appear from the genealogical tree of Mez; this is proved by finding one and the same species on Cyperaceae and Juncaceae (*Dizygomyza luctuosa* Mg. on Carex and Juncus). The *Elachista* species live exclusively on Cyperaceae, Gramineae and Juncaceae. The sub-genus of *Dizgomyza*, *Poemyza*, is restricted to the first two families, while *Dizygomyza* mine on all three families and on the related Iridaceae as well (three species). *Cosmopteryx*, of which several species mine Gramineae, cannot be adduced as indicating relationships; it is known that this genus is extremely polyphagous and occurs in a number of families without any close relationship, such as Urticaceae, Leguminosae, etc. In general, mining species are found relatively seldom on monocotyledons; this may be connected with their primitive position. There are nevertheless several genera, such as those mentioned, which have become specially adapted to them, producing a large number of species. The degree of protein differentiation within this group is relatively small and this is shown by the fact that oligophagy (first degree) and also frequently oligophagy (second and third degree) is found. The close connection which we observed among miners on Glumiflorae, Juncaceae and Iridaceae is based either on the slight protein differentiation of the plants concerned, or the position of these families on the genealogical tree of Mez needs to be revised. On the basis of morphological and cytological similarities, there is little doubt but that the first view must prevail.

Close to the monocotyledon branch are several smaller branches which are of little significance in this discussion; they contain the Lauraceae, whose parasites (*Caloptilia* species) normally occur primarily on the Amentaceae families and the A r i s t o l o c h i a c e a e. The Amentaceae are on the branch above in our diagram and possibly the Lauraceae should more correctly branch off from its base. The Aristolochiaceae are mined by only a single, isolated genus *(Millièreia)*, which is of no importance for us here.

The following branch commences with the R a n u n c u l a c e a e and there is now a radical change in the picture. While we found extensive oligophagy (second and third degree) among monocotyledon miners, in Ranunculaceae a, distinct monophagy becomes the rule. Occasionally cases of oligophagy (first degree) occur, as in Aquilegia and Thalictrum on the one hand, and in Delphinium and Aconitum on the other. Possibly the present distant position of the two first-mentioned genera will have to be revised and included in the system of the Ranunculaceae. Whole genera feed exclusively on these plant family, and monophagy, even first degree, is the normal rule. Although this family is so close to the monocotyledons on Mez's genealogical tree, miners show a much greater degree of specialisation in the selection of their food in the Ranunculaceae and there is little doubt that protein differentiation is much more highly developed than in preceeding families. It must be left to future research to show whether the rate of differentiation has been retarded in some genera, such as Thalictrum and Aquilegia, or whether these two genera should be shown as more closely related than is at present the case. The genus Anemone is an especially fruitful field for study of variations in the degree of monophagy, as has already been pointed out.

R o s a c e a e are the nearest relatives of this family and here the situation is again quite different. While monophagy is the rule with Ranunculaceae, with Rosaceae oligophagy is more normal. Oligophagy (first degree) is extremely common. Species of *Agromyza* mining on Rosaceae accept the most widely varying genera of the family. The same applies to many species of Lepidoptera, even though extensive adaptation and even strict monophagy is found in this Order; this is, however, explained by the high degree of specialisation of the insects concerned. On the other hand, in this case alone we find what is possibly a special form of oligophagy, which clearly indicates a close connection between Rosaceae and the families of Amentiferae, which occupy a very distant position in this system of classification. This connection will be illustrated in detail when discussing the Centrospermae—Amentiferae branch.

In our genealogical tree P l a t a n a c e a e and Leguminosae branch off from Rosaceae. There is little to be said on the first family; it possesses two miners from genera which are also common on Rosaceae. These two miners, however, occupy a somewhat isolated

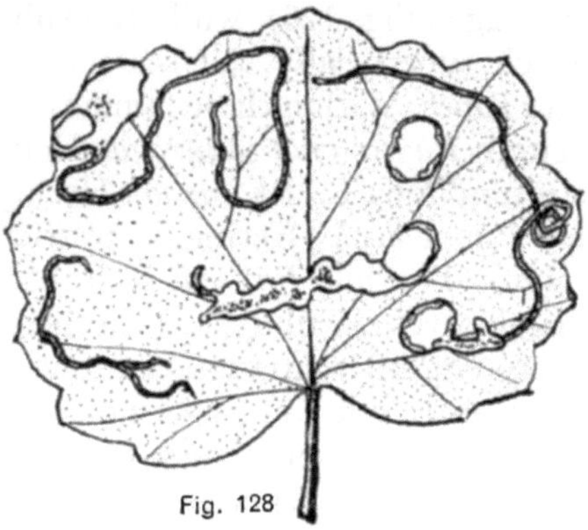

Fig. 128

Fig. 128. Saxifraga: *Incurvaria trimaculella* HS. (also on Ribes as subspecies *quadrimaculella* Höf.).

position among the other members of their genera, parallel to that of the family they mine. *(Lithocolletis platani* Stgr., *Niepeltia).* L e g u m i n o s a e, on the other hand, have a large number of mining species, and whole generic groups on some genera are restricted to this family. Oligophagy (first degree) occurs here also, as for instance in *Liriomyza congesta* Beck. and *Agromyza nana* Mg., and among Lepidoptera in *Stomopteryx*. There are also many clear examples of monophagy, at least third degree, while monophagy (first degree) is found in some species of *Leucoptera*. Of Leguminosae only the sub-family Papilionatae needs consideration here. In this sub-family there are differences in the monophagy (third degree), indicated by the fact that not all genera are attacked; the miners are able to differentiate between the various tribes of the sub-family For instance, certain species of *Agromyza* and *Leucoptera* occur only on G e n i s t e a e, while others are found only on Loteae, Vicieae or Trifolieae. *Agromyza nana* Mg., is very closely related to *A. genistae* Hering but is found primarily on Trifolieae, while the latter is restricted to Genisteae. This tribe also feeds a whole series of *Leucoptera* species, in which monophagy (first degree) is probably the normal rule. Although protein differentiation is not so marked as in Ranunculaceae, it is nevertheless more developed than in Rosaceae.

On the same branch are the families C r a s s u l a c e a e and S a x i f r a g a c e a e. Their close relationship is shown by the occurrence on them of two species of *Chilosia*, which can scarcely be distinguished: *Ch. semifasciata* Beck. lives on Sedum and some other closely related genera, while *Ch. saxifragae* Hering lives on Saxifraga. Saxifragaceae and Ribesiaceae were previously classified as separate families; their identity is suggested by *Incurvaria trimaculella* HS., which mines on Saxifraga (Fig. 128), while the sub-species *quadrimaculella* Höfn, which superficially can scarcely be distinguished from it, is confined to Ribes. Within these families strict monophagy is also sometimes found: *Phytomyza saxifragae* Hering lives exclusively on Saxifraga rotundifolia L., *Ph. aizoon* Hering exclusively on Sax. aizoon Jacqu.; *Ph. sedicola* Hering is restricted to the flat-leaved Sedum telephium L, while *Ph. sedi* Kalt. occurs only on the Sedum with cylindrical leaves (S. album L.). In these two families an extensive degree of protein differentiation can thus also be detected.

Elaeagnaceae and Thymelaeaceae also divide off from the same branch; they support very few miners. The latter family is exclusive host for one species of *Dizygomyza* and the *Phyllobrostis* species; these are so isolated that they are of no value in suggesting plant relationships. At the end of another branch are the Onagraceae, whose principle miners are Momphidae. One case of monophagy (first degree) is to be found here. *Lophoptilus raschkiellus* Z. occurs only on Epilobium angustifolium L., avoiding all other species of the genus. This possibly indicates that this species of Epilobium, which was previously plcaed in a separate genus Chamaenerium, was incorrectly grouped among the Epilobium. Araliaceae which follow, have only very polyphagous miners in the Palaearctic Regions; in other parts of the world, however, for example in New Zealand, mining species are found on Araliaceae, which show distinct monophagy, even first degree.

The adjacent Cornaceae are mined by one species of *Phytomyza* and also by Lepidoptera from the genus *Antispila*. It is particularly worthy of note that *Antispila* species are found exclusively on Cornaceae and Vitaceae, which on our genealogical tree are widely separated from each other. These two families possess a special pigment, "wine-red" distinct from all other related pigments, referred to as "beet-red". The occurrence of this wine-red, together with miners from the genus *Antispila*, which are limited to these two families, possibly suggests that there is a closer relationship between the families than has so far been portrayed in the various botanical systems.

The highest point on this branch is occupied by the Umbelliferae. They are of particular interest to the minologist. The majority of miners on this family are definitely monophagous; frequently monophagy (first degree) is encountered and species of Peucedanum have already been mentioned in this connection. *Philophylla heraclei* L. (also the closely-related North American *Ph. fratria* Lw.) accepts many different genera of the family. The habits of all the Trypetidae, however, to which both species belong, suggest that they have become miners only very recently; Trypetidae (fruit-flies), as their name indicates, are primarily fruit eaters and only very few species have adopted the habit of leaf-mining. Consequently, no extensive adaptations to the mining habit have developed and likewise there has been no clear specialisation on specific food-plants within the family. Apart from this one exceptional case, practically all Agromyzidae which mine Umbelliferae show evidence at least of monophagy (third degree). There is also a whole species group of the genus *Phytomyza*, namely the *albiceps* group, which is almost entirely restricted to Umbelliferae and Compositae; parallel to the *albiceps* group (with yellow frons), the *obscurella* group (with black frons) of the same genus is similarly restricted to Umbelliferae and Compositae. There is every indication that within the Umbelliferae an extensive degree of protein differentiation has occurred, which is not less marked than that which we observed in the Ranunculaceae. Where we find one and the same species mining different genera, it would probably be better to group these genera together. This has already occurred with Angelica and Archangelica, which were previously considered as separate genera; *Phytomyza angelicae* Kltb., which occurs on both but on no other Umbelliferae, suggests that this unification was justified. Extensive protein differentiation might more readily be expected in a family of this kind placed at the top of a branch than in the more primitive Ranunculaceae; this merely shows that the degree of protein differentiation develops independently of the systematic position of the family and that in some cases this development occurs more slowly, in others more rapidly, almost in the form of a leap.

The presence of species of *Phytomyza* from the *obscurella* and *albiceps* groups on the two most highly specialised plant families finds a parallel in the presence of leaf-mining fruit-flies which, in the Palaearctic and Nearctic Regions, only mine Umbelliferae and Compositae. This case is so striking that it suggests some relationship between Umbelliferae and Compositae, which is not at all apparent from our genealogical tree. Morphologists long ago suggested some connection between the two families, and this found even greater support from cytologists; it was confirmed by serum diagnosis undertaken by Blass (1930), who discovered that all reactions between Umbelliferae and Compositae (and the closely related Dipsacaeae) were beyond doubt positive. This deserves special emphasis, since this was not an observation made by Mez or his supporters but was the result of work by a pupil of the Schürhoff school, which considers that serum diagnosis is basically unsuitable for establishing plant relationships. We need thus have no hesitation in accepting the accuracy of their findings. According to this work the Umbelliferae would have to be placed in an entirely different position on the genealogical tree close to the Compositae.

Santalaceae have very few miners and are on a quite different part of the genealogical tree. However, one of these miners, *Liriomyza thesii* Hering is most closely related to certain species which live on Compositae. Furthermore, we also find here the *Epermenia* species, which form a compact genus and mine exclusively on Umbelliferae and Santalaceae. The slight degree of differentiation within the genus *Epermenia* would almost suggest that Santalaceae should be grouped closer to the two highest families on the genealogical tree and should be linked by some degree of relationship to the Umbelliferae.

The next branch of our tree has two families which are of special interest. These are the families Caryophyllaceae and Chenopodiaceae, (grouped together as Centrospermae), and the Amentiferae, of which Corylaceae (Betulaceae), Salicaceae and Fagaceae are of particular interest. This branch commences close to the Berberidaceae, which are almost completely avoided by miners; the only example is one species of *Caloptilia* and this is significant. We know that *Caloptilia* species occur primarily on Amentiferae (using *Caloptilia* in the narrower sense of the term). Considering first the Centrospermae, in which we nowadays no longer include the Polygonaceae, there are a large number of mining species occurring in the various families of this group (although some are merely biological forms). In the previous section we indicated the peculiar habits of *Euspilapteryx phasianipennella* Hbn. which lives on Polygonum and Chenopodium, while also occurring frequently on Lysimachia. Buhr (1937) discovered in the Rostock Botanical Gardens the Anthomyid *Hylemyia (Delia) echinata* Séguy (normally on Caryophyllaceae) on Chenopodiaceae and Amarantaceae, which likewise belong to the Centrospermae, and also once on a Primula; he also found it strangely enough several times on Polemoniaceae, which occupy an entirely different position in the system. Mez's genealogical tree indicates this close connection to the Primulaceae, by placing it in the immediate vicinity of the Centrospermae, on the same branch; other systems normally show the Primulaceae in a quite different position. Mention has already been made of *Dizygomyza flavifrons* Mg. (a typical mining fly of Caryophyllaceae) which also occurs on Beta, a genus of Chenopodiaceae. Morphologically indistinguishable species live on the same families, such as *Pegomyia chenopodii* Rond., *atriplicis* Gour. and *silenes* Hering, whose specific names indicate their host plants. One species, however, cannot be passed over without comment, the true *P. hyoscyami* Pnz., living on Solanaceae; it is difficult to find any points of contact with this species, unless perhaps the

Gnorimoschema species whose relationships have not been closely investigated, may offer some guide. However, if the Primulaceae are placed closer to the Centrospermae (this was done by v. Wettstein without serum diagnosis), a close relationship between them and Solanaceae may well be discovered. Amaranthaceae are closely linked to this related group and in some cases the same miners are found on both; this applies especially to the species of *Microsetia* mining Chenopodium, which may also be found on Amaranthus, although this occurs more frequently in North America than in Europe. In Europe *Microsetia* is restricted to Chenopodiaceae and Caryophyllaceae, thus indicating their relationship. Whether Polygonaceae were separated from Centrospermae justifiably or not must here be left open. When discussing oligophagy (second degree) we quoted several examples of the relationship between Chenopodiaceae and Caryophyllaceae; we might mention here many species of *Coleophora* (not miners but seed-eaters), where closely related species groups affect these two families. No convincing evidence is available on the basis of miner's food selection of a high degree of protein specialisation among the Centrospermae: we frequently find oligophagy (first degree) and (second degree), though this is perhaps less evident in the Caryophyllaceae, which stand somewhat apart, than in the Chenopodiaceae, whose genera appear to vary less in this respect. Protein differentiation appears to be more developed in the Polygonaceae. *Pegomyia nigritarsis* Zett. is very common on Rumex, while *P. bicolor* Wied. is equally common on Polygonum; very seldom does either of these very closely related species feed on the host of the other. *P. bicolor* Wied. has always been bred on Oxyria; this emphasises its close relationship to Polygonum, although botanists classify it as closer to Rumex. *Mantura rustica* L. is found on Polygonum (only P. aviculare L.) and on some species of Rumex. It is strange that none of these species is able to transfer to Rheum, although this is frequently attempted. The mines do not fully develop and the larvae die while young. It must again be left open whether this is due to the extensive protein differentiation in Rheum or whether the reason is that Rheum is not an indigenous species; we have already pointed out the effect on miners of plants which are far removed from their normal habitat.

Santalaceae and Loranthaceae which now follows are on a lateral branch, representing a cul-de-sac, as far as future development is concerned. We have already discussed the former and their presumed relationship to the Umbelliferae. Among Loranthaceae, Viscum and Loranthus are each hosts to one species of miner; however, both these miners occupy an isolated position within their genera and no deductions can be made from their habits. The much discussed question whether extensive differentiation occurs within the species Viscum album L. according to the nature of its host plant finds no confirmation in *Argyroploce woodiana* Barr.; it lives as frequently on mistletoe growing on broad-leaved trees as on conifers. It was originally bred in England on mistletoe parasitic on broad-leaved trees, but in Berlin, for instance, it occurs exclusively on mistletoe parasitic on Pinus.

The amentiferous families require a much more detailed discussion. Their systematic position has been frequently disputed by botanists and they have changed places on several occasions. It is doubtless correct to assume that the peculiarities of this family have developed by adaptation to anemophily or wind pollenization (in Salix pollenization by insects is certainly a secondary development); this has caused an appreciation of the true relationships of this families to be much obscured. We wish to deal here primarily with Salicaceae Corylaceae (Betulaceae) and Fagaceae. These three families are certainly closely related, although this is not clearly indicated by their position on the genealogical tree. For example, Salicaceae should certainly be closer to Myricaceae, as is indicated by the numerous examples of oligophagy, in which the same species of miner occur on both families; this is also not disputed by plant morphologists. The genuine *Caloptilia* species occur on all these families (as also on Juglandaceae). *Phyllotoma* mines in Salicaceae and Corylaceae, on Salix, Populus, Alnus and Betula (also on Acer, although this species group is isolated by its habit of cutting its cocoon out of the leaf epidermis). It would appear that the relationship between Salicaceae and Corylaceae is closer than that of either to Fagaceae. *Agromyza albitarsis* Mg. lives on Populus, the closely-related, only recently separated *A. alni-betulae* Hend. on Alnus and Betula; *Rhamphus pulicarius* Herbst. lives on Salicaceae and Betula, *Lyonetia clerkella* L. on Betula and Salix; however, none of these species nor their nearest relatives live on Fagaceae.

Mining insects provide no guide for the correct placing of the Amentiferae on the Centrospermae branch. The connection, however, can be largely justified even on morphological and cytological grounds. According to the very reliable observations of Brischke the non-mining Tenthredinidae provide a very interesting indication: *Pachynematus albipennis* Htg. lives only on Polygonum and Salix, thus confirming the close connection between the two.

The close connection between the various families of Amentiferae need no special emphasis here. Every specialist in any insect order will be able to quote species from his particular field, which live on different genera of these families. From the point of view of minology, however, there is a strange fact which requires closer examination.

A large number of miners, (as also of free-living species), live on Amentiferae families but also simultaneously on Rosaceae, while avoiding all other families. Such cases are so common and consistent that they cannot be explained on the basis of traditional plant classification. There must be an extremely close connection between these two groups of plants, which cannot simply be explained by suggesting that certain associated substances are present in both and which accounts for this oligophagy. So far none of these substances common to these two, but absent from other families, have been discovered. Furthermore, the species concerned here, (apart from a few exceptions to be discussed later), are not miners. A chance transfer to different plants cannot occur with miners, owing to their habit of remaining permanently on one plant, as explained earlier. However unscientific this may sound from the point of view of the modern student of plant phylogeny, we are nevertheless forced to the prophesy that sooner or later a genuine relationship will be discovered between these two groups of plants, even though at the moment there may appear to be less evidence for it than for that between Umbelliferae and Compositae. The commonest and best-known example of this oligophagy, which I would consider as systematic oligophagy, (second degree), is the occurrence of *Leucoptera scitella* Z. exclusively on the most varied genera of Rosaceae and on Betula. The frequency and consistency of the selection of these food plants show that this is no mere xenophagy, (Fig. 129). Taking into consideration the extremely limited range of food-plants accepted by this genus and recalling the fact that none of these miners will even eat a plant from a closely related family and that many of them are monophagous, (either third degree and sometimes even first degree), then it is difficult to accept that *L. scitella* Z. should form an exception. The assumption that this is a case of disjunctive oligophagy is contradicted by the fact that this species of *Leucoptera* does not represent an isolated exception. *Lyonetia clerkella* L. and *prunifoliella* Hb., *Recurvaria nanella* Hb., *Coleophora siccifolia* Stt., *Coptodisca splendoriferella* Cl., (U.S.A.), the North American Hispid *Baliosus ruber* Web., (which occurs also on Tilia), *Trachys minuta* L., (also on Ulmus and Tilia), species of the genera *Rhamphus* and *Tischeria*, species of the Tenthredinidae sub-family *Fenusinae*, (of which two genera occur on Ulmus and Tilia but nowhere else), all occur on Rosaceae and Amentiferae and only on these families. Also all European species of *Lyonetia* are restricted to Rosaceae and Amentiferae. There are also dual species, pairs of species which are extremely closely related and very difficult to distinguish, of which one lives on Rosaceae and the other on Amentiferae; it may be assumed that they developed into separate species only in recent times. We may mention

Fig. 129

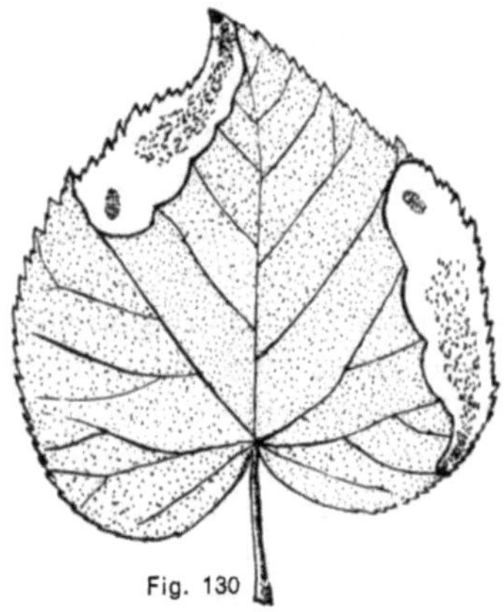

Fig. 130

Fig. 129. Betula: *Leucoptera scitella* Z. (also on many Rosaceae).

Fig. 130. Tilia: *Trachys minuta* L. (also on Rosaceae, Salix, Ulmus).

Lithocolletis corylifoliella Hb. on Rosaceae and on Betula, *Coleophora serratella* L. on Rosaceae and *C. fuscedinella* Z. on Corylaceae, and the species of *Coleophora* of the whole, distinct *palliatella* group, which are only found on these two plant groups.

External feeders which are of less value in substantiating our claims nevertheless possess numerous oligophagous species and genera, which live on Rosaceae and Amentiferae and only on these. The following are a few examples of species which occur on both plant families: among Lepidoptera, *Rhagades pruni* Schiff., *Cochlidion limacodes* Hufn. *Heterogenea asella* Schiff., *Simaethis pariana* Cl., *Pamene spiniana* Dup., *Tmetocera ocellana* Fbr., (but also on Hippophae, a genus of Elaeagnaceae); among saw-flies, *Rhodogaster punctulata* Klg. We may also mention the dual species *Argyroploce profundana* Fbr. and *pomedaxana* Pierce, a pair of species occupying a quite isolated position in the genus and only recently recognised as separate species; the latter lives on Prunus padus, the former on Quercus, but it has not entirely "forgotten" its origin and is sometimes also found on Prunus. The highly specialised *Carpocapsa (Laspeyresia) pomonella* L. lives only on Rosaceae and Juglans. The *Swammerdamia* species of Lepidoptera are confined to Rosaceae and Amentiferae, as well as our true *Caloptilia*, (although one species occurs on Moraceae in the same branch, and another on Ericaceae). *Parornix* species are restricted to Rosaceae and Amentiferae, (with one species on Acer). Among the Tenthredinidae also these two families and no others are affected by the genera *Dianeura*, *Cimbex*, *Trichiosoma*, *Pamphilius*, *Janus* and also the genus *Arge* which, like

Caloptilia, also lives on Berberis; this would seem to provide ample justification for locating the Centrospermae—Amentiferae branch close to that of the Berberidaceae.

Although from the purely morphological stand-point the suggestion of a relationship between Rosaceae and Amentiferae may appear quite improbable, nevertheless the number of cases indicating such a relationship is so large, (and others could be mentioned), that one has to discard the possibility of these merely being examples of disjunctive oligophagy. Taking into consideration the extent to which Amentiferae may have been transformed by their adaptation to wind pollenization, it must none the less be conceded that one is justified in suspecting some relationship, which has so far found no support in any botanical investigations. We are convinced that future investigations, possibly with more subtle methods of serum diagnosis, will confirm the relationship between these two groups of plants.

No clear picture can be formed of the degree of protein differentiation within the families and genera of Amentiferae, since the position varies among the different genera of miners, according to the degree of taste specialisation of the individual mining species. It has already been mentioned that Fagaceae are clearly further removed from Salicaceae and Corylaceae than are the two latter from each other. *Rhamphus pulicarius* Hrbst. occurs even on Betula, (never on Alnus), and Salicaceae, on Myrica but not on Fagaceae. Myrica and Salix are mined by many identical species; both Salix and Populus are host for the species of *Rhamphus* just mentioned and *Rhynchaenus populi* Fbr. and *decoratus* Hbst. It appears that differentiation is most highly developed in the Corylaceae; Betula and Alnus have no species in common, apart from *Agromyza alni-betulae* Hend., which has only recently been separated from the Salicaceae species *A. albitarsis* Mg. It is significant that *Lyonetia clerkella* L. and *L. prunifoliella* Hb., which are equally common on Rosaceae and Betula, are never found on Alnus. Attempts made here to transfer miners from Alnus to Betula and vice versa have never achieved a successful larval development. It is strange that these two species of *Lyonetia* frequently and without difficulty accept Rosaceae, which are alleged to be so far removed systematically, but never accept the genus Alnus, which belongs to the same family! Within the Fagaceae, however, differentiation between the genera seems to be more developed. Individual sections of Quercus, as well as of the genus Fagus, have their own miners; this applies not only to groups such as the hard and soft-leaved oaks, but also to species, such as Qu. pubescens Wlld. Certain less discriminating species, such as the genus *Tischeria* are found on Quercus and Castanea but avoid Fagus, thus indicating the closer relationship between the first two genera;

many species of *Nepticula* and *Lithocolletis* are also common to both genera.

At the end of the Centrospermae—Amentiferae branch are the Moraceae, Ulmaceae and Urticaceae. The close relationship between Ulmaceae and Amentiferae has already been indicated by several examples. Here also we find a further suggestion of the relationship to Rosaceae; *Coleophora badiipennella* Dp. which is normally monophagous on Ulmus, has a sub-species *C. adjectella* HS., (possibly a closely related species), which occurs on Prunus spinosa L. Within the family, the two genera Ulmus and Celtis show a considerable degree of differentiation: the *Lithocolletis* which occur on them are different species and no Ulmus-feeding species of *Nepticula* will accept Celtis. The position is similar with Moraceae, (which now include Cannabinaceae). No miners have yet been found on Morus, while a Cannabis-feeding *Liriomyza* refuses to feed on other genera. *Cosmopteryx* indicates the relationship between Humulus and Urticaceae, and a species of *Caloptilia* the relationship with Amentiferae; two species of *Agromyza*, however, are confined to this genus. There is no very great protein differentiation between the genera of Urticaceae, Urtica and Parietaria, although they occupy the highest position on this branch; *Agromyza reptans* Fall. and *A. anthracina* Mg. occur on both genera. The former is also occasionally found on Cannabis and Humulus, thus emphasising the relationship between Urticaceae and Moraceae. Experiments in which Buhr transferred miners from one genus to another within this family were invariably successful, while they always failed on Ulmaceae; in other respects also Ulmaceae seems to stand rather aside from the two other most specialised families.

Climbing up to the next branch on the genealogical tree we reach the Tiliaceae. There are numerous indications that Tiliaceae and Amentiferae should be closer together and not so isolated from each other. *Trachys minuta* L. lives on Amentiferae and Rosaceae, Ulmus and Tilia, (Fig. 130). Among the saw-flies, as we already mentioned, the *Fenusinae* are found on Amentiferae, Rosaceae, and on Ulmus and Tilia; also, the well-known, conspicuous, snail-like *Caliroa* feed externally on Amentiferae-Rosaceae and on Tilia. The gall midge *Eriophyes rudis* Nalepa, (a species which, like all Cecidozoa, shows a high degree of adaptation to its food-plant), lives on Betula and Tilia as two distinct sub-species. *Atemelia torquatella* Z. which mines on Ulmus as well as on Betula, is no longer found on Tilia. Further similar examples could easily be quoted to support the suggestion that the Tiliaceae should more properly be placed on the Amentiferae branch.

Malvaceae support a whole series of miners; differentiation

within the family seems to be little developed and the majority of miners found on one genus will also be found on other genera of the family.

On the other hand, in Euphorbiaceae which follow, the protein differentiation seems to be very highly developed. None of the Euphorbia species will accept Mercurialis and even within Euphorbia itself distinctions are made between individual species. It might be expected that none of the species feeding on the bushy Euphorbiaceae of the Mediterranean area will accept herbaceous species; however, even among the latter *Liriomyza pascum* Mg., which is commonly found mining E. amygdaloides L., consistently refuses the garden weed E. peplus L.

There is little to say about Oxalidaceae, Zygophyllaceae, Rutaceae, Linaceae and Geraniaceae, which follow on a lateral branch. In most cases they possess miners, which are characteristic of the genera concerned and in Europe are limited to a few genera and species. Their main distribution is in other regions where little is so far known about miners. We are not therefore justified in drawing any conclusions on plant relationships. The Geraniaceae are interesting for two reasons; they possess one species of mining saw-fly, (*Fenella voigti* Hering), while the other members of the genus are restricted to Rosaceae. Presumably, however, this species will have to be separated from the rest of the genus. Also on this family there is one butterfly, (*Lycaena astrarche* Bergstr.), whose larva mines just like those of *Coleophora*. Such a habit is quite abnormal among Rhopalocera. At the end of this branch are the Tropaeolaceae; we have already discussed the importance of this family in some detail in connection with "bridging species."

The families on the next lateral branch also merit little consideration from the point of view under discussion. They have their own, characteristic miners which in each case do not occur on the nearest related family. Only the Vitaceae are worth discussing in greater detail. When discussing Cornaceae we mentioned that there must be some close connection between these two families, which is not apparent from the genealogical tree. The genus *Antispila*, (in Europe and North America), lives on Vitaceae and Cornaceae, but not on the neighbouring Rhamnaceae. The former two families share no miners in common with the latter. Within Rhamnaceae the existing unification of the genera Rhamnus and Frangula is supported by *Bucculatrix frangulella* Goeze, which mines both.

A similar situation is found in the most specialised group of this branch but Ericaceae, (with the closely related Empetraceae), and Aceraceae require separate discussion. When considering the Amentiferae we saw that both the latter families in several respects have close connections with Amentiferae—Rosaceae. It would appear that these families also have not found their correct place on the genealogical tree, but that they should be more correctly on the Amentiferae branch. For instance, *Lyonetia ledi* Wck. occurs on Ledum, (a genus of Ericaceae), while the remainder of the genus are normally restricted to Amentiferae—Rosaceae; it is also able to complete its development on Myrica. *Coleophora siccifolia* Stt. which, as already mentioned, mines Amentiferae—Rosaceae is so similar to *C. uliginosella* Glitz, even in its genitalia, that B a r a s c h, (1934), was inclined to consider the latter as a sub-species. *Parna tenella* Klug on Tilia, supports the demand for including Tiliaceae with Amentiferae—Rosaceae, since it belongs to the Fenusinae which are limited to the two groups; *Phyllotoma*, which occurs exclusively on Amentiferae and Acer, suggests that Aceraceae should occupy a similar position. All European species of *Parornix* live on Amentiferae—Rosaceae and one species also occurs on Acer; the large number of species of *Caloptilia* living on Acer also suggests a similar relationship.

In both families the protein differentiation is clearly highly developed. The genus Negundo, which is possibly incorrectly included in Acer, is not attacked in Europe by a single species of Nepticula, although *N. aceris* Frey and *N. speciosa* Frey are found on all European species of the genus. In this instance, as in others, the more discriminating species of *Lithocolletis* have developed separate species on the different species of Acer, but none will feed on A. negundo L. The various genera of E r i c a c e a e also have own characteristic miners, which in some cases specialise on individual species as, for instance, on Vaccinium; only Calluna and Erica are so similar that they each have the same species of miner.

A number of families which follow on the next branch on the right-hand side appear to be very closely related: these are R e s e d a c e a e, P a p a v e r a c e a e and C r u c i f e r a e. Numerous miners occur on all three families and this is clearly linked up with the occurrence of myrosin cells, which they all possess. Before the Papaveraceae branch off from the Cruciferae one finds the C a p p a r i d a c e a e, which form no exception in their range of miners. The following Cruciferae-miners, which are able to develop normally on all these families, may be mentioned: *Scaptomyzella flava* Fall., *Ceuthorrhynchus contractus* Mrsh., *Phyllotreta nemorum* L., *Plutella maculipennis* Curt. and others.

P a p a v e r a c e a e in fact occupy a slightly more independent position in relation to the Cruciferae, Resedaceae and Capparidaceae, (all grouped together as Rhoeadales), than appears from the illustration of the genealogical tree. Supporting this is the fact that species which are common on the Cruciales, (Cruciferae, Resedaceae and Capparidaceae), are not in the wild state found on Papaveraceae, (although they will accept it in captivity), which lack the characteristic myrosin cells. *Pieris* feeds on the Cruciales and also on Tropaeolum, but refuses Papaveraceae which also belong to the Rhoeadales. The same has been observed in the parasitic fungus Albugo candida (Pers.) O. Ktz., which occurs on Cruciales, but not on Papaveraceae.

The position of Tropaeolum, on which they are also able to develop has already been discussed as a bridging species. Similar examples can be quoted from among the parasitic fungi. Protein differentiation within these families, judging from the habits of their miners, is only developed to a very slight extent; a species which occurs on one genus of Cruciferae, is able to live on practically all other genera, and transfer to Resedaceae causes no difficulties. The same applies to the parasitic fungus Peronospora parasitica (Pers.) Fr., though this has developed a number of biological forms which can be distinguished by statistics of variation. A rather different situation, however, is found in some externally-feeding beetles, and one beetle, (*Psylliodes tölgi* Hktg., which mines in Biscutella), has never been found in any other genus of Cruciferae. A fact perhaps worthy of mention is that, as late as 1902, Krause included all genera of Cruciferae in the single genus Crucifera.

Guttiferae and Cistaceae, which are placed on a separate branch, are not so closely related to the last group, and have no close connection with the Resedaceae, although the two appear to be close together in our illustration. They possess a number of their own characteristic miners, which provide no indications of plant relationships. Differentiation according to the present division of genera within the Cistaceae has been too little investigated, as far as miners are concerned, to permit of any comment. Cistaceae are mined by a large number of insects, especially in the Mediterranean area.

A similar isolated position is occupied by Violaceae, Passifloraceae, (with only a single species of Nepticula), and Begoniaceae. It is of interest to recall the case of xenophagy on the latter, noted in *Pegomyia bicolor* Wd., which is normally confined to Polygonaceae.

The following branch commences with Oleaceae. The close connection between its various genera is strikingly illustrated by *Gracilaria syringella* F. which lives on Syringa, Ligustrum and Fraxinus. The confinement of the Phillyrea parasites to their specific hosts, (for example, *Phytomyza phillyreae* Hering and *Silliana lhommei* Hering), is possibly a result of zoogeographical factors; *Parectopa latifoliella* Mill., however, also occurs on Olea. *Gracilaria syringella* F. is constantly attempting to transfer to Caprifoliaceae at the end of the branch, thus confirming the correct position of the family, although it has not been proved by serum diagnosis.

The families which follow on various lateral branches of the genealogical tree possess only a few miners, which in each case are characteristic of each individual family and provide no further evidence for our argument.

Boraginaceae, Scrophulariaceae and Plantaginaceae now follow, grouped around the Labiatae. Mining insects provide a number of indications of this relationship. We may mention the Labiatae miner *Coleophora albitarsella* Z., which mines in a form *pulmonariae* Rag. on Boraginaceae. The *Apteropeda* species likewise live on these two families and also on Plantaginaceae, which belong to the same group. As a comparison we may recall the externally-feeding larva of the Tenthredinid *Pachyprotasis nigronotata* Kriechb., which also only lives on Plantago and Mentha. Among Coleoptera, *Mniophila muscorum* Koch mines Plantago and Digitalis, a genus of Scrophulariaceae, (according to Kaltenbach). Finally, *Coleophora onosmella* Brahm mines in Labiatae and Boraginaceae. We previously mentioned the isolated occurrence of the fly which normally mines Scrophulariaceae, *Dizygomyza verbasci* Bouché, on Buddleia, a species of Loganiaceae, (Fig. 123). Our illustration shows that both families are on the same branch of the genealogical tree, so that the xenophagy of this species should be more correctly termed systematic oligophagy, (second degree). Protein differentiation appears to have been only slightly developed in these families. Miners on Labiatae, Scrophulariaceae and Plantaginaceae invariably feed on a number of genera within the family and show no evidence even of monophagy, (third degree). Even among normally highly selective Coleoptera, *Dibolia depressiuscula* Letzn. is found mining Ballota and Marrubium and even on Veronica.

At the end of a lateral offshoot of this branch are the Solanaceae which, however, have no particular relevance to the subject under discussion. The occurrence of *Pegomyia hyoscyami* Pnz. on some genera of Solanaceae is of interest, since morphologically it cannot be distinguished from its relatives which live on Centrospermae. The following parallel cases might also be mentioned: Buhr (1937) discovered the fly *Liriomyza solani* Macq., which mines Solanaceae, also on two genera of Chenopodiaceae; the colorado beetle, *(Doryphora, Leptinotarsa)*, is mentioned as occurring on Chenopodiaceae and Amaranthaceae, as well as on Solanaceae; the parasitic Chenopodiaceae fungus Peronospora effusa (Grév.) Rabh. occurs on Amaranthaceae and Portulacaceae, and its var. *hyoscyami* Rabenh. also on Hyoscyamus. There is also little to say on the degree of differentiation between the individual genera; however, the occurrence of *Acrolepia pygmaeana* Hw. on Atropa and on the single species of Solanum S. dulcamara is extremely interesting.

The most highly developed families of this branch are Rubiaceae, Valerianaceae and Caprifoliaceae, each of which has its own characteristic miners, without any obvious interconnection by oligophagy (second degree). This is not surprising in view of the highly specialised position of these families. The slight degree of protein differentiation might not, however have been expected. With practically all species occurring on these families one finds systematic oligophagy (first degree), and monophagy (third degree) is not the normal rule.

Only in the highest family of the Caprifoliaceae does generic monophagy appear to predominate; but Lonicera and Symphoricarpos have the greatest number of mining species in common.

Shortly before this group are the Dipsacaceae which, despite many similarities are not, according to Mez, closely related to the Compositae. However, subsequent serum diagnosis has shown that there is some degree of relationship between this family and Compositae, and *Phytagromyza similis* Bri. (Dipsacaceae) and *Ph. centaureana* Hering (Centaurea), which can only be distinguished by their larvae, provide an interesting confirmation of this. Differentiation between the genera of this family appears to be well developed, although *Trachys troglodytes* Gyllh. mines on Succisa, as well as on Scabiosa and Knautia.

Cucurbitaceae which follow have no characteristic miners. According to more recent sero-diagnostic investigations they should not be placed so close to the Compositae as in our illustration.

Gentianaceae represent a compact group, whose miners have no connections with those of other families. Differentiation between the genera is clearly only slight. *Phytomyza swertiae* Hering, which is characteristic of Swertia and is to be found nowhere else, presumably developed largely as a result of zoogeographical factors.

Campanulaceae also are a compact family with their own characteristic miners. There are no links with other families, apart from the fact that the sub-genus *Calycomyza* of *Dizygomyza* is confined to Compositae and Campanulaceae, thus indicating their relationship. Differentiation within the family is suggested by the habits of its miners to be slight; oligophagy (first degree) is the normal rule. However, there is one species of *Rhinosia* which mines exclusively on Campanula persicifolia L. and refuses all other species of the genus. This doubtless indicates an unusual taste specialisation of the larva concerned.

We have now reached the highest group of the plant genealogical tree, the Compositae. Their special connection with the other most highly specialised group, the Umbelliferae, (although not indicated on the genealogical tree, yet doubtless existing and proved by serum diagnosis), was discussed in detail when we dealt with the latter. Protein differentiation is highly developed in the Compositae, as would be expected from their position as the highest family; this is shown in the monophagy of the majority of their miners. Frequently monophagy (first degree) is found, as in Artemisia vulgaris L. and other species of the same genus. Oligophagy (first degree) can also often be observed, as for example in the occurrence of *Phytomyza conyzae* Hend. on Inula, Buphthalmum, Pulicaria and Telekia. In this case, as in others, the species is able to exist on only a few, closely related genera. *Phytomyza kyffhusana* Hering is restricted to Gnaphalium and Inula; certain species indicate the relationship of Centaurea with Carduus and Cirsium, for which a number of other examples might also be quoted.

The botanist who has followed this discussion will not have been able to avoid confirming many of the points raised. Less frequently he will have shaken his head in doubt and on some occasions, as when we were suggesting the relationship of Rosaceae and Amentiferae, he will have been roused to vigorous protest. The entomologist for his part will have been able to adduce certain cases of food selection from within his particular group which conflict with our suggestions. We therefore once again recall the remark of Fischer & Gäumann that in considerations of this type one must in each case examine the range of hosts of a single parasite and one cannot disprove one's findings by adducing different habits of others. However, as was emphasised at the commencement of this chapter, the selection of food by miners can provide no proof of relationships between the host-plants, but merely gives the botanist indications where such relationships may possibly be found. At best we can merely provide pointers. At all events this subject opens up a field of minology which, although highly controversial, is nevertheless of the utmost interest and it can be anticipated that future research, by both botanists and entomologists, will result in many new and interesting discoveries. It is not sufficient merely to assert that certain associated substances exert such an extensive influence on the choice of food of phytophagous insects. The presence of these substances has been proved in practically no case; on the other hand the fact that the mining insect is so frequently correct when judging plant relationships permits one to deduce that it is also correct in other cases, which the student of plant phylogeny has not yet been able to accept. Much closer cooperation is necessary between the botanist and the entomologist in this border region, in order to achieve a final clarification of the subject.

One final question remains in this discussion and that is to assess the degree of specialisation in relation to the origin of the mining insect. In other words, which is more primitive, which more specialised — monophagy or oligophagy? This leads on to a second question: should we consider monophagous or polyphagous miners as the more primitive?

There is little doubt that in itself polyphagy represents the more primitive type of food selection. This applies not only to miners but to all other phytophagous insects and it appears that parasitic fungi form no exception to this rule. Monophagy only developed later via oligophagy from the earlier polyphagy, through specialisation and differentiation. This view is supported by cases of oligophagy still found in parasitic fungi, which can only be explained by the species, which to-day are able to produce their two generations on plant families far removed from each other and in no way related, being formerly polyphagous. If the transfer to the other family was a secondary development, the species concerned would have chosen a more closely related family, (this appears to be at least a logical assumption). There can thus be little doubt that polyphagy is the primitive, and monophagy the derived, more specialised form of food selection. It would, however, be mistaken to assume that polyphagous miners were more primitive than those in which monophagy is found. The mining habit of insect larvae is itself such a considerable adaptation and specialisation that it can only have developed in most recent times. This is suggested by the fact that only members of the four highest insect orders have adopted mining. Among the more primitive Hemimetabola there are no miners. Among these four orders only the highly developed forms are miners, (apart from a few exceptions: *Eriocrania* among Lepidoptera and the Scatophagidae among Diptera, for example). When polyphagy is found in such species, this represents a secondary development, such as in *Phytomyza atricornis* Mg. and *Liriomyza strigata* Mg., which are the commonest of our polyphagous species. They belong to genera in which practically all other species favour monophagy, or at least oligophagy (first degree). This fact clearly shows their special position and the secondary development of their polyphagy. We are thus justified in considering polyphagy among miners as due to a recent throw-back.

The question must also be considered whether the occurrence of certain miners on primitive plant families is an indication that the mining insects themselves occupy a primitive position, and vice versa. One cannot reliably deduce that such is the case, since the species concerned have not adopted their mining habits in any way parallel to the development of the plant family concerned, but only commenced feeding on it later. There may be cases where such deductions would be justified, as for instance, with the occurrence of the primitive Scatophagidae on the equally primitive monocotyledons. On the other hand one merely has to indicate the highly specialised genus *Hydrellia* of the Ephydridae, which is confined to monocotyledons, to perceive that no rule can be established. Among Lepidoptera, the Eriocraniidae, which are without doubt the most primitive of miners, are linked exclusively to the Amentiferae, which earlier were indeed recognised as a primitive group of families, (they appear as the most primitive flowering plants in the system of Wettstein, 1924, from which all others, including monocotyledons, have evolved); however, this can no longer be accepted, particularly in view of the numerous modifications resulting from wind pollenization. We must therefore be extremely critical and cautious when employing food selection as a criterium for judging the systematic position of insect species.

Food selection by mining insects, and indeed by all phytophagous animals and plants (fungi), as we have seen, represents a valuable aid in establishing plant relationships; we must never, however, forget that it is only one of many aids and we must refrain from any dogmatic expression of opinion. If we have expressed a degree of convinction on a number of doubtful cases, this should merely serve to emphasise the importance and necessity for further research on this remarkable border region between botany and zoology and to stimulate botanical systematists to undertake further intensive investigations into the questions raised.

The last word in assessing plant relationships must naturally rest with scientific botany. The importance of food selection in this respect cannot be proved by the minologist; nor, however, can it be refuted, as was attempted by Buhr (1937), employing methods of experimental minology, (which will be discussed in a later chapter). The experimental transferance of miners onto abnormal food-plants offers no proof and serves but little to clarify the problem. It is then no longer a case of selection by the miner or the ovipositing female, but of compulsion, exercised on the miner. If we learn by this means which substances the larva is able to utilise from other plants for completing its development, this has a physiological significance of a different kind, bearing no relation to the problem under discussion; what is here involved is the economic potentiality of the species and it is known that this can greatly vary, even in closely related species. Interesting as these experiments may be, they do not involve the powers of discrimination of the female nor the stimuli influencing them, which proceed from the plant itself. Our remarks, as has been several times emphasised, are in no way intended as a proof, but merely as an indication of where in the system of classification the botanist may perhaps look for proofs.

13.

MINERS ON AQUATIC PLANTS

This and the previous chapter have only a very general connection. The selection of host-plants by the miners discussed so far was a secondary characteristic, deriving from their family classification. The mining species we shall discuss here are influenced to a far greater extent by their environment, in other words by the occurrence of the plants they select on or in water. Aquatic plants are found in the most varying families and, independently of the family they belong to, they develop similar types of tissue, which are adapted to life on or in water and facilitate their survival. The majority of aquatic plants are characterised by possessing a tissue known as aerenchyma, consisting of a loose agglomeration of parenchyma cells, between which there are wide air spaces. The parenchyma cells are frequently arranged in such a way that often long, continuous air channels are formed which traverse the whole plant. The spongy parenchyma is normally utilised in the formation of this aerenchyma; the palisade parenchyma retains the normal arrangement of layers observed in terrestrial plants, since most plants, (in as far as they are not able to absorb air directly from the water) remain dependent on this tissue for their respiration and assimilation.

The relatively few mining insect larvae found on aquatic plants are frequently in some way adapted to aquatic life, although in many cases a special adaptation does not appear necessary. The species which feed in the aerenchyma of the plant have of course no contact with the water; they are surrounded by air in these air channels inside aquatic plants, even in the submerged parts of the plant. The transfer from terrestrial to aquatic plants is therefore not very difficult and does not require extensive modifications. On the other hand the supply of food in the aerenchyma is not particularly large and one might expect the larva to be compelled to excavate large areas of the plant in order to find sufficient food to complete its development. We therefore frequently find that the larvae feed for a part of their life not in the aerenchyma but in the assimilation parenchyma of these plants. The Agromyzidae (mining flies) must be considered as the least adapted to aquatic life of all insects. However, a few of them have transferred to aquatic plants. A cross-section of a leaf of Triglochin, in which *Phytomyza triglochinae* Hend. has lived (Fig. 131),

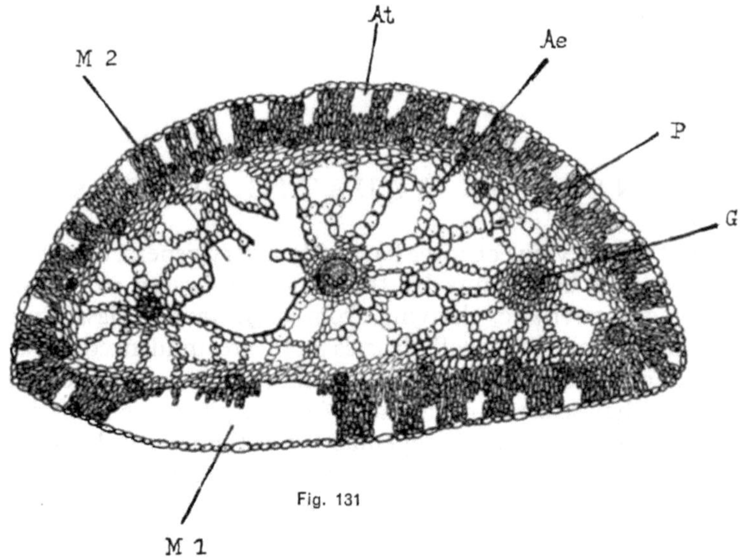

Fig. 131. Triglochin: Cross-section of the leaf with mines of *Liriomyza triglochinae* Hend. (Ae - aerenchyma, At - respiration cavities, G - vascular bundle, P - Palisade parenchyma, M1 - early mine, M2 - final mine.

shows how they select their food within the plant tissue. This plant lives in water but its leaves rise freely into the air, and the larva thus requires no special adaptation to aquatic life. The mine commences immediately below the epidermis of the semi-cylindrical leaf (M1), and to begin with only the palisade parenchyma of the leaf is consumed. Later, when the larva is larger, it proceeds to the centre of the leaf to construct its final mine channel, which now runs solely through the aerenchyma (Ae) of the leaf and is accordingly very long. We also find a similar, at least temporary, preference for the assimilation parenchyma in species which are more adapted to aquatic life, such as the species of *Hydrellia. H. hydrocharitis* Hering mines the leaves of Hydrocharis morsus-ranae L. (Fig. 134), which float on the water. A cross-section of the mined leaf showed that the mine channel ran exclusively through the assimilation parenchyma, which lay towards the air; the aerenchymatous spongy parenchyma on the side of the floating leaf towards the water was quite untouched. Under these conditions a mining larva lives exactly as in terrestrial plants — apart from the fact that the leaf floating on water decays more quickly, resulting ultimately in the penetration of water into the mine cavities. A peculiarity of species which feed primarily on the aerenchyma, is that they frequently cover large distances within the leaf without feeding; in other words there is a change in the relationship between living and feeding quarters within the mine. We shall consider miners in three groups, according to the position on the water of their food-plant.

a) Plants, whose mined leaves rise into the air above the water.

A large number of plants have their roots in water but their leaves are in the air and normally have no contact with the water. Larvae living on such leaves will usually find the same conditions as on terrestrial plants, even though, as we have seen, there may be a larger proportion of aerenchymatous tissue. We may quote as the best-known example *Elachista poae* Stt., which mines on the leaves of Glyceria, an aquatic grass. The mine channels invariably begin at the apex of the leaves and run along the leaves downwards. They are extremely long and usually not limited to one leaf. The larva changes to a second, adjacent leaf, which is usually below water. For this, however, it requires no special modification; the leaf sheaths are close together, so that the larva has no contact with the water when transferring. With other miners, whose feeding channel is not distinguished by such extreme length, the situation is even simpler. A mining weevil, *Hydronomus alismatis* Marsh., lives on Water Plantain. Its mines are long (Fig. 132) but more blotch-like, and with dinstinct channels only when the larva is young. Nevertheless, the larva is sometimes compelled to form a new mine and then wanders around on the upper surface of the leaf until it can find a spot where it can again burrow into the leaf, just as if it was on a terrestrial plant. In this case also the food consists of green parenchyma, as with miners of terrestrial plants. Many species of Hydrellia live in this way, as for instance *H. butomi* Hering and *H. griseola* Fall. on a wide variety of aquatic grasses. The short blotch-mine of such species always remains above water on fresh green tissue of the plant. *Liriomyza ornata* Mg. likewise lives on Butomus, but exclusively on the aerenchyma of the petiole or peduncle. The pale aerenchyma cells clearly do not produce as much food as normal green tissue and the feeding channel is accordingly much larger. In North America *Sphida obliqua* Wlkr., a relative of our *Archanara* species, lives in Typha (Reed-mace), and likewise exclusively in the aerenchyma. It bites through the transverse walls and covers long distances through the aerenchyma, consuming only very little food and thus causing long air channels in the plant. Unfortunately very few investigations have been undertaken into mining habits of our European

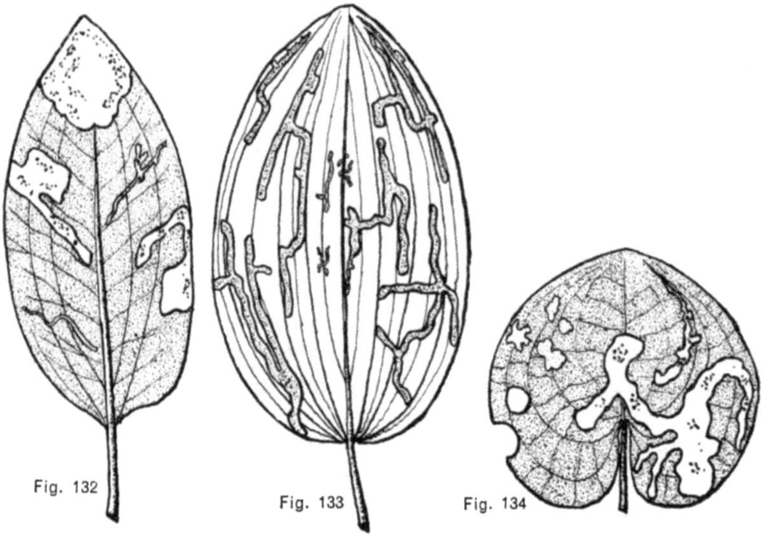

Fig. 132. Alisma: *Hydronomus alismatis* Marsh.

Fig. 133. Potamogeton natans L.: *Cricotopus brevipalpis* Mg.

Fig. 134. Hydrocharis: right, *Hydrellia hydrocharitis* Her.; left, case cut-outs and "window-feeding" of *Nausinoë nymphaeata* L.

species of *Archanara* but there is little doubt that many peculiarities of this type will be discovered among them also. It is known that the North American *A. oblonga* Grote also consumes the vertical walls of the aerenchyma channels so that the apex of the leaf dies and in this way the mine can be more easily found. The channels of *Sphida obliqua*, mentioned above, are from the outside scarcely visible at all.

It is a general peculiarity of mines in aquatic plants, that they cannot be detected as easily as mines in terrestrial plants. An important factor in recognizing a mine is the hollow cavity which stands out so distinctly from its surroundings. When similar cavities, such as those of the aerenchyma, already exist in the normal plant, further cavities produced by mining are so much less conspicuous. Furthermore, a change of colour in the mine is frequently caused by the penetration of air, as will be explained in the following chapter. When air is unable to enter the mine cavity, as in many aquatic plants, any distinct colouration of the mine caused by air is naturally lacking.

Many species of *Hydrellia* live on aquatic plants with only slight modifications, as already described; life near water is characteristic of the Ephydridae, the family to which they belong and their life as miners on aquatic plants in no way varies from that on terrestrial plants. This explains how one of them, *H. griseola* Fall., has without difficulty transferred to terrestrial plants. Like its relatives, it normally occurred on aquatic grasses, showing a preference for the genus Phalaris which grows on banks of rivers, edges of ponds, etc., and then frequently adopted the habit of mining in corn-fields situated in river valleys or other damp places. There it particularly attacked barley (Hordeum) and in many places has become a real pest. Presumably as a result of human cultivation, it has finally penetrated into gardens, which are far from any water. Recently we managed to breed it from Allium porrum L., which gave another interesting and convincing indication of the relationship of these two families of monocotyledons; this represented systematic oligophagy (third degree). The larva of *Schoenobius forficellus* Thbg., which also mines long channels in the leaves of Glyceria, shows in one respect a greater degree of adaptation to aquatic life. It does not proceed from one leaf to the next like the larva of *Elachista* but, when changing leaves, cuts a case out of the end of the mine, just like *Coleophora* species on terrestrial plants, and with this case, allows itself to be carried by the water until it reaches another leaf which it finds suitable; it never returns to the case, and thus, in contrast to *Coleophora*, its frass is not voided through the end of the case but remains in the mine. If it wishes to proceed to a new leaf, it constructs a new case; each case is used as a means of transport only once. The larva shows no other special adaptations to aquatic life. While feeding it utilises the air in the aerenchyma and during its boat journey it has sufficient air inside the case. While in the floating case it frequently protrudes the fore-part of its body to keep a look-out for a suitable new leaf.

Although in general miners inhabiting aquatic plants do not live very differently from those in terrestrial plants, nevertheless a number of special adaptations to life in the proximity of water do occur. Floods and other influences may submerge the host plant at least temporarily and thus cause a change in living conditions. The actual mine cavity is not immediately affected but disadvantages may ensue for pupation and for the pupal site, if this is outside the mine; the water and the current may separate the pupae from their original site and deposit them in places where the normal food-plant is no longer available. Pupae of miners in aquatic plants have actually been found on a number of occasions in river alluvia. The mining larva must, however, desire to remain at the spot where it pupates. Accordingly, puparia of *Agromyza* have been found attached to the leaves of Phragmites with silk, a process which is not usually seen in species which pupate outside their mine. The peculiar extension of the rear end by the posterior spiracular processes in certain aquatic species has already been mentioned; these structural extensions, which serve to anchor the pupa in the plant tissue are very conspicuous (Fig. 55) and it is significant that they have only been found on species of aquatic plants, where they have obvious importance by preventing the pupa falling into the water; *Phytomyza cicutae* Hend. on Cicuta, *Ph. calthivora* Hend. on Caltha and *Liriomyza virgo* Zett. on Marsh-equisetum are appropriate examples. Of two species of *Phytomyza* living on Caltha, *Ph. caltophila* Hering lives on the normal leaves, which are above the water; its puparium has no such extensions. *Ph. calthivora* Hend. occurs only on plants growing in very damp places, so that the lower leaves which it frequents lie either in or on water or mud, and this species possesses these peculiar structures. They indicate a life on aquatic plants, even though the habits of species obtained from puparia of this type, are not yet known.

b) Miners in floating leaves.

When miners occur in leaves which are floating on the water, more extensive modifications may be anticipated. A change of mines will still be possible, not, however, a change of leaves, unless the larva is specially equipped to swim from leaf to leaf. This has not in all cases been investigated. *Tanysphyrus lemnae* Payk. lives in Lemna; the individual plant-bodies are excavated by the larva and the larva occasionally moves across the water with undulating movements, in order to find fresh plant-bodies. When a suitable one has been discovered, it hauls itself up with its mandibles and commences to burrow inside. Pupation occurs on the land. If the larvae become submerged, they perish. We do not know what morphological peculiarities enable them to move across the surface of the water. The closely related *Tanysphyrus callae* Voss encounters no such difficulties; the leaves of Calla which it infests are well above the water; here it produces mine channels, which resemble those of Lycoriidae. Pupation occurs in a somewhat dilated blotch inside the leaf, where it enjoys substantial security, since leaves of Calla are of a relatively firm consistency. The pupa would run far greater danger in the leaves of Lemna, particularly as they decay relatively quickly.

The larva of a Tendipedid, *Cricotopus brevipalpis* Mg., lives in genuine mines, which also comprise feeding quarters, in the floating leaves of Potamogeton natans L. (unlike all other species of its family). It produces linear mines on the upper surface of the leaf (in the palisade parenchyma) and it lives in exactly the same way as miners of terrestrial plants (Fig. 133). It completely avoids the aerenchyma on the lower surface of the leaf on the water, as does *Hydrellia hydrocharitis* Hering with similar habits.

The special danger of early decomposition exists with all leaves floating on water, which of course would allow water to penetrate the mine cavity. This usually occurs at any early stage in the mines of *Hydrellia hydrocharitis* Hering (Fig. 134) and it is rare to find well-preserved mines on this plant. However, the larva has usually already completed its growth and pupated. In other cases the living larva may still be in the leaf. For instance, the larva of the North American *Bellura melanopyga* Grote lives in the leaves of the water-lily (Nymphaea). While mining in the leaves, it finds sufficient air in the aerenchymatous tissues; later it proceeds down into the petiole but meanwhile the leaf blade has decayed and water has entered the mine channel. The larva, therefore, now lives in its channel underwater and since the air in the aerenchyma is no longer sufficient, it is compelled from time to time to come to the surface to obtain a fresh supply of air.

In all these species the adaptations to life near water are not sufficiently developed to result in detectable morphological modifications, which would make it possible for the larva to live underwater. The well-known European Pyrale *Nausinoë nymphaeata* L. has advanced one step further. G. W. Müller (1892) made a careful investigation of its adaptation to life on aquatic plants. He claims that the imago lays its ova on the most varied aquatic plants with floating leaves, such as Nymphaea, Nuphar, Potamogeton and others. On hatching the larva penetrates into the mesophyll of the leaves and lives as a miner. Like the other species we have mentioned, it could in theory obtain its air supply from the aerenchyma of the leaf tissue. However, as Müller showed, it does not do so; the young larva exists in the leaf tissue surrounded by water, which has presumably penetrated through the larva's point of entry into the leaf. It is thus not able to effect normal respiration and in fact the spiracles are closed and the tracheae are not fully developed. It is therefore completely dependent on respiration through the integument, as is found in many aquatic insects. All structures on the surface of the skin are accordingly lacking; it has no bristles, warts, hairs, etc., so that there is nothing to prevent the water uniformly covering the skin. If a terrestrial larva is immersed in water, one finds that air collects between all such protuberances on the skin and prevents a complete covering by water. The *Nausinoë* larva, however, only mines for a short time; it subsequently cuts an oval piece out of the leaf and commences first "window-feeding", and later leaf perforation (Fig. 134, left). According to Müller's investigations it is still completely surrounded by water in its case and absorbs its oxygen apneustically. The case is changed several times before autumn, without any important alteration in its habits. It then finally sinks to the bottom of

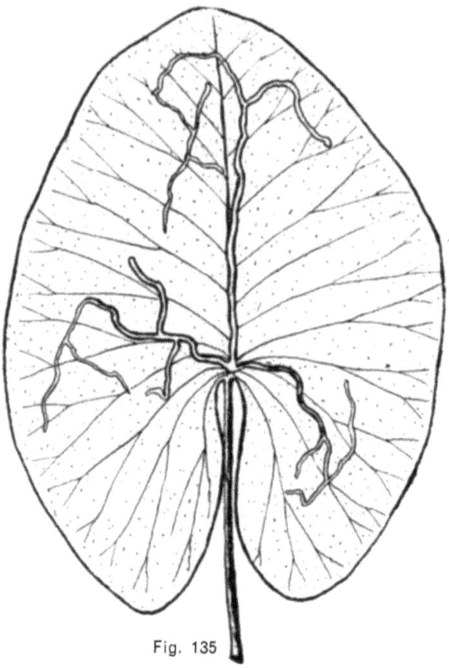

Fig. 135. Nuphar: *Hydromyza livens* Fall.

the water with the rotting leaves to which the case has been attached, where it overwinters. On arrival of spring it crawls up the young shoots of Nymphaea and continues living in the same way. Now, however, it no longer possesses the ability to absorb oxygen from the water through its skin but commences normal respiration. Parallel with this change, the skin loses its former smoothness and develops the surface structures of the normal larva. It can therefore no longer have direct contact with the water. Its case is now filled with air and it lives inside, as in an air cavity. Presumably other species of *Nausinoë* live in a similar way. Blood-gills, similar to those found in the related genus *Paraponyx* (which, however, feeds externally under water, not as a miner) have not been observed in *Nausinoë*. Possibly modifications of this type occur in other larvae having similar habits, during the course of their development. They have never been investigated from this point of view, though this would be particularly valuable for the *Nonagria* species.

The larva of *Hydromyza livens* Fall. (Fig. 135) mining in Nuphar (and sometimes in Nymphaea) deserves special consideration. The female, as already mentioned in the chapter on oviposition, lays its eggs on the underside of the leaf and is thus compelled to go underwater. Usually several larvae are found in one leaf, due presumably to the visits of several ovipositing flies. The mines, which are normally branched, all converge on the uppermost part of the petiole. The larvae do not remain long in the leaf blade but proceed to the petiole, where they pass the greatest part of the time as miners. Pupation also takes place in the petiole. It is particularly interesting that these species form two types of puparia. Schütte (1921), who made a close study of the habits of this species, found both "summer pupae" and "winter pupae", which vary both in their whole development and in their morphology. Before pupation the larva constructs a cavity, a pupal chamber, in the petiole. During this process, the larva which later produces a summer pupa, excavates a channel right up to the epidermis, so that the pupal chamber is enclosed from outside only by a thin layer of epidermis, which prevents the entry of water. The puparium hangs head downwards in the pupal chamber, anchored in the leaf tissue by the posterior spiracular processes. In this way a free space is preserved at its fore-end, ensuring no hindrance to emergence, while it is able to take up air from the leaf tissue with its posterior spiracles. The puparium itself is very thin-skinned. On emergence of the imago the thin covering of epidermis is burst open and the fly crawls up to the surface. Emergence takes place in summer. With the winter pupa conditions are in many respects different. The fly emerges only after overwintering. It does not therefore prepare a chamber for emergence, with a covering of epidermis. The petioles rot and disintegrate in the autumn and the enclosed pupa becomes free. It sinks to the bottom of the water and only rises again in the following spring for the fly to emerge. Why it rises in this way is not known; presumably gaseous products of metabolism provide the puparium with the necessary buoyancy. In this winter pupa the ventral wall of the puparium undergoes considerable dilation after pupation, which presses the rear spiracular processes firmly into the leaf parenchyma; owing to the longer time spent in the petiole, the provision of air for breathing is even more important than for the summer pupa. The winter pupa therefore possesses an air-filled space between the actual pupa and the last larval skin, into which metabolic products are presumably given off during the winter and which, together with the subsequent rise in temperature of the puparium, produce the necessary buoyancy. There is a marked deposit of calcium in the puparium wall of the winter pupa, which is absent from the summer pupa. It is not difficult to imagine the origin of this calcium deposit; an accumulation of calcium in dipterous puparia is encountered very seldom. In the stems of Nymphaeaceae there are peculiar structures composed of calcium oxalate, the trichoblasts. They resemble battle-axes of ancient warriors and are covered with long thorn-like processes; they occur in large numbers among the cells of the aerenchyma. Examination of the larva's intestine shows that the trichoblasts are no longer to be seen; they have thus been broken down in the intestine. From this it can be assumed that the larva assimilates the calcium oxalate during digestion; it then undergoes a change and, under the influence of the body's carbon di-oxide, is transformed into calcium carbonate which is deposited in the walls of the puparia, presumably first passing through the Malpighian tubules. In this way the calcium oxalate of the trichoblasts (which are assumed to be a deterrent to insect feeding) is rendered innocuous; superfluous carbon di-oxide in the body, liberated by respiration, is retained in compound. This finally results in an effective strengthening of the puparium, which must be considered as a valuable protective aid, especially in view of the long duration of the winter quiescence.

It has been possible to breed the larvae of *Hydromyza livens* Fall. away from water, in closed, air-tight jars, which prevent any drying of the plant tissue; in this way imagines have been produced both from summer and winter pupae. This tends to show that no really extensive adaptations to aquatic life have occurred, which the specialised habits of the species concerned might have led one to expect.

c) Miners in submerged leaves.

Mining insect larvae are also found on the underwater leaves of many aquatic plants, such as Potamogeton and Stratiotes. Once again it is the genus *Hydrellia*, which possesses a number of species living in this way. Their mines are all very similar and resemble the examples illustrated in Fig. 136. They consist of channels, which are frequently branched and irregularly traverse the leaf; they usually end up towards the centre of the leaf base or near the mid-rib, where pupation also takes place. The larvae vary as little as the mine channels, so that it has not so far been possible to distinguish the individual species either from the mines, or from the larvae or puparia and for this reason it is necessary to breed the imagines. Since the puparium is found in the mines, this is not difficult. The mined leaves need not necessarily be kept underwater; the flies emerge satisfactorily if the portions of the leaves containing puparia are kept in air-tight jars with an adequate degree of moisture. Species of *Hydrellia*, which normally live submerged, often emerge surprisingly rapidly, sometimes only a few hours after they have been removed from the water. Possibly the fully-formed imago remains for some time in the puparium to await a drop in the level of the water, which would be more favourable for its emergence.

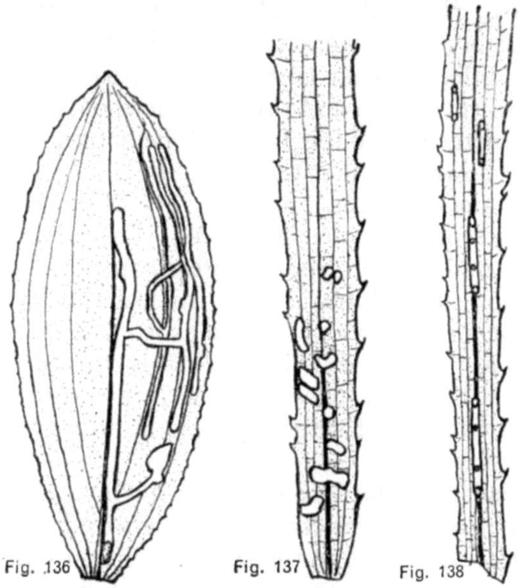

Fig. 136. Potamogeton: *Hydrellia fascitibia* v. Ros.

Fig. 137. Stratiotes: *Hydrellia stratiotae* Hering with main portion of the mine in the mid-rib.

Fig. 138. Stratiotes: Tendipedidae mines with two identical openings.

The larvae of *Hydrellia*, like those of all other Ephydridae, but unlike the rest of the Acalyptrata, are characterised by having the whole body surface covered with small setae. When submerged, they are thus surrounded by a layer of air, which adheres to the setae and protects them against any deleterious effects of the water. In all species of *Hydrellia* the posterior spiracles are extended into two long points (Fig. 34), with which the puparia are firmly attached to the inner leaf tissue. This applies not only to miners but also to species feeding in root stocks, etc., which leave these plant parts for pupation and transform anchored in this way and projecting out into the water. Even though the *Hydrellia* species appear to live surrounded by air as if they were in terrestrial plants, the channels in the submerged leaves nevertheless very rapidly decay and there is always the likelihood of water entering the mine cavity; the importance of the spiracles being imbedded in the plant tissue, as a means of ensuring a supply of air is thus obvious. Although a wide variety of species occur in similar mines on similar plants, there is little difficulty in distinguishing them on emergence. For example, from similar Potamogeton mines found at the same spot, we obtained successively *Hydrellia fascitibia* v. Ros., *H. cochleariae* Hal., *H. viridescens* R.-D. and *H. maura* Mg. Possibly future workers will be able to discover sufficiently perceptible differences on larvae and puparia to make possible determination of the mines, and thus obviate the necessity of having to await the emergence of the imago. The mines, moreover, frequently extend down into the stem.

Hydrellia stratiotae Hering produces very different mines in the submerged leaves of Stratiotes (Fig. 137). The larva lives most of the time in the mid-rib and cannot be seen from the outside; from here it eats short, lateral tracks into the leaf and immediately retires again to the mid-rib, where pupation also takes place. Both larva and pupa can only be found if the mid-rib is opened up. The larva lives here under the same conditions as in a terrestrial plant and owing to the abundance of aerenchyma is in no way influenced by the water.

With the peculiar Pyrale *Acentropus niveus* Oliv., only a part of the larvae bore into the leaf on hatching; the others feed externally on the Potamogeton leaves. Mining in the leaf or its mid-rib only continues for a few days, whereupon the larva leaves its mine and feeds under a leaf-fold or under portions of the leaf which have been bitten off and spun together. Later, whole leaves are spun together and pupation also occurs under this web. This all takes place under water. Unfortunately the larvae have not yet been examined with the same thoroughness as those of *Nausinoë*. Probably, however, one will discover a greater degree of adaptation to life under water; this is already suggested by the strange reduction of the wings in the female.

d) Miners with truly aquatic habits.

In the foregoing sections we discussed species which, although living on aquatic, even submerged, plants, did not substantially differ in their habits from species living on terrestrial plants. However, there is one whole family of miners which is not able to live out of water — that is the Tendipedidae (Chironomidae). Only they have true aquatic habits: they are characterised by the possession of blood-gills, with which they can absorb air direct from the water. We already mentioned above the unique position of *Cricotopus brevipalpis* Mg., (Fig. 133). All other mining species of this family construct a characteristic linear mine, which is completely straight, does not broaden out, but remains the same width throughout and possesses two openings. The Tendipedidae construct their living tubes in all possible places, even among colonies of Bryozoa, and mining represents an exception limited to relatively few species. This takes place not only on plant parts submerged in a shallow covering of water; *Endochironomus nymphaeae* Will. lives up to 1½ metres below the surface of the water in the centre of the petiole, still feeding on plant matter. When the tube becomes too small, it is extended; Gripekoven (1914) who spent much time studying the mining Tendipedidae (although he did not consider mining in the strict sense applied by us), claims that this extension occurs according to a consistent mathematical formula: the increase in length bears the same proportion to the old tube as the old to the new.

A hole is constructed at each end and the larva closes the earlier holes by spinning. A leaf mid-rib is favoured as a site for these channels (Fig. 138), in which the larva spends its whole life, including pupation; occasionally, however, the larva changes both its channel and the leaf it is in. This presents no difficulty, as it is able to move rapidly through the water by undulating movements.

The mines of the Tendipedidae resemble each other very closely and cannot be distinguished. Thanks to the detailed research of Thienemann and his assistants, exact information on the characteristics of the larvae and pupae of the various species are now available. It is thus possible to determine the different species without great difficulty from the larva or pupa (see Figs. 32 and 33 where the most important parts for purposes of determination are illustrated).

The larvae spend sometime mining in fresh, green leaves and are then pale reddish or green *(Phytochironomus)*, or they show a preference for rotting leaves *(Glyptotendipes)* and are then usually dark red or brownish red. However, there is no fundamental difference in their food in either case.

The peculiar method of feeding of the Tendipedidae has recently been made known by the work of B. M. Walshe (1947). She bred the larvae in glass tubes, the width of which equalled that of the tubes they themselves normally construct and this permitted detailed observation. The interesting discovery was then made that the larva constructed a net-like web with its spinneret. While doing this it anchored itself firmly by its posterior prolegs and rotated its forebody. The resulting web, which narrows to the rear and gives the impression of a conical net, is constructed in an extremely short space of time. The larva then hangs from its mouth parts from the front of this net and by rapid undulations (about 2 per second) sets up a current in the tube which carries along minute organisms and single-celled plants (Algae) which are caught up in the fine mesh of the web. Everything larger than 17 μ, and the majority of particles with a diameter larger than 12 μ, is retained in the net, including all possible detritus and pollution of the water, in addition to living organisms. The larva now ceases its undulating motion and eats the whole net and everything it contains, whereupon it constructs a new net in the same fashion. This all occurs so rapidly, that two successive nets will be constructed in the space of $1\frac{1}{2}$—2 minutes. From time to time the larva interrupts this process and eats the lateral walls of its tube, which are of course covered with silk and where a number of organisms are caught. Presumably on these occasions parts of the leaf tissue are also consumed and the tube is thus enlarged. Occasionally the larva eats tissue outside its tube, by attaching itself with its posterior prolegs and stretching its fore-body out of its tube. The fore-legs are made great use of in constructing the net work. With true miners, as opposed to the species living in special tubes, the larva turns round and while washing in the current of water turns the end of its body towards the net. These are truly pantophagous species, which eat everything which may appear in their net, whether of animal or plant origin.

Some species among the Tendipedidae also construct false or artificial mines by gnawing a narrow excavation in the leaf surface, which they close on its upper surface with silk mixed with detritus of all kinds and then live inside in the same way as other species.

Our observations have shown that apart from the last family, the connections of miners with aquatic life are of a very limited nature and there is relatively little evidence of any special adaptations. These species almost without exception, have adopted habits linking them with water only as a secondary development. In many cases they live on aquatic plants very little differently from normal terrestrial-living species. Certain factors must, however, exist, which we are at the moment unable to appreciate, which cause them to adopt such habits, independently of the systematic relationships of their food-plants. They represent a fruitful field of research for the minologist. Much remains to be done; it is necessary to investigate the differences between individual species, as regards larva and puparium, as well as the connection of the different species of *Hydrellia* to specific plants. We know especially little concerning the host plants of the various species of *Hydrellia* and the consistency with which they are affected. Most of what has been achieved so far has been based on chance discoveries. Many years of work will be necessary but valuable results can be anticipated, which will enrich our knowledge of minology. Intensive work on this group will also probably lead to the discovery of greater evidence of adaptation to aquatic life than is available at the present time.

14.

COLOUR AND DISCOLOURATION OF MINES

See Colour Plate 1

The great interest shown in mines by entomologists for many years past is to a large extent due to the distinct variation of colour between mines and the surrounding leaf. Feeding channels in other parts of the plant, such as roots, pith of the stem and fruits, are less conspicuous and do not show the peculiarities which are characteristic of each species, as clearly as is the case with mines. It is not always the mine itself which stands out so distinctly as a result of its different colouration; in some cases the mine is less obvious but its surroundings are discoloured in a characteristic fashion due to the influence of the mine. Both the colour of the mine and the discolouration of the surrounding leaf are frequently of great value for determining the species of the mine-producer and they therefore merit detailed consideration. Furthermore, other problems of great interest arise in this connection, characteristic of minology as a border region between zoology and botany; in many respects this is a virgin field, offering boundless possibilities and promising substantial results to future research.

I. Colour of the Mine.

The colour of the mine, which often deviates so strikingly from that of its surroundings and emphasises so clearly every detail of its shape (and also makes mines obvious to anyone with only a general interest in nature study), is usually the result of the fact that parts have been eaten out of the plant tissue. Air normally penetrates the resulting cavities and this produces a different colour from that of the rest of the leaf. We know from the early chapters of this book that the mining larva feeds on various tissues of the plant; the colour of the mine may also vary according to which parts of the leaf have been eaten away. Once fully conversant with every detail of this problem, it is possible to deduce which tissue has been removed from the leaf, merely from the differences of colour of the mine and without having to make a microscopic examination of the inside of the leaf. The colour also varies according to the type of light in which the mine is examined. In certain cases the appearance of a mined leaf when viewed from above against a dark background (direct light) is quite different from its appearance when the light passes through the leaf, in other words when the leaf is viewed with transmitted light, against the light or in front of a bright source of light. The variations in the colour of the mine with direct and transmitted light are particularly noticeable in the epidermal mines produced by sap-feeders. When giving an exact description of the colour of a mine, it is therefore necessary to give details for both types of light.

Further factors influencing the colour may also be present. In many cases it is not only the removal of plant tissue which determines the striking colour, but changes in the tissue which remains may also have an effect. These changes may take the form of withering and degeneration of the remaining cells under the influence of the activity and excreta of the larva, or the cells may be impregnated with certain substances and thus assume a peculiar colouration, from which the species of the mine-producer can be determined. In both these cases the colour of the mined part of the leaf differs so substantially from the undamaged part, that the mine becomes easily visible. Below, the various types of mine colouration are described and explained in relation to the histology of the leaf.

1. Light green in transmitted light, grey-green in direct light. This is the most commonly occurring mine colour and is found frequently among the representitives of all Orders. It is caused by the complete or partial removal of the palisade parenchyma lying on the upper surface of the leaf, while the epidermis remains undamaged. Air then enters the cavity formed in this way which gives it the somewhat greyish tone in direct light, while with transmitted light this is absorbed far less than in the intact layers of the leaf, which are also far thicker. In other cases the spongy parenchyma of the lower surface of the leaf is consumed and the palisade parenchyma is left, and this causes a similar appearance. This light green is especially prevalent in mines of Diptera but is also found with Lepidoptera and Coleoptera. The common linear mines of *Phytomyza cirsii* Hend. on Cirsium and Carduus may be mentioned as examples.

2. Yellowish green in both direct and transmitted light. This applies to inter-parenchymal mines. The larva eats a layer of parenchyma but not one adjoining the epidermis; the layer consumed lies deeper and the layer adjoining the epidermis remains intact. In the majority of cases this also is not entirely unaffected but withers and undergoes some degeneration; the granules of chlorophyll, the

chloroplasts, become smaller and the whole tissue layer appears lighter and paler. This yellowish green colouration of the mine which results even in direct light, enables the mine of *Phytomyza affinis* Fall., which is as common as that of the previously mentioned species and occurs on the same plant, to be easily distinguished (Plate 1, Fig. 7 and Fig. 5 of the text). In the early instars when the parenchyma above the mine cavity is still unchanged, such interparenchymal mines are difficult to detect, as for example with *Phytomyza heringiana* Hend. on Malus. However, once the process of undernourishment has set in, leading to the impoverishment of these cells in chlorophyll, the mine soon assumes a distinctive appearance and one can detect which tissue the larva has singled out for preference.

3. Darker green than that of the rest of the leaf arises not from the production of a cavity but from the filling up of this cavity. We previously mentioned that the mining larva does not digest the granules of chlorophyll but that these are again voided and comprise a substantial part of the ensuing excrement. The chlorophyll does not remain unaffected by the influences of the digestive ferments but is transformed into the darker green chlorophyllan. In species in which the mine cavity is largely filled by the excrement and when this change in the chlorophyll has taken place, the mine channel appears a darker green than the rest of the leaf and stands out through its darker colouration from the undamaged parts of the leaf. This is found particularly clearly in early spring in the linear mines of *Phytagromyza xylostei* R.-D. on the leaves of Lonicera, and later in the mines of *Liriomyza congesta* Beck. on a wide variety of Leguminosae. The manner of the formation of this dark green colour results in it being not particularly durable. With time, the frass undergoes more and more decomposition and is washed away by rain, damp, etc., so that it tends to become lighter and lighter; as more air penetrates, the oldest mines of these species appear greyish and finally quite whitish. In order to detect this dark colouration, it is necessary to examine fresh mines, in which the larva is still feeding or which have only recently been abandoned. The nature of the formation of the colour makes it self-evident that the colour does not affect all parts of the mine; where the larva is still feeding or where it has only recently been feeding, the mine always appears lighter than the rest of the leaf, owing to the fact that no or very little frass has as yet been deposited at these spots. For this reason the blotches produced by *Chylizosoma* species on Liliaceae and Orchidaceae are darker green than the rest of the leaf only in the centre, and round the edges they remain light. The transformation of chlorophyll into chlorophyllan does not take place in all mining insects; in many cases no

recognisable change occurs to the chlorophyll and the frass line is the same green as the rest of the leaf; the colour of the mine channel therefore scarcely differs from that of the rest of the leaf and when the mine is completely filled with frass, it remains very difficult to detect until the frass decomposes and so causes the change in the mine's colouration. This is the reason, for example, why the completely frass-filled channels of *Nepticula continuella* Stt., on Betula and of *N. basiguttella* Hein. on Quercus are so difficult to discover while the larva is still feeding in the leaf.

4. P u r e t o y e l l o w i s h w h i t e in direct and transmitted light. This applies to mines in which the whole parenchyma between the upper and lower epidermis has been eaten away, so that only the epidermis remains. The colour of the mine is determined solely by the colour of the epidermis. If this is quite colourless, the mines appear pure white; if it has a yellowish tone, the mines are lighter or darker yellow according to its exact tone; in this case, however, all trace of green is absent. The mines of *Coleophora*, in which almost without exception the mine cavity has been carefully excavated, leaving no remnants of the parenchyma, are characteristic in this respect. The colour of the epidermis in *Coleophora* mines is therefore particularly apparent. Uniformly coloured mines are also produced by *Microsetia sexguttella* Thbg. (*stipella* Hbn.) on Chenopodiaceae and by numerous species of *Nepticula*. These full-depth mines occur with all insect Orders, although they are commonest among the Tenthredinidae; these, however, frequently leave remains of the parenchyma in the mine, causing a more or less marked greenish tint. This is especially apparent among species of *Fenusa*, whose initial mine is always somewhat shallower than the later portions. Owing to the small degree of light absorption in the mined area of the leaf, these mines may be readily photographed in transmitted light.

5. S i l v e r y w h i t e in direct light (in transmitted light the mine is scarcely seen or may be quite invisible). In this case the larva has fed exclusively in the epidermis cells and air has entered the resulting cavity. The very thin skin above this air, consisting only of the outer walls of the epidermis cells, produces the characteristic colouration (Fig. 4. and Plate 1, Fig. 8.). Epidermal miners can be recognised by this silvery white, which is especially striking in upper surface mines, as for instance in species of *Phyllocnistis* and in the early instars of the Gracilariidae. As epidermis cells contain no chlorophyll, the cavities can scarcely be distinguished in transmitted light. This peculiar silvery white colouration frequently remains even when the mine has become deeper and is visible in transmitted light, as in the mines of *Acrocercops brongniardella* Fbr. on Quercus; the thin skin lying above the air in the mine cavity continues to give the

mine its characteristic silvery white colour; this can also still be seen in the blotch mine, although it is usually found primarily in linear mines. Where *Phyllocnistis* channels continue into the twig cortex, air does not penetrate as easily as into the cavities in leaves; the mine, in which the thin skin of the outer walls of the epidermis cells lies directly against the parenchyma (due probably to the greater tensions present in the twig), then assumes a yellowish-green colouration in direct light, for instance in the cortex mines in the middle instars of *Phyllocnistis saligna* Z. on Salix (Fig. 16).

6. Silvery white with patches of rust-brown occurs in the same way as in the previous group; in transmitted light the mine is scarcely visible or even invisible. The cause is again the activity of sap-feeders but either certain substances are deposited in the thin outer skin or a process of withering develops, which gives it its rust-brown tone. This colour is found especially often in the early mines of *Callisto denticulella* Thbg. on Malus, when as is normal, the mines lie on the upper surface. When, as commonly occurs in some years, the early mines are formed on the lower surface, this colour never arises. Possibly this is connected with the numerous hairs on the lower surface. The rust-brown colouration is also frequently found on the upper-surface tentiform mines of *Lithocolletis stettinensis* Z. on Alnus; lower surface mines of the same genus on the same plant also sometimes show traces of this colour but only in individual patches. In tropical mines and in the early instars of Gracilariidae it is extremely common. The cause of this variation in colour has not yet been investigated; it is not known whether it is due to substances deriving from the plant itself or which are excreted by the mining larva. Possibly tannic acid derivatives are to some extent responsible.

7. Greenish-white mottling is particularly common and striking in mines of the genus *Lithocolletis*. This mixed colouration results from the larva not consuming the whole parenchyma below the epidermis but merely eating holes extending to the epidermis into the uppermost layer of parenchyma, which otherwise remains intact; these holes appear lighter and whitish in the midst of the green layer of parenchyma and produce the effect of mottling. This is characteristic of the genus concerned and of some other Gracilariidae but it is not limited to Lepidoptera; it also occurs with Diptera, especially with species which produce inter-parenchymal mines, in other words which feed on central layers of the leaf but which drive out short channels to the epidermis, which then appear lighter than the remainder of the mine. This occurs in *Phytomyza heracleana* Hering on Heracleum and in some related species of this group in the leaves of Umbelliferae.

Fig. 139

Fig. 139. Rhamnus: Serpentine mines with blackish-violet discolouration and "window-feeding" of *Bucculatrix frangulella* Goeze.

Fig. 140. Rhamnus: Part of a leaf cross-section of the serpentine mine of *Bucculatrix frangulella* Goeze, showing the dark impregnated tissue. Inside the mine the large pellets of frass.

8. A reddish-brown colouration of the mine may be based on two separate causes. This may be merely the result of the natural colour of the epidermis, as we find especially in the leaves of Alnus. Consumption of the parenchyma leads to the colour of the epidermis standing out clearly; this is found with many Alnus miners such as *Fenusa dohrnii* Tischb., in the *Rhynchaenus* mines occurring on the same species and in many others. The mines of the species of *Coleophora* which live on Alnus and which are also coloured in this way show clearly that this is really due to the colour of the epidermis.

On the other hand a discolouration of the epidermis may ensue from certain substances originating either from the plant or from the mining larva. The characteristic blotch-mine of *Tischeria dodonaea* Stt. (Plate 1, Fig. 3) serves as a good example of this and it may always be easily recognised by its beautiful brick-red colour. Frequently this reddish-brown discolouration occurs when the epidermis cells die. This is confirmed by the fact that in some mines the oldest portion already shows a reddish-brown discolouration, while the end of the mine on which the larva is still feeding is of a fresh green colour. This is seen especially clearly in species of *Leucoptera*

PLATE I.

PLATE I.

Colour and Discolouration of Leaf-Mines.

Fig. 1. Leaf of Quercus: "Green islands" resulting from mines of *Nepticula subbimaculella* Hw. in the angles of the veins in a leaf showing autumnal discolouration.

2. Leaf of Betula: Interference with growth on one side of the leaf through "ringing" of the more lightly coloured right half of the leaf, and the withered marginal area, caused by the mine of *Phylloporia bistrigella* Haw.

3. Leaf of Quercus: Mines of *Tischeria dodonaea* Stt. with brick-red discolouration.

4. Leaf of Succisa: Accumulation of violet anthocyanine under the influence of the mine of *Phytomyza succisae* Hering.

5. Leaf of Taraxacum: Pustules (intermediate stage between gall and mine) with accumulated anthocyanine of *Cystiphora taraxaci* Kieff.

6. Leaf of Senecio nemorensis: Ringing with accumulation of violet anthocyanine caused by the mine of *Vidalia cornuta* Sc.

7. Leaf of Cirsium: Inter-parenchymal, and therefore yellowish green, linear mine of *Phytomyza affinis* Fall.

8. Leaf of Populus tremula: Epidermal linear mine of *Phyllocnistis labyrinthella* Bjerk.

9. Leaf of Rumex: Serpentine mines of *Nepticula acetosae* Stt. with accumulation of red anthocyanine.

10. Leaf of Populus tremula: "Green islands" in leaf showing autumnal discolouration, caused by mines in leaf-base of *Nepticula argyropeza* Z. developing from the gall-covered petiole.

1.
2.
3.
4.
5.
6.
7.
8.
9.
10.

and in the common *Rhamphus pulicarius* Hrbst. on Salix and Betula.

9. Brown to deep black. The mine takes on this characteristic colouration largely as a result of the decay of portions of tissue, though in some cases the impregnation of certain areas of tissue with substances produced by the larva may exert some influence. The latter view is supported by the fact that this darker colouration occurs in many species of *Leucoptera*, such as *L. scitella* Z. on Rosaceae and Betula, only in the vicinity of the frass deposits (Fig. 129). The ornate serpentine mines of *Bucculatrix frangulella* Goeze on Rhamnus have a deep blackish-violet colouration (Fig. 139). A cross-section of one of these serpentine mines, as illustrated in Fig. 140, indicates a considerable impregnation of the plant tissue; this applies not only to the intact parenchymal walls between the spirals of the mine but also to the epidermis cells and this has resulted from some dark substance probably originating in the larva's excrement. The early channels of *Argyroploce woodiana* Barr. on Viscum in late winter and the first days of spring show a brown discolouration arising in exactly the same way, even though the inside of the mine channel is lined with silk. *Liriomyza virgo* Zett. which mines in the cortex of the shoot of Equisetum also has channels with a dark brown discolouration. Wulff (1902) showed that the brown colouration in Equisetum is due to transformed chlorophyll and this presumably applies also to this case. It may be assumed that the process of transformation of the chlorophyll developed under the influence of the excretory products of the larva. The deposit of black pigments can be seen very conspicuously in the mines of *Zeugophora* and certain species of *Rhynchaenus* on Salix and Populus. It will be recalled how especially plants of this genera are very subject to blackening when preparing herbarium specimens and it is tempting to seek the cause of this deep black discolouration solely in the nature of the host-plant itself. It must not be overlooked, however, that numerous other mining insects are found on both these plant genera (such as many species of *Nepticula* and *Lithocolletis*), without any discolouration of the mine ensuing. There can thus be little doubt that the parasite itself is at least partially responsible. Possibly anthocyanines also have some influence in producing these very dark colours.

10. Red to blue anthocyanine colourations. Anthocyanines are plant pigments not linked to specific pigment bodies but which are in solution in the cell sap or the protoplasm. Their occurrence is widespread throughout the plant kingdom and they are commonly responsible for colour in fruits, and are also extremely common in perianths of flowers and in gall formations. With an alkaline cell sap reaction the anthocyanines appear blue, with an acid reaction they are red; with a more neutral reaction between these extremes one

finds a wide range of colours from reddish-violet to bluish-violet. The development of pure blue anthocyanines is found in mines very rarely; most frequently certain reddish-violet colourations occur, very striking examples of which are found in the pustule mines (frequently considered as galls) of the genus *Cystiphora* on various Compositae (Plate 1, Fig. 5). Frequently a purplish-red is found, which may be somewhat clouded by the green of the chlorophyll. With the same species on the same food-plant the development of this colour may distinctly vary; for instance, blister mines of *Leucospilapteryx omissella* Stt. living on Artemisia vulgaris L. are found with a beautiful red colouration and side by side with them, admittedly in a minority, are mines of a clear green with no trace of anthocyanine formation. Light rose is very often found in the mines of *Lithocolletis tremulae* Z., which produces lower surface tentiform mines on Populus tremula L. This case derives particular interest from the fact that with this species only the outer walls of the epidermis cells remain above the mine cavity; the anthocyanine therefore no longer exists in the epidermis cells but has merely in some way impregnated their outer walls. Pure red is as rare in mines as pure blue. The best-known and most striking example of this is represented by the ornate serpentine mines of *Nepticula acetosae* Stt. on Rumex (Plate 1, Fig. 9). It is interesting that in this case the formation of anthocyanine is largely confined to the parts of the mine in spirals. The subsequent, almost straight, length of mine channel has the normal yellowish colour of Rumex mines. Even some distance from the mine anthocyanine bodies may arise and this will be discussed later. Probably, after completing a moult, when the larva ceases producing spirals, a change has occurred in its digestive tract, which is not without its effect on the formation of red pigment. The leaves of Rumex possess a general tendency for developing anthocyanines, as can be confirmed from the effects of occasional injuries or from the linear mines of *Mantura* species or of *Pegomyia* species in their early instars. The regularity of this extensive formation of red pigment in the mine of these species of *Nepticula* clearly suggests that the larva itself is to some extent responsible.

There is so far no convincing explanation of the causes of the occurrence of anthocyanines beyond their normal development in perianths and fruits. Various authors mention as factors in anthocyanine formation excessive sunlight, a damp atmosphere, accumulation of the products of assimilation and an abnormally large supply of oxygen. It is difficult to decide which of these factors should be held responsible for the formation of anthocyanine in leaf-mines. We know that large quantities of undigested products of assimilation are present in the excrement voided by the larva, since carbo-hydrates are not absorbed by the larva's body. Also the air in the mine channel is especially damp (this will be discussed in a subsequent chapter in relation to the formation of hyperhydric cells, which develop particularly well in damp air). Moisture is unable to escape from the completely enclosed mine channel and the degree of saturation is increased still further by the many cells which are not consumed by the larva but merely lacerated, resulting in the evaporation of their moisture content into the mine cavity. According to Schürhoff (1924), there is a lesser degree of translocation of the products of assimilation in damper air, so that one factor may support the other. Air penetrating into the cavity provides a larger supply of oxygen for the plants and thus all these factors may play a part in the formation of anthocyanines. Katič (1905) also discovered that all factors which promote the formation of anthocyanine cause a reduction in the number of chloroplasts. May not the opposite also apply? We have frequently noted a reduction of the chlorophyll granules due to miners; the possibility cannot be excluded that this has some connection with the formation of anthocyanines. Küster (1925) also discovered that tissue which has little chlorophyll is able to form anthocyanines particularly readily.

In the occurrence of these pigments in mines one other factor must be considered; that is the influence of the mining larva resulting from its own excretory products. It could not otherwise be explained why anthocyanines arise with one species, while on the same plant they do not arise with another species; furthermore, it would be difficult otherwise to account for the regularity of their occurrence in certain species. We know that certain tannic acid derivatives play some part in the development of anthocyanines in galls and it is natural to assume that similar substances contained in the larva's frass will at least assist in the development of anthocyanines in mines. At the same time one finds that some plant families are particularly disposed to anthocyanine formation; for example, they are most frequently encountered in Compositae and Rosaceae.

Zweigelt (1931) distinguishes two groups of anthocyanines: the constitutional and autumnal, as he terms them. The constitutional anthocyanines develop primarily in the epidermis and arise during the plant's period of most rapid growth; they are reduced in autumn. These can be contrasted with the autumnal anthocyanines which develop only in mature tissue and in the mesophyll of the leaves. It would appear that the anthocyanines found in mines can be included in neither of these two groups, since they occur in both epidermis and mesophyll but only in the older plant tissues; they are never found in mines which are formed in early spring.

2. Discolouration in the vicinity of the Mine.

1. Ringing. In the most varied species of mines the leaf is frequently seen to be lighter and more yellowish-green in the vicinity of the mine channel, or it may even sometimes appear a quite light yellow. In such cases the neighbouring cells are clearly undernourished, resulting in an impoverishment in chlorophyll, which is expressed in the yellow colouration. If the infested leaves are still young, this phenomenon is usually combined with disturbances in growth. It occurs with great regularity with the mines of *Phylloporia bistrigella* Hw. on Betula. (Plate 1, Fig. 2). This phenomenon is known as "ringing". The assumption is that nutrient materials can only reach the affected areas through narrower anastomosing vessels, which can only carry considerably less food than is required for full development. This phenomenon is particularly striking when the larva starts by mining for some time in the mid-rib, which contains the main conducting vessels. We find a ringing of this type in the leaves of Alnus, when the larva of *Heliozela resplendella* Stt. is still mining in the mid-rib and two lateral veins, in other words before it has constructed the terminal portion of its mine. The area between the two main lateral veins and the tip of the leaf are then usually much lighter than the rest of the leaf (Fig. 141a). One can thus detect from the ringing, which leaves are inhabited by larvae and they may be collected for breeding; the terminal portion of the mine is constructed so rapidly, that the mine is invariably found to be abandoned by the time it can be easily recognised. In smaller leaves the larva is frequently unable to find adequate sustenance in the mid-rib and then constructs an abnormally long channel in the leaf outside the mid-rib (Fig. 141b). Not so many conducting vessels are destroyed in this case and therefore one does not find the lighter, yellowish-green colour, which might be expected. In some cases it appears as if the ringing and the ensuing reduction in chlorophyll were deliberately produced by the larva, so regularly does it occur. Thus, *Vidalia cornuta* Sc. always constructs its early mine in such a way that the tip of the Senecio leaf in which it is mining is "ringed"; the ringing, however, is frequently obscured by a distinct anthocyanine development (Plate 1, Fig. 6), and the final, sizeable blotch-mine is always made in this ringed portion of the leaf, never in the undamaged part. It would almost appear that this ringed tissue is either essential or at least preferred as a source of food. Since the supply vessels are ruptured, not only is there reduction of chlorophyll in the ringed area but also an accumulation of assimilated products and it need therefore cause no surprise to find that there is often extensive anthocyanine formation in ringed areas. Phenomena of ringing have been

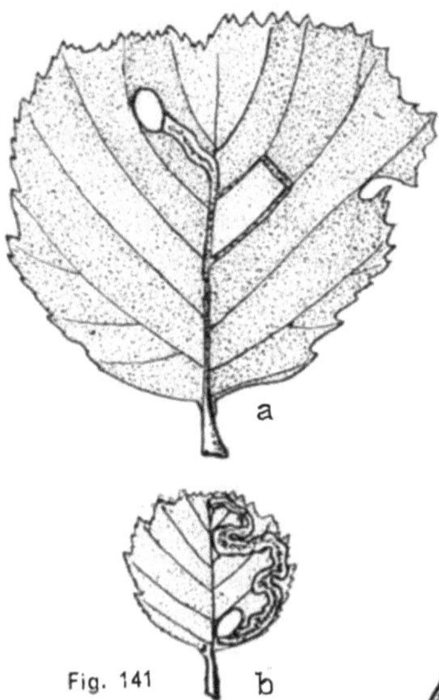

Fig. 141

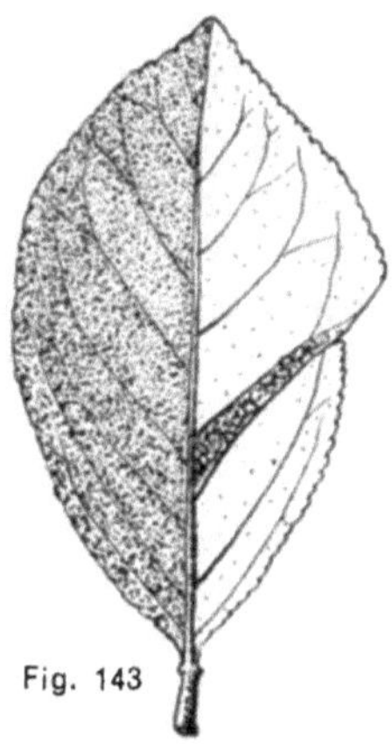

Fig. 143

Fig. 144

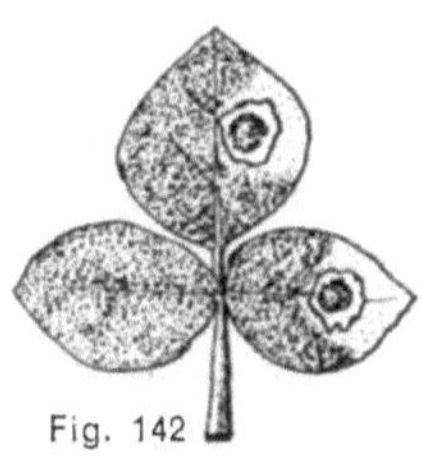

Fig. 142

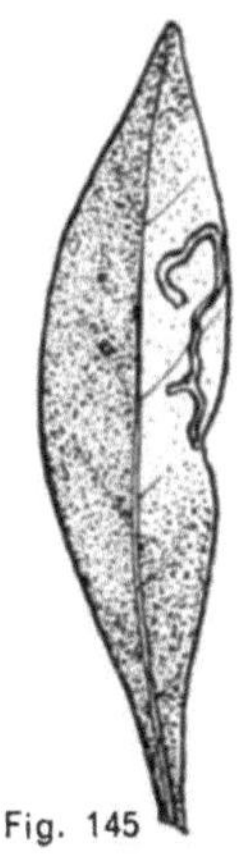

Fig. 145

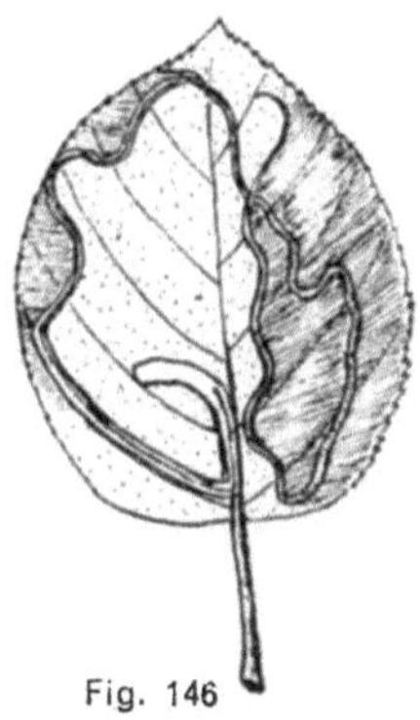

Fig. 146

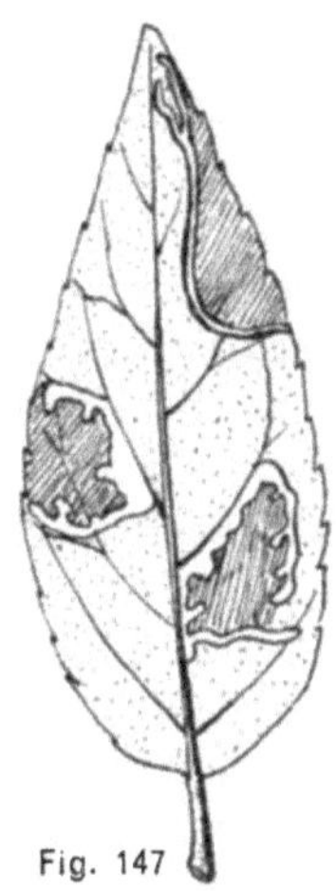

Fig. 147

Fig. 141. Alnus: *Heliozela resplendella* Stt. a. normal channel, b. abnormal channel on an unusually small leaf.

Fig. 142. Cytisus: *Leucoptera phyllocytisi* Hering. This very superficial mine causes "ringing" towards the margin of the leaf.

Fig. 143. Prunus cerasus L.: *Lithocolletis cerasicolella* HS. Although the mid-rib is undamaged, one side of the leaf is "ringed".

Fig. 144. Prunus cerasus L.: *Nepticula prunetorum* Stt. Although the serpentine mine has avoided the vessels, one side of the leaf has developed "ringing".

Fig. 145. Olea: *Prays oleellus* Fbr. "Ringing" of one side of the leaf.

Fig. 146. Amelanchier: *Lyonetia clerkella* L. Necrotic areas caused by the mine in the living leaf.

Fig. 147. Phillyrea: *Cacophyia permixtella* HS. The small channels enclose necrotic areas.

observed not only in mines but also in galls, which occurred on the mid-rib or one of the stronger lateral veins; the space between two lateral veins or a strip extending to the margin of the leaf is then a paler green. Czapek also noticed considerable development of anthocyanine in conjunction with ringing, as described with *Vidalia.*

This explanation of the phenomena of ringing, as first outlined by Schneider-Orelli (1909), apparently however, only satisfactorily deals with a part of the question. There can be little doubt that the larva itself is to some degree responsible for reduction in chlorophyll. That the rupturing of conducting vessels provides the main cause in many cases cannot be disputed but this by no means deals with the root of the problem. For instance, one frequently finds ringing where a rupture is not sufficiently vital to explain the ensuing cell degeneration. Ringing has been noted in *Leucoptera phyllocytisi* Hering even when no substantial vein had been severed; furthermore, the mines of this species are so superficial that the damage to the vascular bundles could not produce such marked results. A similar situation can often be found in *Lithocolletis cerasicolella* Hs. on Prunus. In Fig. 143 we see how one whole side of a leaf appears to be ringed, without the mid-rib having been in any way affected by the mine. Extensive evidence of ringing is also found on the same plant under the influence of the mines of *Nepticula prunetorum* Stt. (Fig. 144). The shape of the mine, which in this case is a distinct spiral, immediately suggests that no substantial vessels have been damaged. As we saw earlier, the formation of heliconomes and visceronomes is a process to enable the larva specifically to avoid vascular bundles. When nevertheless in such cases parts of the leaf definitely turn yellow, it must be presumed that the "ringing" is not caused by a purely mechanical severing of conducting vessels, but rather that certain substances produced by the larva itself result in the impoverishment of the surrounding tissues in chlorophyll, substances which in all probability are contained in the larva's excrement. The negligible damage to the leaf caused by *Prays oleellus* Fbr. in Olea (Fig. 145) allow of no other explanation of this process. On the other hand there are numerous species which cause extensive damage to the mid-rib or main lateral veins, without any suggestion of ringing, especially in species which spend their whole life in the mid-rib, only driving out occasional short lateral channels. Thus, the mine of *Ceutorrhynchus contractus* Mrsh. in radish leaves in early spring causes large areas of the leaf to turn yellow, without any of the larger vascular bundles having been severed. Ringing is particularly characteristic of many species, such as *Lyonetia clerkella* L. which lives on Rosaceae and Betula and was the first object of study for this process, and also *Phylloporia bistrigella* Hw. described above. The fact that the parts of the leaves of Quercus lying nearest to the summer-miner *Nepticula basiguttella* Hein. assume the autumnal colouration in autumn more rapidly than the rest of the leaf, shows that certain analogous processes are at work in the leaves before they become visible externally.

2. Necrobioses (necrotic areas). In certain species one quite frequently finds that the area of the leaf surrounded or enclosed by conducting vessels completely dies. These portions of the leaf assume autumnal colouration in the summer, while the remainder of the leaf is alive and fulfilling its normal functions. Such areas of the leaf killed by the activity of the miner are known as necrobioses. They are found particularly often in *Lyonetia clerkella* L. (Fig. 146). It is reasonable to assume that in this case destruction of conducting vessels is so extensive that no food at all can reach the isolated areas of the leaf and that they consequently die. Meanwhile, one also finds that in some cases the main vessels are not injured as extensively as might be expected, as in the common necrobioses caused by *Cacophyia permixtella* HS. in the leaves of Phillyrea (Fig. 147). Severing of conducting vessels is even less responsible for necrobioses, as found with *Phyllocnistis suffusella* Z. in the leaf of Populus nigra L. As early as September one side of a leaf traversed by numerous mine channels was seen to be dead, while the other side was a normal green. As the larva only mines the epidermis cells, no vessels at all are injured. These necrobioses can thus in no way be considered as a "ringing" in the literal sense of the term.

Presumably some substance is produced by the mining larva, probably in its excrement, which results in a "poisoning" of the cells in the vicinity. This explanation is supported by the fact that it is always the same particular species which cause necrobioses in the leaf as a result of their mining; these are frequently the same species as produce ringing; in fact, necrobioses may be considered as a more advanced stage of ringing. They are also frequently found in heliconomes and visceronomes, such as in the early mines of *Eriocrania sparrmannella* Bosc. and *Nepticula continuella* Stt. on Betula, and also in *Nepticula fulvomacula* Skala on Ulmus. In some cases one finds these phenomena in conjunction with anthocyanine formation.

3. Red or blue caused by anthocyanines. Not only the mine itself but also its immediate or more distant surroundings may appear extremely conspicuous through the occurrence of anthocyanines; however, in some cases of this red colouration, the disintegration of chlorophyll, which has a red component which is particularly evident in the autumnal colouration of the leaves, is a decisive factor. In general, the same causes and considerations apply here, as in the colouration of the actual mine. In the vicinity of the

mine pure blue is very rare. The best examples are found in *Phytomyza succisae* Hering in the leaves of Succisa but not, however, in *Trachys troglodytes* Gyllh. or *Phytomyza olgae* Hering, which mine on the same plant. The blue colouration in this case is often so extensive that the insignificant mine channels can scarcely be recognised at all (Plate 1, Fig. 4). A similar colour usually develops in conjunction with *Phytomyza glechomae* Kltb. on Glechoma, and is usually especially evident in the vicinity of the puparium, which remains in the leaf. We previously mentioned that in dipterous larvae a particularly large quantity of frass is voided immediately prior to pupation and this provides the obvious explanation of the localisation of colour in the vicinity of the puparium. A reddish-violet colour of various shades is not unusual in the linear mines of *Liriomyza strigata* Mg. on various Compositae; with *Micrurapteryx pavoniella* Stt. on Aster the whole dilated mine and the surrounding parts of the leaf appear a reddish-violet. In the ringed portion of the Senecio leaf infested by *Vidalia cornuta* Sc. anthocyanines of various shades are also common. With *Nepticula acetosellae* Stt. a pure red occurs not only in the mined area of the Rumex leaf but frequently in the vicinity of the mine also. The beautiful colour in this case is certainly formed by anthocyanines and not by disintegrated chlorophyll. The surroundings of the mines of *Lophoptilus raschkiellus* Z. on Epilobium are a similar bright red. In some cases the red colour in the vicinity of the mine is a characteristic feature of the species, as with *Nepticula erythrogenella* de Joann. on Rubus, which has been named after it. In exceptional cases, Hermann Schulz noticed that in September a similar red colour develops in the vicinity of the mine of *N. rubivora* Wocke, although it is not usually found with this species.

As with the colour of the mines themselves, it is reasonable to assume that in these cases also the formation of anthocyanine is induced by certain substances produced by the larva. When the early linear mines of *Acidia cognata* Wied. on Tussilago do not develop to the normal blotch owing to the prior death of the larva, the formation of anthocyanines can be noted, although this is never found when the larvae produce their usual blotch mine. Probably certain substances arising from the decomposition of the dead larva promote the formation of anthocyanine. The discolouration arises either in the epidermis or in the upper layer of the palisade parenchyma; for instance, according to B. Fletcher the larva of the Indian *Acrocercops hyphantica* Meyr. lives in the veins of a species of Nephelium, which thereby become red and this dark colouration of the veins makes the mine extremely difficult to detect. In our European *Phytomyza succisae* Hering the mine and puparium are also difficult to see in the middle of the large blue patch it causes.

At the same time the formation of anthocyanine depends to a certain extent in the food-plant of the miner. We already pointed out in the first part of this chapter that such discolourations are particularly common in Compositae and Rosaceae. However, on this point differences may exist with in the same family and even within the same genus. For instance, A. Braun (1917) reports that the vicinity of the mine channel of the North American *Nepticula bifasciella* Cl. is discoloured red when the species mines on Prunus serotina Ehrh. but that this colour is absent when the larva lives on P. americana Marsh. Probably individual variations within the same species of host-plant have some effect, which might explain why anthocyanines may be either present or absent in the blister mines of *Leucospilapteryx omissella* Stt. on Artemisia vulgaris L. Definite answers to all these queries will only be obtained when the exact factors influencing anthocyanine formation have been thoroughly investigated and confirmed experimentally.

4. "Green islands" in the autumnally discoloured leaf. One can frequently find leaves, which have fallen or may still be on the tree and which have assumed their brown or reddish autumnal colouration, showing areas which have remained bright green, in other words possess the same colour as the leaf during its period of active life. If these leaves are examined more closely, it will be seen that this green patch contains a mine of some insect larva, which is frequently still inhabited by its producer. These "green islands" in the autumnally discoloured leaf long ago attracted the attention of entomologists and botanists. It was discovered that the larva continues feeding in the dead and frequently fallen leaf and is able to continue its normal habits unchanged for a varying length of time. Attempts have been made to explain this peculiar phenomenon and two main, largely conflicting, theories have been advanced. Wood (1894) assumed that some substance was produced by the larva in the green island, which preserved the chlorophyll and continued to render this part of the leaf palatable to the larva, which was thus able to complete its development. On the other hand Trägårdh (1913) maintained that feeding by the larva severed vascular bundles which effectively prevented any translocation of substances from this part of the leaf; in other words normal autumnal reduction of the leaves could not be followed by any translocation of the products of reduction. This would represent an analogy with the "ringing" process described above and would explain why these islands remain green in the leaf which has turned yellow. In attempting to judge which of these explanations most closely accords with the facts, we must recall exactly what takes place during the development of autumnal colouration in leaves.

Throughout the period of active life a continuous process of reduction of the plant's chlorophyll is taking place under the influence of light and oxydation. This process is not normally visible externally since parallel with it chlorophyll is being constantly renewed. This renewal ceases in autumn and the products of reduction become apparent. Thus the autumnal xanthophyll which causes the yellow colouration is likewise an isolated component of the former chlorophyll, as also the red pigment, which appears in many autumnally discoloured leaves. There is thus never any question of translocation of chlorophyll but only of its disintegration, and only certain nitrogen compounds are translocated. Severance of conducting vessels which may in fact have taken place in varying degrees in the vicinity of the green islands has no bearing on the retention of chlorophyll in them. The absence of any translocation of chlorophyll can also be shown experimentally. If oak leaves with mines of *Nepticula albifasciella* Hein. or the last generation of *N. betulicola* Stt. in the leaves of Betula are collected and the individual leaves are kept in an air-tight jar, one finds that after a time the leaves turn yellow. (*N. albifasciella* Hein., in contrast to its close relative *N. subbimaculella* Hw. mines in the green leaf and thus lives about a month earlier, before the leaves have become discoloured). The surroundings of the mines in the leaves, however, retain the bright green colour originally shown by the whole leaf, and do not turn yellow at all or only much later. Since no translocation is possible in these picked leaves and there is obviously no movement of sap towards the roots, this experiment shows that translocation has nothing to do with the formation of the green islands. One also finds that fallen leaves, which may still be green, only turn yellow on the ground and, if a mining larva is present, develop green islands, after some time has elapsed.

This shows that the opinion of Wood, that the larva produces certain substances which result in the conservation of chlorophyll is in every respect more accurate. This view is supported by the fact that in certain very superficial mines, such as those of *Tischeria ekebladella* Bjerk. on Quercus and many species of *Lithocolletis* in their earliest instars, the formation of green islands can be observed, although no large or essential vascular bundles have been severed. Green islands are found under analogous circumstances in galls: frequently the gall alone remains green, when the whole leaf has already turned brown. It is known that in many cases the development of a plant gall is induced by the oviposition of the female; there is thus the possibility that the formation of green islands may also be caused by substances transmitted by the ovipositing female. However, this is not the case with mining insects, as can be seen when the larva dies in a very early instar. The formation of green islands is then either entirely absent or is checked at its very beginning. In some species, such as *Scolioneura nana* Klg. on Betula, the development of the green island occurs only after the construction of the last stretch of the mine. It can thus be assumed with some degree of certainty that products of metabolism produced by the larva must be considered as the basis of this phenomenon and presumably the essential substances are to be found in the larva's frass. It should not be difficult to produce experimental proof of this also. Interesting, although indirect, proof has already been obtained by G. Voigt (1946). He observed the presence of green islands on autumnally-discoloured leaves when bird droppings were on the leaf and where the leaf substance had not been attacked by any insect larva. After this report had come to my notice, I was able to confirm it by many examples. This appears to suggest that no specific substance in frass produces the striking colour but general substances which are present in the excrement of all animals. When they penetrate to the inside of a leaf, they naturally produce this phenomenon with greater regularity and more intensity than mere bird droppings, which clearly cannot penetrate the leaf so easily.

The similarity with the processes causing "ringing" of the leaf can be seen clearly in species which live in two generations, of which one only appears in late autumn. In the first generation pale green areas known as "ringing" develop in the leaf, while the typical formation of green islands occurs with the autumn mines of the second generation. It can be seen from this that both phenomena can be ascribed to the same cause.

The green islands are in many cases extremely conspicuous, especially when they occur with the greatest regularity, as for instance, with the channels of *Nepticula turicella* HS. in the leaf of Fagus or with *N. subbimaculella* Hw. on Quercus (Plate 1, Fig. 1). It can therefore also be assumed that they are of great ecological importance for the mining larva. O. Richter (1917) showed with the iodine test that in the green islands in fallen leaves on the ground starch is still being formed. This proves that in all the cells and tissues of the plant which have remained green, it is not a question of conservation, comparable to fixing a specimen for purposes of microscopic examination, in which the tissue is maintained in a certain condition but only by killing it; rather is it a question of maintaining alive cells and tissues in the leaf which is otherwise dead. Thus, the expression "chlorophyll conservation" which we have used for this phenomenon is not really accurate. We usually find this occurrence in species, whose larvae appear very late in the year: the larvae would not be able to complete their development owing to the early death of the leaf, unless they were able to continue normal feeding in the

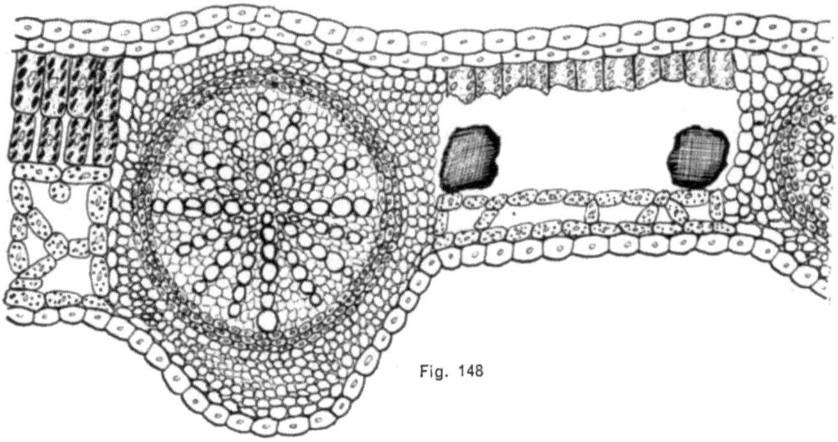

Fig. 148. Populus tremula L.: *Nepticula argyropeza* Z. Cross-section of the mine in the leaf-base.

green island until their pupation. The larvae of *N. argyropeza* Z. are often found as late as December continuing their feeding in fallen leaves covered by snow and completing their normal development; this of course would be quite impossible without the formation of green islands.

A rather more complicated and exceptional example of the formation of green islands is found in certain species of *Nepticula* living on Salicaceae, especially in *N. argyropeza* Z. on Populus tremula L. (Plate 1, Fig. 10), *N. hannoverella* Glitz on P. nigra L. and *N. turbidella* HS. and *N. populi-albae* Her. on P. alba L. These species are considered by cecidologists as gall-causers; before producing its mine channel in the leaf-blade, distinguishable by the double line of frass, the larva lives in the petiole, which becomes thickened as a result of its mining activity. Since the larva is in no way dependent on this thickened area and has no further contact with it, the species should be more properly considered as miners than as gall-producers. Two things are of particular interest in the green islands caused by these species: firstly, they are formed in the leaf, before the larva has itself entered the leaf-blade and secondly, they appear almost without exception at the leaf-base, in the vicinity of the petiole; only more rarely is a green area found between two veins nearer the apex. Both peculiarities have the same cause, which can only be discovered after examining a number of cross sections of the mine channel, made at various different places.

The cross-section of the mine channel in the leaf-blade, as illustrated in Fig. 148, shows no unusual features. It is predominantly

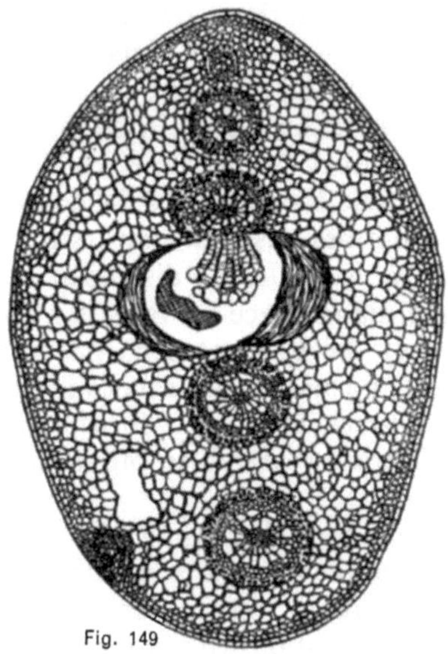

Fig. 149. Populus tremula L.: *Nepticula argyropeza* Z. Cross-section of the petiole with the mine in two different instars.

an upper surface mine, in which remains of the uppermost layer of palisade parenchyma are left intact. The cross-sections of the mined petiole (Fig. 149) show something remarkable. The mine channel commences beneath the epidermis and runs initially only through the abundant parenchymatous tissue of the petiole but is then directed to the vicinity of a vascular bundle. This is always the bundle which supplies either the left or the right side of the leaf-base. Presumably in this way the substances contained in the frass, which result in the formation of the green island, reach the leaf-base via this vascular bundle in substantial quantities and with a strictly localised effect; the future mine will be made in just this area. It is particularly remarkable that the larva always locates one of the two vascular bundles which lead to the leaf-base, and only very rarely does it make a mistake and attack the parenchyma near the wrong bundle. This results in the green island developing some distance from the leaf-base and the larva then has far further to travel before finding the area where it can feed to maturity. The fact that the infant larva lives in the neighbourhood of a specific vascular bundle of the petiole, results in the formation of the green island even while it is still in the petiole, and also ensures that the area of the leaf in which it will complete its development remains alive; this is of particular importance, since it is one of the species which feeds latest in the year. Furthermore, this area develops at a place necessitating the shortest possible journey from the petiole; the larva would otherwise be compelled to cover a considerable distance through areas of the leaf which had already died.

Our observations have shown that in all probability the widely varied colours in the vicinity of the mine which we have described, can be ascribed to a single factor, probably contained in the larva's frass. The chapter on frass disposal and its importance for determining mines which we have already discussed in detail, can now be supplemented by the significance of frass for the larva's ecology. From the regularity of these phenomena one can deduce without hesitation that ringing, formation of anthocyanine and development of green islands are of great importance for the life of mining larvae; it will be the task of future research to prove this importance in its individual aspects.

The problems we have been discussing here are of a general nature; they probably find their counterparts in other cases of parasitism and are not limited to mining insects alone. In this connection it is interesting to study the situation with parasitic fungi, as expounded by Fischer & Gäumann (1929); one finds a number of interesting analogies. With leaves infested by fungi one can detect an increase in assimilation and respiration, and a reduction of chloroplasts known as hypoplasy has also been noted. In the vicinity of affected areas the authors observed holonecrotic (dead) tissue, just as we noted dead areas with leaf-mines (*Bucculatrix frangulella* Goeze, Figs. 139, 140), while in other places impaired but still living tissue (termed plesionecrotic) is found. With fungal parasites, as with others, injuries are caused which lead to increased acidity of the cell sap, which explains why we so rarely find blue anthocyanine and so much more frequently reddish-violet. The formation of "green islands" can also be observed in fungal parasites; the best known and commonest example may be seen in the leaf blotches occurring in autumn on Acer, caused by Rhytisma acerinum Pers., which provide an interesting analogy to those of miners. Phenomena found with fungal infection of leaves comparable to the ringing we have described, are considered by these authors not as a blocking of the conducting vessels by the hyphae of the fungus and even less as a severence or destruction of the relevant vascular bundle, but as a toxic effect of fungal excreta. The causes of retention of chlorophyll as a result of fungal infection is still unexplained; the reasons postulated may equally be applied to leaf-miners, so that the remark of Tubeuf (1916) seems justified in both cases: "It appears to me that galls and areas infected by fungi represent areas of increased metabolism under the stimulus of the parasite and that they form centres of attraction for nutrient substances owing to the increased respiration or supply of food-stuffs to the parasite . . . . The infected parts of the leaf are under the influence of the parasite; they form together, one might almost say, a foreign body in the leaf . . . ." The mysterious position of this "state within a state" represented by a mine in a leaf will doubtless still produce many surprises as more detailed investigations are undertaken.

15.

GROWTH CHANGES IN THE MINED LEAF

MINE AND GALL

Attention has already repeatedly been drawn to the fact that the connections between the mining parasite and the infested leaf are particularly close. When discussing the colour and discolouration of the mine and the surrounding parts of the leaf, we noticed a connection between the two and saw that the plant reacted to the attack of the parasite. We shall now find that inside the leaf along the mine channel, much further reaching changes take place, accompanied by deviations in the growth of the plant, which are quite unknown in a normal leaf. Striking analogies occur in such cases with the growth abnormalities which were discovered in the investigation of plant galls; it can be shown that we are now entering a border region between the study of mines and galls, which appears particularly suitable for focussing attention on the development of gall producing habits and showing how this at least partially has been derived from mining. These changes are in many cases apparent externally, in others they can be detected by a microscopical examination of the histology of the mined leaf; in many cases the two go together, with the external signs indicating that changes have also taken place inside the leaf.

1. External changes.

The influence of oviposition or of mining often results in deformities and rents in the leaf, which are frequently characteristic of the mining species concerned and can be useful for purposes of identification. This shows that the attacks of the parasite are clearly of a more fundamental nature than would at first sight appear. Thus, the leaf develops with a lack of symmetry; holes appear or parts are missing from the normal leaf, or certain parts, particularly the apex of the leaf, are deformed in a characteristic manner, which can completely alter the normal appearance of the leaf. Two factors in particular are responsible for such pathological growth abnormalities.

1. The degree of the leaf's development at the time of oviposition or during construction of the mine.

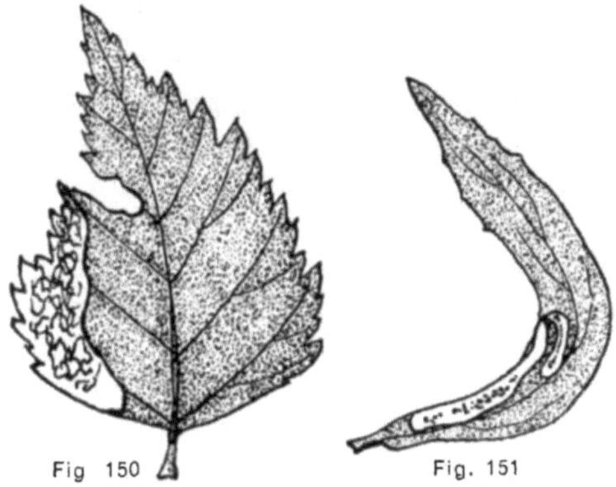

Fig 150 Fig. 151

Fig. 150. Betula: *Eriocrania* sp. Tear in leaf at beginning of mine.

Fig. 151. Epilobium: Distortion of apical leaf caused by mine of *Anybia epilobiella* Roem.

If oviposition has occurred when the leaf, which is later to be mined, is still very young, in other words when the leaf is still extremely small, the act of oviposition represents a serious injury to the leaf. So many of the young cells are thereby damaged, that sufficiently serious abnormalities occur during the development of the young leaf to result in fairly striking deformities during later growth. A characteristic example of this is found in the mines of the Eriocraniidae, which are the earliest to appear in the year. In a large number of cases the egg is laid actually on the edge of the leaf. As the leaf grows and the mine develops, marginal gaps occur on the spot where the egg was laid, which distinguish this spot and are characteristic of the members of this family (Fig. 150). It is of interest in this connection that there are some species, which appear later than the majority in the family (i.e. E. sparrmannella Bosc.); in these species the egg is laid more towards the centre of the leaf and gaps of the type described are never to be found; they can in this way easily be distinguished from the other members of the family.

Such results can be produced without the whole plant having been necessarily in an early stage of leaf development, when it was attacked; they also appear in more fully developed plants, when the youngest leaves at the extremities of the shoots are attacked. *Anybia epilobiella* Röm. normally lives in the fully developed leaves of Epilobium somewhat later in the year. No externally visible abnormalities in the leaf development can be detected. When, however, in exceptional cases one of the youngest leaves at the tip of the shoot is mined, this also produces remarkable distortions in the whole leaf

(Fig. 151). Such distortions are often found in conjunction with an asymmetrical development of the leaf; one side of the leaf remains smaller than the other and assumes a quite different shape. Parallel with this one finds evidence of undernourishment in the more undeveloped part of the leaf, which recalls the "ringing" already referred to.

The mine of *Phylloporia bistrigella* Hw. may be mentioned as the commonest example of this type and which occurs by preference in the youngest leaves at the extremities of the shoots, even when the leafage has reached a state of considerable development. The regularity with which the mine is found in the deformed and undernourished leaf leads to the supposition that the caterpillar finds conditions in the incompletely developed parts of the leaves more to its taste than in the normal leaf (Fig. 152). It is reasonable, therefore, to consider this type as a "weakness parasite", a destignation coined by mycologists for such an occurence and which is discussed further in the chapter dealing with the damage caused by miners to plants.

Distortions are also frequently found particularly on the margin of the leaf, such as may easily be recognised in early spring in the serpentine of *Phytagromyza xylostei* R.-D. on Lonicera and of *Phytomyza vitalbae* Kltb. on Clematis. While all these cases refer to parenchymal miners, with which such developments in the leaf are readily understandable, they are also to be found in conjunction with sap-feeders. The purely epidermal mines in the leaf of Populus nigra L. caused by *Phyllocnistis suffusella* Z. produce the same abnormalities in a leaf at the tip of the shoot, as can be found in this case predominantly on the edge of the leaf. One would scarcely believe that the slight destruction of leaf substance caused by this attack solely on the cells of the epidermis would have such a considerable effect; however, it seems clear that with the elimination of the protective skin, desiccation inside the leaf must be appreciable, thus accounting for these abnormalities in development. It has already been pointed out that the tendency for necrotic processes to appear in such leaves, is far greater even than in leaves infested by other species of miners. Some of these species doubtless cause a genuine ringing process, that is a severance of conducting vessels, which results in the undernourishment of the affected parts of the leaf.

The preference for such undernourished parts, which we have just noticed in the case of *Phylloporia*, is by no means an infrequent exception; on the contrary, it can very often be observed that miners tend to affect the parts of the leaf "ringed" in this way. One should recall the regularity with which *Vidalia cornuta* Sc. affects the "ringed" side of the leaf (Plate 1, Fig. 6).

Fig. 152

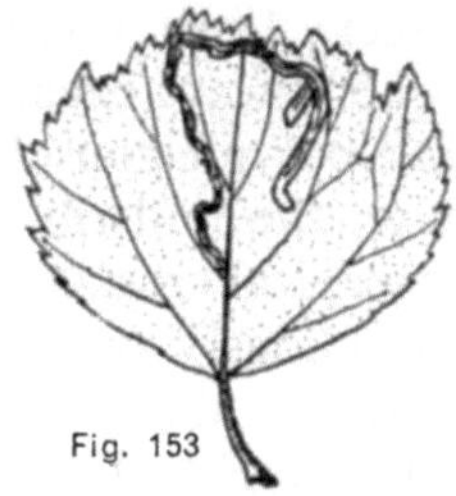

Fig. 153

Fig. 152. Betula: *Phylloporia bistrigella* Hw. Asymmetric distortion and "ringing" on left side caused by the linear mine.
Fig. 153. Betula pendula Roth. Linear mine of *Anoplus plantaris* Naez. has impeded development of tip of leaf.

2. The position of the mine in relation to the vascular system of the leaf.

Even in cases where the ova of mining species are laid on leaves which are no longer extremely young, very frequently extensive growth changes can be observed, particularly if the egg has been laid on or in a main or important subsidiary vein. It is quite obvious that in such a case symptoms of undernourishment will develop particularly readily in the parts of the mined leaf which are supplied by the vascular bundles passing through the affected areas. Thus, whenever the egg has been laid in the mid-rib, subsequently varying degrees of tearing are regularly encountered. This is specially characteristic of many species of the genus *Rhynchaenus*, especially with *Rh. quercus* L. on Quercus and *Rh. fagi* L. on Fagus. The close connection between the site of oviposition in relation to the mid-rib and the appearance of tears or rents is shown by the very different behaviour of two species on the same plant. *Rh. saltator* Geoffr.-Fourcr. is always distinguished by conspicuous rents near its mine; in this case the egg is laid on the leaf mid-rib; the mine is made invariably towards the apex of the leaf. In rare exceptions oviposition takes place on one of the more important lateral veins and this also leads to serious deformities. The mine of *Rh. rufus* Schrk., which also occurs on Ulmus as in the previous species, is always made directly along the edge of the leaf; no important part of a vein is affected during oviposition and in this species damage to the leaf is far less extensive. The mines of *Anoplus plantaris* Naez. (Fig. 153) are very characteristic, as a result of the deformities to the leaf which they cause. In this case also the egg is almost without exception inserted into the mid-rib of Betula pendula Roth. The result is that the elongated leaf tips, which are so characteristic of this species, do not develop normally, so that in outline they appear very similar to those of Betula pubescens Ehrh. Oviposition or the course of the mine on the mid-rib do not always cause such deformities. For instance, with the common and polyphagous *Liriomyza strigata* Mg., which mines predominantly on the mid-rib, results of this kind are never to be found. A microscopical cross-section of these mines shows, however, that the larva fastidiously avoids the vascular bundles and constructs its mine only below or above or somewhat to the side of the conducting vessels (Plate 2, Fig. 1).

In certain cases the destruction of important tissues does not produce a weakening but on the contrary a strengthening of the mined areas of the leaf. For instance, in *Nepticula* species of the *argyropeza* group one finds a thickening of the petiole, although an essential part of the parenchyma near the vascular bundles has been destroyed (Fig. 149). This is, however, explained by the fact that callus tissue develops in such quantity in connection with an enlargement in the parenchyma cells that the result is an increase in size of the petiole, with a marked expansion of its sides. Phenomena of this type will be discussed in greater detail below. Oviposition in the mid-rib is rarely found except with Coleoptera and *Phytomyza ilicis* Curt.; even with species whose mine is linked predominantly with the mid-rib, the egg is inserted into the leaf some distance away and only after hatching does the larva proceed to the mid-rib by the shortest possible route. The reason is possibly that in the vicinity of the conducting vessel in the vein there are extremely tough strengthening elements (sclerenchyma), which would put a particular strain on the ovipositioning female; species with a relatively soft ovipositor thus find it impossible to lay in the mid-rib itself.

II. Changes inside the leaf.

Every wound on a plant results in changes to the injured tissues. In the majority of cases these changes are limited to the damaged cells but in other cases cells further removed from the point of injury may be affected. It should therefore occasion no surprise that with the extensive injuries represented by mines in a leaf tissue, changes are induced which under given circumstances cause a characteristic reaction of the plant. The mined leaf is endangered in two ways; firstly, a process of excessive evaporation ensues which may lead to desiccation even in undamaged cells inside the leaf; secondly, the destruction of certain cells makes translocation at least more difficult, if not quite impossible for certain parts of the leaf. When the plant is still in an early period of growth, it reacts to the injuries caused by the mine by attempting to bridge over the cavity which has been produced. It employs for this purpose cells or tissues which never occur in the normal leaf and are only formed for this special purpose. In some cases, the result is a hypertrophy, that is, the enlargement of cells through increased growth. In this way hyperhydric cells arise, which are distinguished from normal cells of the leaf by their greater water content and increased turgidity. Their formation presupposes either considerable moisture in the mine cavity or some distinctive internal condition of the cells which prevents their normal transpiration of water, as may occur when they are impregnated with poison. The enlargement or hypertrophy of cells is not all that occurs in the processes we are describing. One frequently also finds hyperplasy, a process whereby abnormal tissue develops through an increased rate of cell-division.

The commonest type of reaction by the plant to the wound caused by the mine consists of the formation of callus tissue in the mine channel. A callus is a loose layer of parenchymal tissue without any specific shape, which sooner or later adapts itself to the shape of the mine cavity which it fills. Callus tissue may be formed not only by the meristem, the cell-producing tissue of the leaf which has remained in an embryonic stage of development and is usually attached to vascular bundles, but also all fully developed tissue, which may be present in the leaf in various forms has the same capacity. Naturally, meristematic tissue is able to produce callus more easily and in greater quantities than fully-developed tissue. Callus of both these origins is found in mine cavities. The formation of callus is not limited to the inside of mines; such cells are formed with every type of injury, sometimes assuming a very beautiful form, as for instance in the so-called frost blisters in the leaves of Buxus. Callus tissue plays a particularly important role in the formation of plant galls.

Callus cells have various characteristics which enable them to be easily distinguished from normal plant cells. We previously mentioned the increased turgidity of the cells as a result of their higher water content. Furthermore, they are often enormously enlarged by comparison with the cells of the tissue from which they have arisen. The cells are frequently linked together in chains as in the formation of hairs or fungal hyphae and traverse the mine cavity like threads. They have either only few minute chloroplasts or these may be entirely absent. Such cells are therefore always pale and when they replace assimilation parenchyma, they are never able to play any important part in the process of assimilation. They have extremely delicate walls and any special thickening of the cell walls, such as is sometimes found in callus, is never found in mine channels except

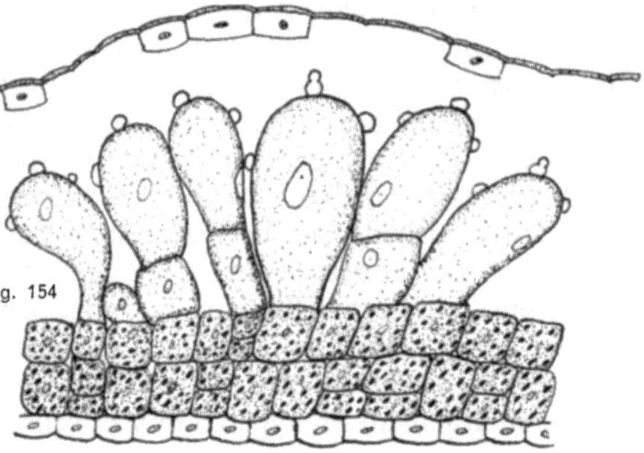

Fig. 154. Clematis: Cross-section of mine channel of *Phytomyza vitalbae* Kltb. with formation of callus, on which pectine granules have formed.

some little pectine granules on the cell walls (Fig. 154). Having such thin walls makes them specially desirable for certain micro-organisms, which can easily colonize the mine channel; we shall discuss these in a later chapter. This is also the reason why callus cells do not retain their initial shape but change it under the influence of pressure from their own and other tissues and usually become distinctly flattened (Fig. 160). They appear to be no richer in cytoplasm than other plant cells; however, owing to their greater water content, it is more diluted and takes up a greater space.

We have already mentioned the most important conditions under which callus is formed. The greater dampness of the air in the mine channel is the biggest single factor. However, that dampness is not the only factor inducing the formation of callus is shown by the fact that it is never found in mines in submerged leaves. It never occurs in the mine channels of *Tendipedidae*, in which moreover the larva gnaws at the walls of the channels at regular intervals, nor in the mines of *Hydrellia* species, in which the larva never returns to preceding areas of its channel. Presumably the greatly reduced absorption of oxygen underwater precludes the formation of callus in this case. This would suggest that oxygen is an appreciable factor in the formation of these cells. We have already pointed out the importance of the proximity of vascular bundles; an abundant supply of food thus seems to be a factor assisting the growth of these cells. Further possible influences favouring the development of callus will be examined below. The main factor, and this should not be overlooked, is moisture in the mine channel and it should cause no surprise that we find a particularly extensive development of callus in mines in plants, whose leaves contain particularly large quantities of moisture, such as members of the Crassulaceae.

From the facts set out so far, it is apparent that no complete reconstitution of the tissues eaten by the mining larva is possible; no genuine regeneration in fact occurs. However, this applies only to certain cases where we can show that no or only the most limited assimilation can take place in callus tissue. On the other hand, callus cells presumably do have some special functions and do not exist merely to fill up gaps which have arisen; thus in a certain limited sense they may be considered as regenerated cells. Furthermore, there are certain cases where true regeneration does occur, where in fact similar tissue to that destroyed is re-formed, although this is found only very rarely. Below we outline the various possibilities of bridging over mine cavities, as far as they have been observed to date.

1. Formation of callus by hypertrophy of cells in the palisade parenchyma.

The cells of the assimilation parenchyma in the vicinity of the mine channel become enlarged and this leads to the filling up of the cavity. Gertz (1918) described such structures, although we have only exceptionally been able to find them; this apparently represents an isolated and rare exception.

2. Callus formation from within the spongy parenchyma.

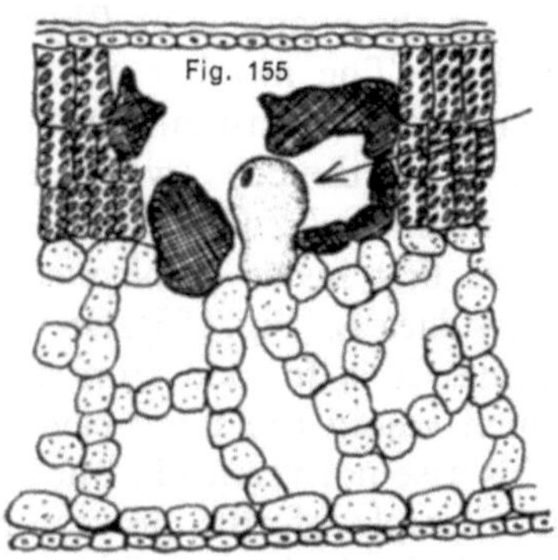

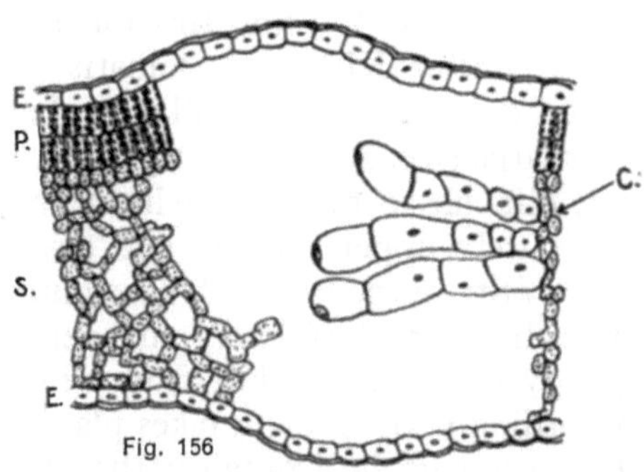

Fig. 155. Vaccinium vitis-idaea L.,: Cross-section of commencement of mine of *Nepticula weaveri* Stt. A callus cell has penetrated between the lumps of frass.

Fig. 156. Malus: Cross-section of mine channel of *Lyonetia clerkella* L. with threads of callus growing horizontally (C). (E - epidermis, P - palisade parenchyma, S - spongy parenchyma).

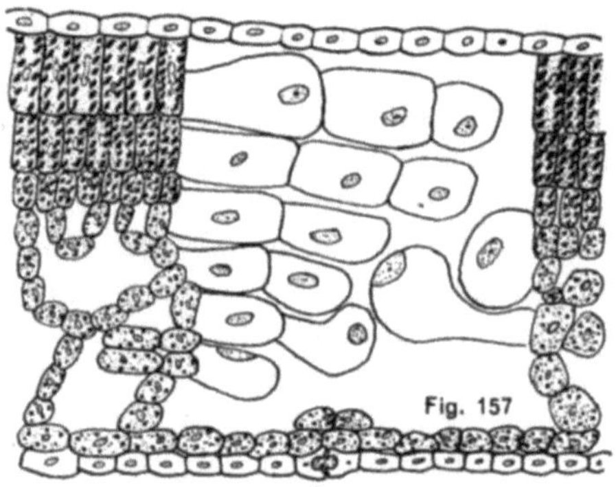

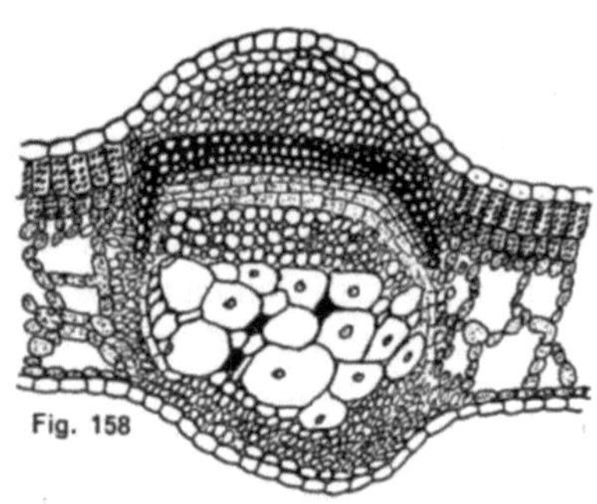

Fig. 157. Malus: Cross-section of beginning of mine channel of *Lyonetia clerkella* L. Threads of callus practically fill the mine and appear to proceed from the palisade parenchyma.

Fig. 158. Quercus: Cross-section of leaf mid-rib. The lower portion of the vascular bundle has been eaten away by *Rhynchaenus quercus* L. and filled out again by callus, with grains of frass between.

This is the commonest type of callus formation. When the larva has eaten away the assimilation parenchyma, callus cells are cut off from the remaining spongy parenchyma and grow out into the mine cavity. One can notice a difference between upper surface and lower surface mines in the way this occurs. In the case of an upper surface mine, the callus cells arose primarily from the remaining cells of the spongy parenchyma below the mine cavity, in other words they grow vertically. We find this for instance in the channels of *Phytomyza vitalbae* Kltb. on Clematis (Fig. 154). The number of cells involved is relatively small but they tend to be distinguished by their large size. A similar origin can be noted in the cells (which are very difficult to find) in the linear mines of *Nepticula*, while these are still purely upper-surface, and also in the first instar of *Nepticula weaveri* Stt. in the leaves of Vaccinium vitis-idaea L. (Fig. 155). On the other hand in completely full-depth mines it is formed from the spongy parenchyma, but from cells which adjoin the mine cavity. Accordingly, the growth takes place in this case not vertically but for the most part horizontally; the classic example of this process is found in the channels of *Lyonetia clerkella* L. in the leaves of Malus. The cells are usually grouped together in large numbers and form distinctive tubular formations (Fig. 156). When the callus is in a stage of greater development these tubes irregularly fill out the entire mine cavity and are then no longer purely horizontal; in optical section the impression frequently arises that their point of origin is the assimilation parenchyma (Fig. 157). However, if they are examined with varying adjustments of the microscope and with varying depths of the cells themselves, one can see clearly that the callus cells which have replaced the assimilation parenchyma have in fact originated in the spongy parenchyma. Variations can also be noted in the quantity of callus, as can be seen from Fig. 156 of the end of the mine and Fig. 157 of the beginning of the mine, with the same species on the same plant; these variations do not depend on the amount of time available for the formation of callus but in the varying width of the mine channel; the narrower the channel, the greater is the quantity formed.

Callus cells essentially resemble the parenchyma cells which give rise to them but are larger and with a higher water content, and their chlorophyll content is much reduced. Originally they are cylindrical or ovoid but as soon as they appear in larger numbers and begin exerting pressure on each other in their restricted space, they become polygonal and may even appear completely flattened. They vary considerably amongst themselves; some are very large, while others scarcely differ in size from normal spongy parenchyma cells and can only be distinguished by their other characteristics.

3. Callus formation in the vicinity of meristematic tissue.

Young, undifferentiated tissue found in or near vascular bundles is naturally best suited to callus formation. These cells are therefore particularly well developed in mines running through or near vascular bundles and they may also be found in channels, whose substantial width normally militates against their formation. As an example we may mention the linear mine of *Liriomyza strigata* Mg. (Fig. 12), which follows the leaf veins. Plate 2, Fig. 1 shows the beginning of the development of callus in the extreme corners of the mines. Tissue formed in this way penetrates into parts of the mine channel which are no longer so close to a vein and frequently fills them out so completely that the epidermis bulges and the result is thickening of the leaf in this area (Plate 2, Fig. 3). In such cases the mining larva has not attacked the vascular bundle itself and no formative tissue has been injured. When this occurs, the development of callus is even more active. This can be found especially in species which construct their mine channels exclusively inside the mid-rib, such as the *Rhynchaenus* species (Fig. 158 and Plate 2, Fig. 2). A cross-section of the mid-rib of an oak leaf in which *Rhynchaenus quercus* L. has been living shows that the lower half of the vascular bundle has been completely eaten away. No cavity has been formed, however, since the whole mine channel has been filled up by an abundant growth of callus tissue. The grains of frass left in the mine have been compressed by the callus cells and completely flattened; they are com-

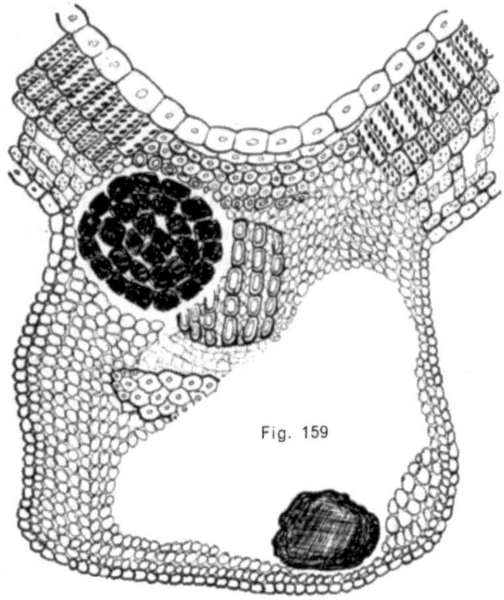

Fig. 159. Alnus: Cross-section of the leaf mid-rib with two stages of the mine channel of *Heliozela resplendella* Stt. (second generation). No callus cells.

pletely surrounded by the newly-formed tissue. When discussing "ringing", we referred to the strange habit of *Heliozela resplendella* Stt. which as a characteristic way of mining first in the mid-rib and subsequently in two lateral veins of an Alnus leaf (Fig. 141). This species is very common in the autumn generation and does not easily avoid detection. If an infested leaf is examined, the mine, as illustrated in Fig. 159, will be found in the mid-rib. The beginning of the mine is to be found in the upper part of the mid-rib and is distinguished by substantial quantities of frass; this is probably connected with the fact that the larva eats here far more thick-walled cells (strengthening elements (sclerenchyma) and portions of the vascular bundle). Since there is more indigestible matter in this food, the quantity of excrement is also larger. The later channel runs for the most part in the parenchyma of the mid-rib, a softer, more nutritious tissue and the grains of frass are accordingly smaller. In these mines, which are to be seen in September and October, there is no trace of any callus cells, although the vascular bundle has been seriously injured. The situation with the rarer first generation which mines during the plant's most active period of growth, is quite different. If mines found in July are examined, the mine channel is completely similar. One sees that the same portions of the leaf are eaten as in the second generation but the reaction of the plant to this

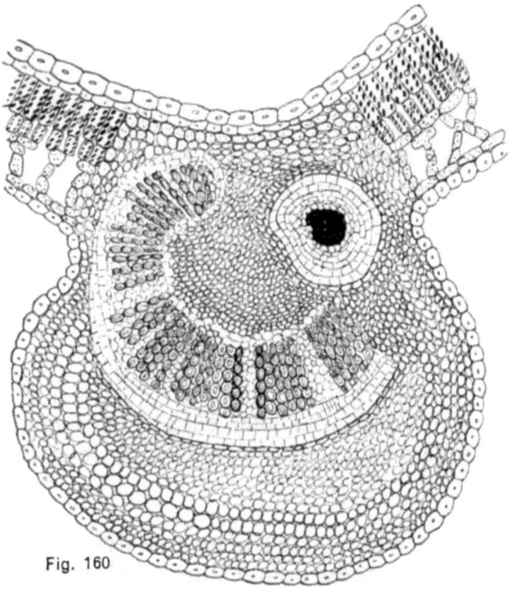

Fig. 160. Alnus: Cross-section of the leaf mid-rib with mine of *Heliozela resplendella* Stt. (first generation). Mine channel completely filled up by callus.

injury is quite different. Callus tissue has grown in such abundance that it completely fills the mine channel; the individual cells have lost their customary shape, clearly under the substantial pressure to which they have been exposed and the grains of frass have been compressed to a thick lump, which is still visible in the centre of the channel.

4. Regeneration by the palisade parenchyma.

This phenomenon is the least frequently encountered and has only been found in the mines of *Agromyza alni-betulae* Hd. on Betula (Fig. 161). The plant reacts most vigorously to attack by this parasite, as is shown by the numerous distortions and deformities of the leaves under the influence of the mine channel. In the majority of cases the mine cavity is filled up by callus, as in the cases just described. In addition, however, individual instances of genuine regeneration have been observed; the tissue which has been removed is replaced by new tissue of the same type and in this case, as it is an upper surface mine, by assimilation parenchyma. It looks as though new palisade parenchyma has developed from the spongy parenchyma; it is also possible, however, that it has been produced by neighbouring palisade parenchyma. The regenerated cells are identical

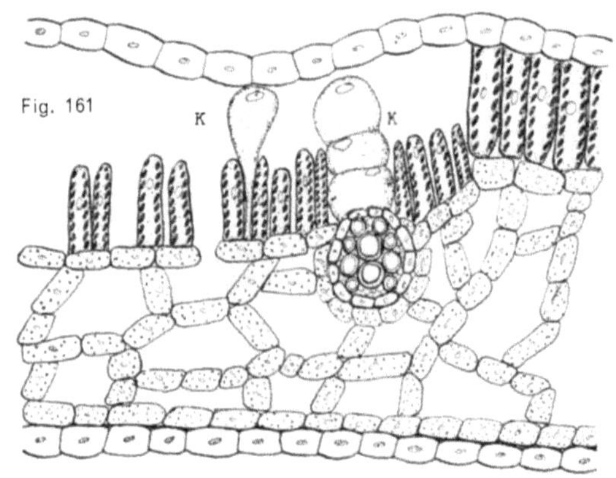

Fig. 161. Betula: Cross-section of mine channel of *Agromyza alni-betulae* Hend., with newly formed palisade parenchyma, interspersed by callus cells (K).

with the others, especially as regards normal development of chloroplasts, (which are always reduced in callus cells); at the time we made our examination they were merely somewhat smaller. It is impossible to say whether, with further growth, they would have reached the epidermis and thus even more closely resembled normal upper surface tissue. As it was, they formed the impression of a layer of palisade parenchyma, which had slipped down and become somewhat smaller. Furthermore, normal callus cells grew among them (Fig. 161), of a type familiar from other mine channels. Judging from their external appearance, it can be assumed that these new cells are capable of assimilation. They cannot, therefore, be considered as callus; they represent a newly-formed tissue, which apparently fully replaces that which previously occupied the area. This is not, as with callus, merely a wound tissue, whose sole purpose is to fill up the cavity caused by the injury, but a genuine replacement of tissue which has been removed, with all the morphological and functional characteristics of the originally existing healthy tissue. Moreover, in the channels caused by this mining fly the callus cells found together with this new tissue were shown to be much richer in chlorophyll than is normally the case, so that presumably they also to some slight extent participate in assimilation.

It would be useful if systematic observations of this type could be carried out; information so far available is largely the result of chance discoveries. The formation of such regenerated parenchyma has been observed in the following mines; observers were Gertz (G.), Hering (H.), Küster (K.) and Schneider-Orelli (S.):

FOOD-PLANT	MINE-PRODUCER	OBSERVER
Aegopodium podagraria L.	*Phytomyza obscurella* Fall.	G.
Alnus incana DC., glutinosa Grt.	*Heliozela resplendella* Stt.	H.
Betula pendula Roth.	*Agromyza alni-betulae* Hend.	H.
Brassica sp.	*Scaptomyzella flava* Mg.	K.
Clematis vitalba, L.	*Phytomyza vitalbae* Klt.	H.
Dahlia variabilis (Willd.) Desf.	*Liriomyza strigata* Mg.	H.
Ilex aquifolium L.	*Phytomyza ilicis* Curt.	K.
Lamium album L.	Spec.	G.
Lonicera periclymenum L., xylosteum L.	*Phytagromyza xylostei* R.-D.	G.
Malus silvestris Mill.	*Lyonetia clerkella* L.	S.
Prunus cerasus L.	*Lyonetia clerkella* L.	H.
Phaseolus vulgaris L.	*Liryiomza strigata* Mg.	H.
Quercus robur L.	*Rhynchaenus quercus* L.	H.
Sedum telephium L.	*Phytomyza sedicola* Hering	K.
Vaccinium vitis-idaea L.	*Nepticula weaveri* Stt.	H.
Valeriana officinalis L.	*Liriomyza valerianae* Hd.	H.

This list does not include cases where mere traces of largely ephemeral callus tissue has been found in the corners of *Lithocolletis* and other similar mines. It does not really provide a basis for judging the frequency of this phenomenon in mines; for this, more extensive investigations are necessary. It is as yet impossible to say just where regenerated tissue may or may not be formed Gertz was never able to detect it in the linear mines of *Phytomyza ranunculi* Schrk. on Ranunculus or of *Nepticula floslactella* Hw. on Corylus. In 9 species of *Nepticula* we examined, not even traces of it could be found; only in the case already mentioned — *N. weaveri* Stt. — was it in evidence in small quantities. This must appear somewhat strange. The mine channels of *Nepticula* are so extremely similar in their whole appearance to those of *Lyonetia clerkella* L. that an inexperienced minologist can easily confuse them, unless they have already reached a stage of full development. We know, however, that callus tissue develops readily in *Lyonetia* mines but not, however, in the so similar *Nepticula* mines. This clearly suggests some influence on the part of the miner; in other words this new tissue does not represent a one-sided reaction of the plant to its injury. Structures of this type have never been discovered in the common and polyphagous *Phytomyza atricornis* Mg. They have also never been found in species mining in monocotyledons, although it is known that callus tissue sometimes occurs in abundance, for example in orchids.

We already noted that m o i s t u r e is one of the main conditions for the formation of regenerated tissue of all types in mine channels. We must also mention here the higher temperature in mines, which we discussed earlier; the two factors together give the inside of the mine cavity its "hot-house atmosphere", which is especially favourable for the growth of such tissue. Callus is accordingly found particularly in narrow mine channels, while in the larger and more airy mines it occurs only in the corners, where the mine adjoins undamaged tissue. Callus has also been formed in "artificial mines", produced by the prick of a needle. The development of this tissue, as already shown, is favoured by the proximity of v a s c u l a r b u n d l e s, which are able to supply the abundance of food required by the newly-developing cells. The larger and more extensive are these bundles, the more successfully does the callus grow; the best examples are thus found when the mine runs either in or in the proximity of the mid-rib. Similarly, the mine cavity is seen to contain less callus, the greater the distance of the mine from the mid-rib. The state of the infested plant's growth is an important factor in the formation of regenerated tissue. The younger the infested parts of the plant are, the greater is the development of callus. From a certain time onwards the plant is unable to form new cells in mines or it can do so only to a very limited degree. This time is naturally largely dependent on climatic factors but varies with individual species of plants. Species, whose period of main development is in autumn, will produce new cells in mines, when this is no longer occurring in other species. For instance, structures of this type may still be observed in Dahlia and in Ilex in autumn, at a time when they can no longer be found in any other plants. The extent to which variations in this process may occur even with the same species of plant and the same miner has already been shown in the case of the second generation of *Heliozela resplendella* Stt.

G e r t z finds two other factors which are responsible for callus formation in mines. He accepts that genuine "ringing" results in this proliferation of certain of the cells. In support of this view Gertz quotes experiments undertaken by H a b e r l a n d t (1902), who isolated mesophyll cells of plants and established them on nutrient solutions on which they remained alive for some time and were capable of assimilation. Immediately products of assimilation had accumulated in these isolated cells, they began to develop similar proliferations. However, we have previously noted that the phenomenon known as "ringing" is frequently not caused by any severance of conducting vessels; we also noted that callus formation is most marked in the vicinity of vascular bundles and deduced from this that in such cases the specially abundant supply of nutrient products must be the cause. It is impossible to assume that an abundant supply of nutrient materials takes place to areas where all removal of products is impeded. Furthermore, one finds that the callus formation is not restricted to the side of the leaf which is "ringed", that is to say cut off, but cells grow into the mine channel from the opposite side also. It is not certain whether carbo-hydrates are the substances which accumulate in ringed areas; they are in any case the most important substances for the growth of callus. It is possible, however, that other substances which are not accumulated in these cases are of greater importance for the formation of these new structures. Accumulation of products of assimilation in the cells cannot therefore be considered as an essential condition for proliferation.

Gertz further maintains that an additional factor is the increased counter-pressure in the mine cavities. He suggests that the surrounding, intact parenchymal tissue exerts a certain pressure and some cells escape this pressure by growing out into the mine cavity. The main objection to this idea is provided by the fact that formation of callus continues even after the mine cavity has been filled up; it has even been discovered that abnormal pressure exists in the mine cavity, which may cause the walls of the mine to bulge out into either the upper or lower surface of the leaf. Furthermore, we frequently find that cavities inside a part of the plant are not influenced by pressure from neighbouring tissues, as we already saw in the construction of the mine of *Liriomyza triglochinae* Hend. on Triglochin (Fig. 131). Purely mechanical influences of pressure can therefore be completely discounted when seeking an explanation of the causes of callus formation.

Probably one will be forced to the conclusion that the parasite itself exercises some influence, which at least assists in callus formation. Just as we agreed, when discussing "ringing" and anthocyanine formation, that probably some influence is exerted by frass or substances it contains, with regeneration of tissue there is little doubt that the same thing applies. It will be difficult to prove this but it can be assumed with some degree of certainty.

The fact that newly-formed cells are frequently encountered with some species of parasite, while they may be completely absent with others clearly suggests that the parasite itself exerts some influence. For instance, callus has never been found in the mines of *Phytomyza atricornis* Mg., no matter what plant it infests, while the equally polyphagous *Liriomyza strigata* Mg. causes the formation of these cells in the most varied plants. The fragmentary nature of the data available permits of no conclusion as to whether or not individual species or families of plants have a predisposition for the formation of these new cells. It is justifiable to assume, however, that one food-plant is more, another less inclined to their formation. This view is supported by the fact that callus is found with some consistency in the mines of *Lyonetia clerkella* L., when they are constructed in Malus, but on the other hand when they occur in Prunus cerasus L. this new tissue is not found at all or only in very slight quantities. However, it appears that all these other influences occupy a secondary position in relation to the importance of moisture in the mine channel and proximity to vascular bundles.

The question of the function of these newly-formed cells is again by no means clear. First of all they fill in the gaps caused inside the leaf by the mining of the larva, they thus serve to bridge over the mine cavity inside the tissue. This results in alleviating the danger of desiccation, at least in some of the affected spots. Their increased turgidity enables them to be considered to some extent as a strengthening element; this applies of course not to the individual cells, which are characterised by their thin walls, but to callus as a compact tissue area. They are not able to replace the assimilation parenchyma, which in so many cases is attacked by the larva. We already pointed out that they suffer from a distinct reduction in chlorophyll, which enables them to undertake assimilation only to a very modest degree. We noted only one exception to this, where regeneration of assimilation tissue also occurred. It seems natural to conclude that they represent a factor assisting in translocation. Confirmation can be seen in the fact that they originate largely in the spongy parenchyma, which itself is a tissue facilitating the transportation of substances. The investigations of Schneider-Orelli showed that transportation of products of assimilation (carbo-hydrates) cannot be undertaken by callus cells. He also showed by the iodine test that there was no removal of carbo-hydrates from "ringed areas", independently of whether the mine channel was filled with callus cells or not. The main function of callus cells, however, is the transport of water into parts of the leaf, which have been cut off by the mine from the remaining conductive tissue of the plant. The surprisingly thin walls of these structures itself suggests transportation of water as a principal function. Schneider-Orelli also produced direct proof of this claim: on comparing leaves in which substantial vascular bundles had been destroyed, it was found that the parts of the leaf isolated in this way retained their turgidity and remained alive, provided that callus had been formed in the mine channel. When this was not the case the affected parts of the leaf dried up. The main function of these new cells therefore appears to be to maintain distribution of water throughout the leaf. This of course does not exclude the possibility of other substances being transported by the same means. It has been found that the newly-formed cells are particularly susceptible to the pigments which have been introduced into the plant for staining the protein-conducting tissues, such as the sieve tubes (Phloem). It can possibly be concluded from this that some part of the transportation of proteins is effected by callus tissue. When such a colouration is present, some of the undamaged cells in the vicinity of the mine channel are tinged, which suggests that a general accumulation of protein occurs in the various cells in the surroundings of the mine cavity; this probably represents a "wound reaction", which is a normal development in the plant after injuries and which enables the affected parts of the plant as far as possible to make good all damage caused.

If we accept the possibility of the mining larva exerting some influence on the plant, leading to the formation of the peculiar regenerated tissue, we have already entered the "noman's land" between minology and cecidology, the study of galls. An essential characteristic of gall-producing organisms is that they cause various new structures to arise in the plant, on which they are dependent for their food. We are thus in a definite border region and it will sometimes be difficult to determine whether a given structure is a mine or a gall. The term gall used in a loose sense can be applied to every reaction of a plant to oviposition or feeding by a living organism; the stricter, more scientific sense of the term, however, implies that the tissue formed as a result of this reaction is also eaten by the larva, in fact that its only function is to serve as food for the larva. We must adhere to this stricter definition of the gall (the cecidium), if we wish to be able to distinguish at all between gall and mine. In a mine this particular significance of the newly-formed tissue is entirely absent; the larva never returns to the parts of its mine which are filled with callus, while in a gall the larva remains within the newly-formed tissue or returns to eat it; normal tissue is usually not suitable as its food. However, there are certain species of mining insects which can definitely not be included in either category. A transitional stage is represented by *Liriomyza strigata* Mg., which constantly returns to the channel when these new structures have developed and is in many cases, therefore, forced to eat them away. A similar situation is found with the species of *Nepticula* of the *argyropeza* group on Populus, which we have also previously mentioned; these latter indeed are considered by cecidologists to belong to their sphere, even though the thickening of the petiole they produce is very slight. In both these cases, however, it must not be overlooked that the larvae only return to the mine cavity produced earlier, during pauses in feeding. Both feed again later on the actual leaf-blade and visit the mid-rib or petiole only when they interrupt their feeding. If they in fact eat the tissue which has subsequently developed in the mine cavity, they do so not for any compelling biological reason but only in order to keep their channel open. We see, however, in these cases how a gradual transition between miners and gall-producers is taking shape. It is easy to imagine how in the course of development over thousands of years a special adaptation of the larva to this particular tissue might take place and that after a certain period mining larvae might become transformed into gall-producers. There are to-day certain species in which this phenomenon occurs with considerable regularity and to a striking degree, so that it is difficult to be sure that one is dealing with miners and not with gall-producers. *Phytomyza cecidonomia* Hering on Hypochoeris and *Ph. cardui* Hering on Cirsium and Carduus construct completely similar mines on the plants in question, mining in the mid-rib of leaves, usually those nearest the root. These mid-ribs become considerably swollen (by enlargement of the parenchyma cells) and appear just like a gall, so that in both these cases the two species might be considered as gall-causers. The extreme adaptation to the gall tissue has not, however, developed to the extent which is usual in cecidogenous species. Both species may also mine considerable lateral tracks in the rest of the leaf-blade outside the thickened area of the swollen mid-rib. This makes it very clear that we have entered a border-region between the two ecological groups, galls and mines; certain forms will constantly be occurring, for which a definite classification as either mine or gall must be left to the judgment of the individual worker. In such cases one can be guided by the systematic position of the producer and for example, in cases of doubt, include all structures produced by mining flies (Agromyzidae) as mines, while doubtful structures produced by the Cecidomyidae will be considered as galls. It must be constantly realised, however, that one is dealing here with ecologically intermediate forms, which might be considered as mines equally as well as galls. A relevant example is provided by the well-known gall-midge on Buxus *Monarthropalpus buxi* Lab. which produces small pustules in Buxus leaves; also by the various species of *Cystiphora* which also produce round pustules in Compositae leaves and are invariably characterised by the conspicuous anthocyanine formations they induce (Plate 1, Fig. 5). The newly-created tissue is found only in very small quantities, so that the deformities to the plant caused by these species have not the normal appearance of galls but look more like mines; they are included in the classification of mines as also in that of galls. A microscopical cross-section of these structures shows only callus cells, which are eaten by the larva but no other type of newly-formed tissue; the larva appears merely to bite into the parenchyma cells and to imbibe their sap; probably a constant flow of sap is directed to this spot, which itself provides adequate food for the larva. That gall-causing habits may in many cases be derived from mining is shown by the fact that there are certain genera of insects in which, of closely related species, some produce mines and others galls. For instance, the Melanagromyza species live partly in normal leaf-mines, others *(M. pulicaria* Mg., *M. beckeri* Hend.) are limited to the mid-rib of leaves, others again live in the leaf mid-rib and stem parenchyma, while yet others feed only in the pith of the stem, until we finally reach species, such as those of the *schineri* group, which cause genuine shoot galls in the twigs of Salix and Populus. Further, there are certain species in the normally leaf-mining family *Nepticulidae*, which have developed into miners of the cortex of the twig, such as the European *Zimmermannia* and the North American *Ectoedemia*. The genus *Psacaphora*, which in our region lives in leaf-mines, has a North American species which also occurs in the cambium of Salix. We thus find the most varied instances of intermediary stages between mine-producing and gall-causing habits.

On assessing the significance of these varied habits, we reach the conclusion that:

Leaf-mines are the most primitive forms of leaf-galls, while leaf-galls represent a highly specialised type of leaf-mines.

Leaf-galls will not, of course, have been derived from leaf-mines in all cases; gall-causing habits are just as polyphyletic as mine-producing habits; however, this rule can certainly be applied to a large number of these structures.

Finally, it is necessary to point out one peculiarity of mines in relation to the new cell growths which can be found inside them. They practically never contain a genuine cork tissue. We know that when a plant is injured, it frequently reacts by the formation of cork in the wound. Only Lüders (1900) claimed that cork tissue is formed in the portion of the mine channel of *Phyllocnistis saligna* Z., which runs through the cortex of the twig. The cork normally formed when parts of a plant are injured, does successfully enclose the injury but, as with all cork tissue, isolates the area from its surroundings. In mine channels this is never the case; what is necessary here is to fill out the cavities and to link up the tissues which remain. This can be achieved by the thin-walled callus cells but never by an isolating cork tissue.

The close relation between mines and galls gives added interest to an isolated case, which has only once been observed, of the occurrence of a mine on gall tissue. The branching linear mine of *Liriomyza strigata* Mg. was found on the lepidopterous gall of *Gnorimoschema cauliginella* Schmid on a stem of Silene (plate 2, Fig. 5). When we recall that callus cells are found particularly frequently in the mines of just this species, it need cause no surprise, if in this case *L. strigata* found the gall tissue, which in many respects resembles callus cells, specially palatable. The mine was confined to the area of the gall and did not continue into other parts of the stem. Consequently it remained much shorter than is normal for this species. Buhr (1941) also observed the occurrence of the mine of *Ophiomyia alliariae* Hering on the galls of *Ceuthorrhynchus chalybaeus* Germ. on Alliaria. A preference for gall tissue by mining species may well be much commoner than has so far been observed.

It will also be of interest to mention certain combinations of galls and mines, as can be seen in *Chirosia parvicornis* Zett., mining on Pteridium. The larva mines in the tip of a frond and does not mine only in the actual leaf-blade but also feeds in the mid-rib, on its lower side. This vein is unable to continue its growth on its lower side, while the upperside, which has remained intact, grows normally. This causes the upper part of the vein to become longer than the lower side, with the result that the tip of the frond curls downwards. The larva continues its mining in this curled portion of the frond of the fern and enjoys there greater protection; it is not so easily visible in this structure, as if it was mining openly in the surface of the leaf. This leaf-roll caused by the feeding of the larva is considered as a genuine gall, and inside it the larva mines in the same way as those of all other mining flies. In this case mine and gall are strangely enough combined. Such combinations of gall and mine have so far only been discovered in very few cases and represent a particularly high degree of specialisation by miners.

16.

THE SUBSEQUENT FATE OF THE MINE

Our earlier definition of a mine showed that it represents both the living and feeding quarters of an insect larva. When larval development is complete and the larva has ceased feeding, at least one function of the mine comes to an end; it now no longer serves as feeding quarter. In all cases where pupation takes place inside the leaf, its second function as living quarter remains; however, when the larva leaves the infested part of the plant in order to pupate, the mine ceases to have any further connection with the insect which produced it. It now frequently undergoes fundamental changes, which may considerably alter its appearance. This is especially important for the determination of species. An abandoned mine may alter so radically that the identification of its producer causes some difficulty. Considerable experience of minology is essential to undertake the necessary identification. It is therefore important for anyone wishing to identify mines to know what changes occur in mines after the larva has completed its feeding. Two different possibilities must be taken into account.

1. The larva pupates inside the mine; the mine cavity is only abandoned on the emergence of the imago. In such cases it would appear natural for the mine to be more strongly constructed than in species which transform outside. This is effected primarily by the use of the silk inside the mine cavity. Many species, especially among Lepidoptera, utilise silk throughout the greater part of their lives. Spinning is frequently intensified prior to pupation and a special pupal site may be constructed largely of silk, in which the pupa is better protected than in the simple mine cavity. An analogous process is the special overwintering cocoon constructed by some species, which often has a very characteristic shape, such as the circular cocoon of *Atemelia torquatella* Z. (Fig. 58). In many species of *Lithocolletis*, which are otherwise difficult to distinguish from their mines alone, the structure and colour of the cocoon in which the pupa rests, is important for identifying the species. The imago emerges either through the upper or the lower surface of the leaf. The empty pupa case can frequently be found projecting through the leaf epidermis in mines from which the imago has only emerged a short while before. Later, it becomes detached and it is necessary

to search the mine for the actual exit hole. In some mines the early channels become exposed as a result of growth changes in the leaf even when pupation has taken place in the mine; this, however, has no harmful effect on pupal development, as can be seen especially in *Rhynchaenus* species. The younger the state of the leaf when it was first mined, the less remains of the original linear mine. The torn areas are in such cases a characteristic feature of the leaf and the minologist who is familiar with them will be able to link them up without difficulty with the species of their producer, even if the mine itself was never completed.

2. The larva leaves the mine and pupates outside. In such cases it is natural that the larva takes less trouble to ensure the preservation of the mine. This applies to the majority of mines, in which during the course of time changes occur, making exact identification extremely difficult. The thin epidermis above the mine, which is no more than a fine skin, or maybe the outer wall of the epidermis, which is all that remains with sap-feeders — these are soon destroyed. Dampness especially leads to the rapid decomposition of this outer wall of the mine and as soon as it is broken at any point, damp air and rain penetrate inside the mine and produce further disintegration. In the first instance, the frass line becomes less distinct and thus characteristics important for identification vanish. The beginner in the field of minology will at first find it far more difficult to identify mines in which the frass has been washed away, until he has developed the necessary experience. The actual time of construction of the mine is important in cases of this type. Mines constructed in early spring usually have a conspicuous and striking frass line when they are fresh; however, the frass contains relatively little hard and durable matter, since the cells which the larva has eaten have only become slightly thickened and woody. In such cases the frass is washed away more quickly than in autumn mines, when the excrement contains much indigestible matter of a more lasting nature. For instance, the linear mines of *Phytagromyza xylostei* R.-D. (see Fig. 57) which can be found commonly in early spring on Lonicera and Symphoricarpos are distinguished by their broad frass line which is a beautiful dark green, and is made up of pulpy lumps of frass. However, after only a short time the mine loses its striking dark-green colour and appears much more whitish; the darkest grains are preserved somewhat longer but then they too disappear. Similarly, the mines of *Liriomyza congesta* Beck. which are so common on almost all Leguminosae very soon lose the bright green colour, which is so characteristic of their fresh condition. Another peculiarity which causes difficulties in identifying abandoned mines is the process of curling and withering which develops in leaves with so many species. This occurs especial-

ly in cases where the larva mined the leaf in a very early stage of its development and in which no real strengthening of the remaining cells has occurred. These soft parts of the leaf which surround the mine collapse as soon as the larva leaves the mine, since the existing turgidity is lost through evaporation. A collapse of the mined areas of the leaf can be seen especially in species of Tenthredinidae mining in early spring, in *Fenusella* on Populus, in *Hinatara recta* Thoms. on Acer platanoides L. and in *Parna tenella* Klug. on Tilia. In species mining later in the year tissue is already so much stronger that a collapse of this sort no longer occurs, as in *Phyllotoma* species on Acer, even though in this case a large portion of the leaf substance is eaten away. In general the mined areas of the laef are more likely to collapse, as the size of the mine increases, producing a correspondingly greater loss of water; it follows therefore that more blotch-mines than linear mines become unrecognisable in this way. However, if very young parts of the plant have been attacked, even very smal mines such as the blotches mined by species of *Coleophora* break open or are covered in by the collapsing parts of the leaf. The linear mines of *Agromyza flaviceps* Mg. of both generations in leaves of hop (Humulus lupulus L.) remain visible in the leaf in a good state of preservation for a considerable time; *Agromyza igniceps* Hend., which likewise mines on Humulus, however, and occurs much earlier, is characterised by the way the mine channels break open, since in this case the tissues are still much more delicate. Frequently the earliest stretch of the mine is completely obscured by such tears, while the larva is still mining in the leaf. *Elachista luticomella* Z. also mines in the earliest days of spring in the leaves of various grasses, especially in Dactylis. The long, narrow linear mine runs first along the whole length of the margin of the leaf and only later does the larva proceed to the centre of the blade. The original mine channel completely breaks away from the leaf, the edge of which now appears to have been milled along its whole length. These milled edges on leaves of Gramineae show the micro-lepidopterist in which plants the larva should be looked for. In this case we see that even the absence of the mine from the leaf may serve as a guide for correct identification of the species in question. Frequently the whole mine falls away from the leaf in other species, especially when the mine is very small and when necrotic areas have developed in the leaf as a result of the mining; these areas then break away from the plant in their entirety. This is frequently seen in the early mines of *Bucculatrix crataegi* Z. on Rosaceae, especially species of Prunus; on the same plants the incompletely developed serpentine mines of *Nepticula prunetorum* Stt. fall out of the leaves in the same way and can be recognised by the characteristic hole which remains. Larvae

Fig. 162

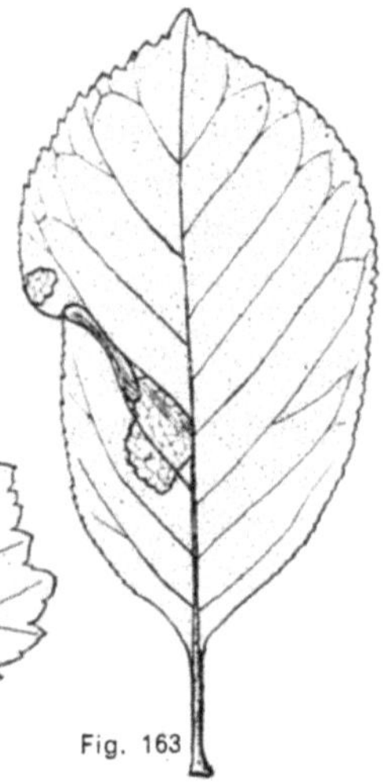

Fig. 163

Fig. 162. Betula: *Agromyza alni-betulae* Hd., larva eaten out from within the mine by bird's beak-cut.

Fig. 163. Prunus cerasus: *Lithocolletis cerasicolella* HS., than inhabited by *Capua reticulana* Hb.

in mines are also hunted by natural enemies such as ants and birds, which eat out the whole area of mine containing the larva. Generally this is done only at the exact spot where the larva is located. Since the larva is most easily visible when approaching maturity, cut out areas of this type are usually found at the end of the mine and anyone commencing the study of mines might be inclined to assume that the larva itself had constructed this cut out area, as is done by species of *Antispila* and *Heliozela* and by some species of *Rhynchaenus*. Such mines are often very conspicuous with an irregular area cut out of the leaf, especially in *Agromyza alni-betulae* Hend. on birch (Fig. 162). When ants eat out the larva of *Agromyza nana* Mg. on leaves of Medicago, it can frequently be found that the whole mine channel which expands into a blotch is eaten away, so that the resulting hole preserves an outline of the complete mine.

The callus tissues, which were discussed in detail in the previous chapter, also often cause extensive modifications, deformities and rents in the mined leaves, which radically alter the normal appearance of the mine. They develop in such profusion that they split apart the remaining parts of the leaf and thus produce the most varied deformities. Uncritical observers then believe such structures to be galls, while they are in reality mines which were long previously abandoned by the insect larva, before the deformities developed.

The abandoned mine subsequently offers a welcome hiding-place to numerous other insects which have no direct connection with the mines but which show a liking for these protective retreats. We shall discuss this further in the chapter on inquilines and parasites; here we need merely mention that these later inhabitants can substantially alter the original appearance of the mine. This occurs especially when the larvae of Tortricidae or Gelechiidae settle in the mine and from here construct a web over the surface of the leaf, beneath which they commence "window-feedings" (Fensterfrass).

There are some miners which have such habits, so that in such cases one might expect to find these mining species. For instance, *Gypsonoma oppressana* Tr. lives in the autumn in a very short mine, from within which it later constructs a frass-studded web over the lower surface of the leaf, beneath which it commences window-feeding. This method of feeding, illustrated in Fig. 163, cannot, however, be considered as mining. *Lithocolletis cerasicolella* HS. constructs the familiar lower surface tentiform mine in the leaf. Long after the moth has emerged and left the leaf, the larva of *Capua reticulana* Hb. takes up residence in the empty mine and from here it feeds in the same way as is normal for the *Gypsonoma* larva on Populus. The *Capua* larva does not thereby become a miner; it is extremely polyphagous and lives not only in the most varied leaf habitations which it webs together with silk but also in green, as well as ripe berries of various species of Ribes; it is thus not in any way selective. The minologist who later finds such a remarkable tentiform mine, with window-feeding taking place under a web of silk, may be caused a quite considerable headache in attempting to identify it. He must therefore constantly bear in mind what may subsequently become of the mine, after it has been abandoned by its producer.

17.

INSTINCT CHANGES IN MINERS

The mining habit represents a degree of specialisation, which among phytophagous insects is only exceeded by gall-producers, and also by parasitic insects. We have seen how this specialised habit is linked with a whole series of morphological and physiological adaptations, and that modifications have occurred to entire organs, rendering any other type of life impossible for the insect concerned. Despite these extensive modifications, the mining habit has not with every species become rigidly automatic, permitting of no flexibility in accordance with changes in external influences. Even among miners we find that the life patterns assumed by each species may vary when new external factors arise. Detailed study is constantly bringing forth fresh evidence to prove that within certain limitations the fixed instincts of mining larvae may be changed. A wide field of activity lies open here for the minologist and if the variations in mines of the same species are carefully observed, there will be constant opportunities for noting adaptations to changed environmental influences even in the living conditions of miners.

Naturally, instinct changes of this type will be most frequently encountered among species whose life in mines has not yet become as essential as in the most highly specialised forms of this ecological group. The most frequent deviations from the customary habits of mining will be found particularly among optional miners. It will also be possible to discover numerous examples of transitions towards mining, which have as yet remained undetected. These may be able to throw some light on the varied origins of mining as a type of existence. There will be a number of cases where a transition to mining may be anticipated, even where it has not yet occurred. The larvae of the Noctuid *Eublemma noctualis* Hb. produce a feeding pattern on the leaves of Helichrysum which resembles a mine, although it is not actually one. They consume the whole mesophyll of the leaf but leave the hairy appendages of the leaves, so that outwardly these retain their normal shape. However, they do not produce genuine mines, since they also eat the epidermis, only leaving its covering of hairs. It is reasonable to assume that on occasions they will start leaving the epidermis cells which bear the hairs and so become genuine miners. The mining habit is of such relatively recent origin with case-bearers that especially here many cases of transitions to other types of existence may be discovered. For instance, the *palliatella* group of *Coleophora* show particularly clearly how the transition occurred step by step from leaf perforation to "window-feeding" and then to genuine mining, both in individual species and also within single species. In *Scoparia illota* Philp. (New Zealand) the final adoption of the mining habit has not yet been effected. According to Chappel (1929) this species mines at night in the leaves of Polypodium; sometimes, however, it spins leaves together and beneath this protective web eats the epidermis also. We know that species of *Scoparia* live primarily on mosses and lichens; this species has transferred to vascular plants and adopted mining but with only such a slight degree of adaptation that it does not always adhere to its newly acquired habits. *Apterona crenulella* Brd. has already become a definite miner, producing its blotch mines in the same way as the genus *Coleophora.* If the characteristic cases shaped like snail-shells are not found, the mines can scarcely be distinguished from those of *Coleophora;* however, the opening in the epidermis through which the larva penetrates into the leaf, which is exactly circular in *Coleophora,* in this other instance is less regular. Feeding areas of *Apterona* have been found on the downy lower surface of one of the leaves of Onychocarpus in which no well-defined point of entry had been formed; the larva had stripped off lengths of the epidermis and then consumed the parenchyma below. Such areas can no longer be considered as actual mines. In this case the insect returned to what was probably its original method of feeding, presumably as a result of the hairy texture of the plant. Thus, on the one hand one finds transitional stages towards mining as a method of feeding in species which have not adapted themselves irrevocably to the mining habit, while on the other hand there is a return to the more primitive external feeding in cases where mining represents the normal rule.

Where adaptation has not developed so extensively, it is possible to find examples where larvae visit "natural" mines and construct "artificial" mines. We term artificial mines the structures resembling mines, in which the epidermis of the leaf is also consumed but where the larva attempts to achieve a similar degree of protection as in true mines by constructing a web in place of the missing epidermis, beneath which it lives in the same way as in a mine. In the chapter on mines on aquatic plants we already mentioned that some Tendipedidae gnaw channel-like furrows above which they weave silk and beneath which they then live in the same way as the many other Tendipedidae whose habits we have described. Meanwhile the original true mining habit may be continued by such methods of feeding, as for instance with *Gypsonoma oppressana* Tr., which is so common in autumn on Populus, and whose larvae first live in a short linear mine but then emerge from the inside of the leaf and commence window-feeding beneath a frass-studded covering of silk on the underside of the leaf. In this case the early genuine mine is continued as an "artificial" mine, which can only be recognised as such on close examination. On the other hand among the species which are not so rigidly adapted to mining, there are some which feed in "natural" mines. They seek out natural cavities in the plant, in which they feed and sometimes remain for longer periods than would be expected from their normal life cycle. Such habits have been observed especially among Thysanoptera. H. Buhr discovered these insects in considerable numbers in clefts of the tissues in leaves of Digitalis. These gaps which are always present in the normal leaf were not visited merely temporarily for protection, but Buhr also found regular deposits of frass which indicated that genuine feeding had been taking place in these "natural" mines. The occurrence of two species of Tortricidae in a "natural" mine which was once discovered on Tragopogon is especially significant. The epidermis above the mid-rib in the leaves of this plant frequently does not tightly enclose the tissues, so that a cavity exists. The larvae of *Cnephasia chrysantheana* Dp. and *Argyroploce lacunana* Dp. developed in this cavity and they only left these natural mines when they were full-grown, in order to pupate outside. This does not represent an individual case, for numerous leaves containing larvae were found in the area. The mines formed in this way were scarcely visible externally, since the larvae had only fed in the neighbourhood of the mid-rib itself but not to either side of it; they merely appeared somewhat lighter than the remaining parts of the vein and betrayed their presence by the ejection hole for frass (Plate 2, Fig. 6). While this habit may not appear particularly surprising in the case of the *Cnephasia* larva, since it normally always lives in a small mine while it is young, it is nevertheless very strange with the *Argyroploce* larva, which from its infancy onwards is an external feeder. With both these larvae, however, it is most unusual that they should become life-long miners. Apparently the new type of shelter suited them so well that they gave up the rather more difficult work of feeding between the spun together leaves. Since the minute larvae were extremely difficult to detect in these mines, it can be assumed that such changes of instinct are even commoner than has so far been discovered. We should at this point also mention feeding in secondary mines, which can also be found among Thysanoptera, when the larvae enter mines which were constructed by other species and have already been abandoned by them. In this way they were found in old *Dizygomyza* mines on

Gramineae. In these they constructed small, circular openings through which they ejected their frass; the presence of these openings in the epidermis, which appeared strange in a dipterous mine, attracted the attention of a minologist.

The necessity for a change in instinct occurs in the numerous cases of "faulty" oviposition. We know that in many miners the type of ovipositon remains constant and we also know how this may influence the life of the future mining larva. In most species of *Rhynchaenus* the egg is laid in the mid-rib near the apex of the leaf, where the largest part of the mine is also usually formed. In this area the leaf tissues are less interspersed with strengthening elements than at the base of the leaf, and the leaf substance is therefore softer. It is found that the larva's development in *Rh. quercus* L. and *Rh. fagi* L., for instance, becomes all the more difficult, the further away from the apex of the leaf the egg was laid. Mines situated in the immediate vicinity of the leaf-base usually have the beginning of a channel but the larva is unable to complete the mine; it is unable to cope with the tough consistency of its food. The intermediate areas indicate the possibilities of adaptation which are open to the larva, with a more or less complete development of the mine even with the different type of food. Similarly, if the egg is deposited on the wrong surface of the leaf this may prove a decisive factor in inducing different habits in the larva. For some species this is without importance, for instance, for the polyphagous *Phytomyza atricornis* Mg., whose larva develops equally well whether the egg was laid on the upper or lower surface of the leaf. For other species this is quite impossible. *Dizygomyza posticata* Mg., is able to develop only on the upperside of the leaf. If an egg has been wrongly inserted into the tissue of the underside, an unnatural looking mine arises, which is never completed; the larva dies as a result of the unaccustomed food. We know that especially for the genus *Lithocolletis* the leaf surface on which the mine is formed is always consistent and characteristic. Even with Populus on which mines of other insects are able to develop indiscriminately either on the upper or lower surface, as was described in the chapter on food selection, it can be observed that *Lithocolletis* occurs only on the lower surface, even when in profusion, with 50 or more mines of this species on a single leaf. As an exception one sometimes finds that the female makes a mistake in the side of the leaf when ovipositing. This is found most frequently with *L. emberizaepennella* Bché. on Lonicera and Symphoricarpos. When this occurs, an unusual-looking blotch-mine is formed on the upper surface of the leaf, which the minologist finds it difficult to account for. The folds in the epidermis, by which the mine can be recognised with certainty, are spun relatively late.

The larva invariably perishes. This also sometimes occurs in the leaves of Alnus with *L. rajella* L. It normally lives on the lower surface; occasionally, however, upper surface mines are also found, which recall those of the normally upper surface feeder, *L. stettinensis* Z., but can be distinguished by their greyish-white, not so yellowish-green, colour. In this instance also, development to the pupal stage has never been observed. In none of these cases are the insects able to develop the change in instinct necessary for finding their normal food in the leaf. *Callisto denticulella* Thbg. behaves quite differently in this respect. Its early mine is frequently found on the upper surface of leaves of Malus; later, after leaving the mine, the larva lives beneath a folded-over leaf-edge. In some years and in various localities one may find considerable numbers of mines which are constructed on the lower surface of the leaf. These mines also have a very different appearance from normal ones and are difficult to recognise as belonging to this species. In this case, however, development continues normally and it appears that no disadvantages ensue for the larva from the abnormal habits in the first instar. This is possibly connected with the fact that the larva does not need to spend its whole life as a miner; its adaptation to mining has not developed in such a specialised form, as is apparent from the difference in leg structure from the larva of *Lithocolletis*, which we discussed earlier.

With the same type of oviposition on the respective leaf surface, some significance must be attached to whether the egg is laid on the margin of the leaf or more towards the centre. For many species the selection of the oviposition site is constant and characteristic; in some cases and with certain species, however, one finds that the place chosen for oviposition is not constant, causing a variation of the mining instinct in the resulting larva. When *Nepticula atricollis* Stt. has laid its egg on the margin, the mine channel largely follows the edge of the leaf and is relatively straight. However, when the egg has been deposited more towards the centre of the leaf, the mine has a very different aspect. The mine channel lies in a tight spiral like the coils of an intestine, and only later when the larva is larger, is this characteristic given up. The explanations of these variations in methods of constructing the mine is provided by the leaf venation. On the margins the veins are small and weak, so that the larva has no need to avoid them. In the centre of the leaf considerable difficulty is encountered in crossing the much stronger veins and this is therefore avoided. Thus, in this species a change in the feeding instincts of the larva enables it to adapt itself to the consequence of variations in the site of oviposition.

There are numerous cases in which changes in the feeding instinct in the same individual during the different stages of its development are very considerable. The most striking examples can be seen in the species of *Coleophora* of the *lixella* group, which grow in the receptacle or seeds of Labiatae but then after hibernation, and sometimes even previously, adopt the quite different habits of miners on Gramineae. Here, living conditions on the various food-plants are very different indeed; that normal development is possible at all results from the fact that this change of host has a firm ecological basis going back for untold generations and has become a regular part of the life cycle. This is analogous to the different forms of the same species of parasitic fungus which can be found on such very different host-plants. The change of instinct which is necessary in this case is already firmly implanted and is fixed by heredity. A similar thing can be found in some other species of *Coleophora.* The minute larva of *C. serratella* L. mines on various Rosaceae, producing small blotches in the leaf in autumn and then firmly attaches itself to the twigs for hibernation. In early spring it frequently no longer mines (especially on Malus) but eats out the buds. Only when the buds have sufficiently developed, does it again commence mining. It apparently has no difficulty in changing its instinct in this way. The internal tissues of the buds appear to provide it with food which so closely resembles that of normal leaf parenchyma that the change of instinct creates no problems. In other species the same sort of thing is frequently found after completion of hibernation. Often of course mining is given up altogether and replaced by a different method of feeding. This is what occurs with the species of the *palliatella* group of the genus *Coleophora.* As an example, we may quote the habits of *C. betulella* Hein.-Wck., as described by Sich (1921). The young larvae live in autumn, mining in leaves of Betula. However, the frass does not remain in the mine, as with other members of the genus in their first instar but is ejected outside by the larva, where it is retained by a few threads of silk. In this way a small cone of frass is formed, which is made into the first larval case, which the larva later enters and carries around with it, now mining frass-free blotches with a circular hole in other areas of the leaf in the normal manner of *Coleophora.* After hibernation the mining habit is entirely abandoned and the larva commences window-feeding in the leaf, in which only a single epidermis remains. In an exactly similar way about 60% of the young larvae of the saw-fly *Hoplocampa testudinea* Klg. are miners in their first instar. Subsequently they bore inside the flesh of the apple. In this species the mining instinct is not so firmly established that no further changes are possible.

The specialised type of existence may be subject to change within certain limits even within a single species. As an example *Lithocolletis klemannella* F. which forms lower surface mines on Alnus may

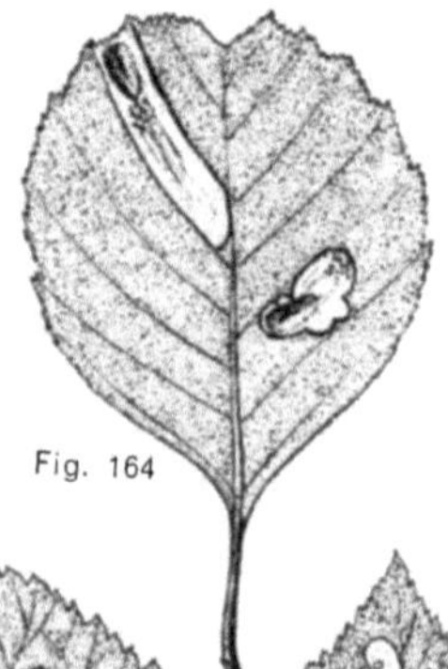

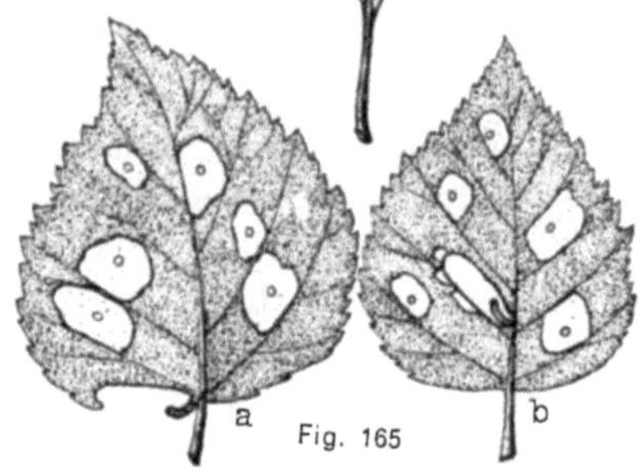

Fig. 164. Alnus: *Lithocolletis klemannella* Fb. Left, a lengthy, shallow mine; right, a short, completely transparent mine; each has a pupal chamber at the end.

Fig. 165. Betula: *Coleophora fuscedinella* Z. with leaf cut-out for production of new case (old case attached). a. normal cut-out on margin of leaf. b. Abnormal cut-out in the centre of the leaf.

be mentioned (Fig. 164). The mines of this species are usually relatively short and of considerable depth and when held to the light they are therefore very transparent. Sometimes abnormal forms are found which extend to great length between two veins; these are shallower and in transmitted light appear more greenish. They then have an extraordinary resemblance to the normally lengthy mine of *L. froelichiella* Z. and can only be distinguished by the expert by the fact that they do not commence in the angle of the lowest vein, as in the last species. Clearly the parenchyma of the lower surface is preferred by the larva; this is not eaten exclusively, merely because the larva desires to avoid having to cross the tougher veins. Near the apex of the leaf, where the veins are no longer so strongly developed and where all the leaf tissues are softer, the mine naturally develops in a quite different way. It is quite understandable that the differences may be even more marked with optional miners. On hatching from the egg the young larva of *Laspeyresia molesta* Busck first seeks the softest possible tissues into which it burrows, showing a special preference for young shoots. If these are not available in the immediate vicinity, it mines in the petioles, in order by this means to reach the shoot. Some degree of instability in the feeding instincts also occurs in the genus *Coleophora*, which we have assumed on a number of occasions to have adopted the mining habit only in recent time. *C. serratella* L., *C. hemerobiella* Scop. and *C. anatipennella* Hb., which live in leaf-mines on Malus, have all been found in mines in fruits. Voigt even noticed the former of these eating seeds, in the same way presumably as its ancesters once did and as is found in numerous species of the genus even today. The manner in which the larva's case is cut out also frequently shows considerable flexibility. In *C. fuscedinella* Z. this normally occurs as follows: the initial case is attached to the edge of the leaf, an area of the leaf is mined and then this mined portion is cut out to form the new case, leaving the old case still attached to the leaf (Fig. 165a). Sometimes the larva manages to effect this whole operation in the centre of the leaf (Fig. 165b). The species adopts strange habits when it occurs on Betula nana L. The leaves of this plant are particularly small and when the larva wishes to leave its hibernation case, which was constructed in the autumn, it is unable to find a sufficient number of large leaves in the following spring to prepare its final case, as it does normally after hibernation. It therefore interpolates an intermediate stage (Fig. 166, 2), a short case which it carries only until the leaves have become large enough for one of them to suffice for the construction of the final case. The production of such a case, which must be considered as falling outside the normal plan of the insect's life,

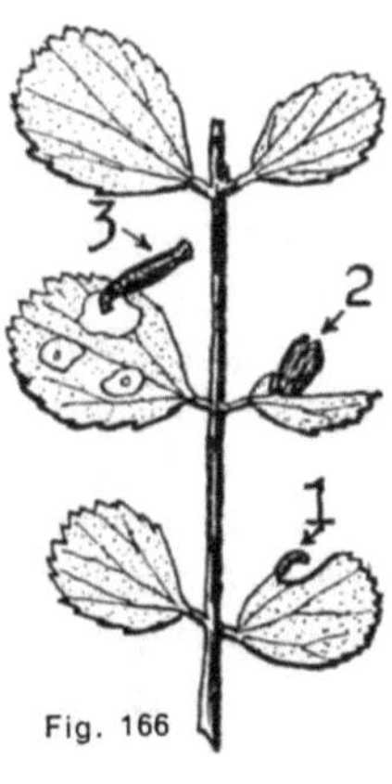

Fig. 166

Fig. 167

Fig. 166. Betula nana: Owing to the small size of the leaves the larva of *Coleophora fuscedinella* Z. has constructed after its hibernation case (1), not as usual its final case (3), but first of all an intermediate case (2).

Fig. 167. Betula: Forked channel of *Rhynchaenus rusci* Herbst.

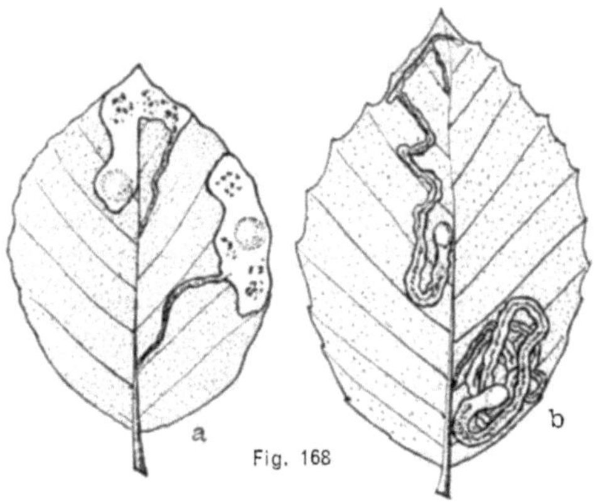

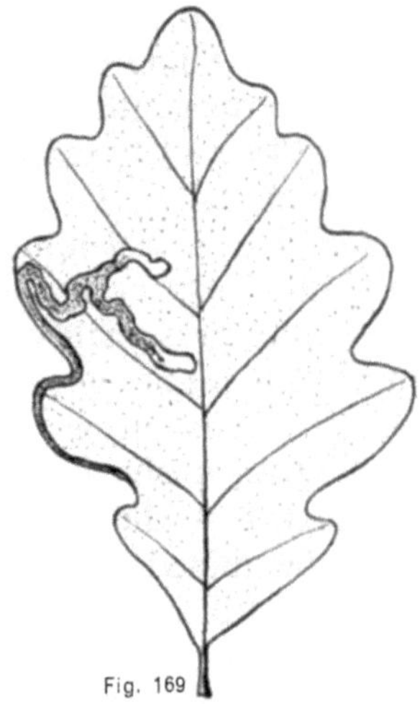

Fig. 168. Fagus: *Rhynchaenus fagi* L. a. Normal linear-blotch mine. b. Abnormal linear-mine which has been formed while pressing for preservation.

Fig. 169. Quercus: *Nepticula samiatella* Z., with abnormally forked channel.

indicates a capacity for changing its instinct which is only rarely encountered elsewhere.

A number of changes in the instincts of mining larvae may occur under the influence of external stimuli. We mentioned previously that some species, which normally mine leaves, become stem-miners if the leaves — especially around the apex of the shoot — are too small to ensure adequate nutriment for the larva. This does not appear to present great difficulties to a species which is as polyphagous as *Liriomyza strigata* Mg. and Buhr in fact found it stem-mining in this way on a wide variety of Compositae. However, even species with more specialised forms of adaptation, such as *Phytomyza gentianae* Hend. possess a capacity for change in this respect. While they normally construct long blotch-mines beginning in the shape of a star in the leaves of Gentiana, when occurring in the small leaves of Centaurium they find the space inadequate; they are then able to develop as complete stem-miners. The minologist discovers a particular form of instinct change when drying and pressing between sheets of paper mined leaves which still contain larvae. In such cases remarkable changes in the type of feeding occur, which sometimes make the mines unrecognisable. For instance, with *Rhynchaenus rusci* Hrbst. on Betula, whose channel always uniformly follows the edge of the leaf, one finds that the channel forks (Fig. 167). The mines of *Rhynchaenus fagi* L. and of *Rh. quercus* L. become quite unrecognisable;

in this case the mine does not extend at its end, as is normal, into a large blister but remains linear throughout, with numerous branches or coils, and at the terminal portion of the channel a pupation cocoon is even constructed (Fig. 168); a mine pattern is thus formed which can only be ascribed to the correct species after considerable experience. In the same way blotch-mines of *Pegomyia nigritarsis* Zett. when drying under pressure turn into long linear mines. A plausible explanation of this change in mine construction appears difficult to find; we are faced with the basic problem as to what even under normal circumstances leads to a differentiation between linear and blotch-mines. Branching of this type occurs also in the wild state, without any cause being immediately apparent. While the species of the genus *Nepticula* normally construct mines which never fork or branch, very rarely a single individual of some species will be encountered which produces a forked channel, as is illustrated in Fig. 169.

Finally, certain cases of deviation in normal instinct occur, for which any external influence can apparently be excluded. As a rule one never finds a *Nepticula* crossing the mid-rib of a leaf, except possibly at the extreme tip of the leaf. We know that the numerous strengthening elements contained in the mid-rib make it very difficult to cross. However, as a rare exception we do find these species sometimes crossing the mid-rib. It was once observed in *N. microtheriella* Stt. in a leaf of Carpinus and also in *N. ulmicola* Hering in a leaf of Ulmus. In both cases the mid-rib was crossed by the very young larva, which makes it all the more remarkable. What induced the larva to take this action remains obscure; its development nevertheless followed its normal course. In this connection it might perhaps be mentioned that the mines of *Lyonetia clerkella* L., which closely resemble those of *Nepticula* species, very frequently cross the midrib when the larva is in its first instar. Sometimes, overcrowding in a limited space takes place in species where it does not normally occur. Thus, on one occasion a distinct serpentine mine was found in *Nepticula luteella* Stt. on Betula, thereby rendering it very similar to that of *N. hübnerella* Hbn. on Crataegus; in this case, however, the spiral did not coil around externally from the inside but internally from the outside, so that the coils became shorter and shorter as the size of the larva increased. In this instance the larva had been parasitised and was in any case unable to reach maturity.

The possibilities of instinct changes are also frequently expressed in the selection of food; only in this case not the miner itself but the ovipositing female is involved. In this connection we wish to refer once again to the organoxeny and xenophagy, which have already been discussed in detail. Disjunctive oligophagy especially presupposes extensive possibilities of instinct change. A change in host-plant can in its turn lead to changes in the construction of the mine. *Nepticula ulmivora* Fol. in the Berlin Botanical Gardens also accepts the genus Zelkova which belongs to the same family. When it occurs on this genus of Ulmaceae, which is nevertheless very different from its normal food-plant, one finds very frequently that the mine channels are forked. It can therefore be assumed that the differing nature of the food-plant was detected by the larva, which attempted by branching its channel to escape from these areas of unaccustomed food.

If we survey the whole field of miners, we find that their ability to react is substantially greater than that of gall-producers and parasites, which are limited to a narrowly circumscribed type of existence. However, even with miners their ecological potentialities are insufficient to guarantee the development of the miner to the adult stage when the ovipositing female has made real mistakes.

18.

EFFECTS OF TWO MINES IN ONE LEAF. OCCURRENCE IN PROFUSION

While the female of external feeders can lay a considerable number of eggs close together without thereby endangering the life of the young larvae, since on hatching they scatter over the whole of the plant or even further, it is generally not possible for miners to deposit eggs close together in this way. The young larvae rarely if ever undertake any considerable movement on hatching but usually bore into the leaf straight from the egg. If a large number of eggs were deposited side by side on a leaf, this would not provide the numerous young larvae with adequate possibilities of development. Therefore, in mining insects the eggs are usually laid singly, one to each leaf or at least on different parts of a large leaf. There are some species where the female deposits several eggs together but in these the larvae subsequently live communally in one large common mine; they are mostly species which feed on large leaves. For instance, the Anthomyidae when ovipositing usually lay several eggs together, which are attached close to one another on the underside of the leaf and not inserted into the leaf; in this way the beginning of a mine of an Anthomyid can easily be distinguished from that of an Agrymyzid, which inserts its eggs into the leaf. A similar type of oviposition is found with the species of *Chylizosoma* which mine on Liliaceae and Orchidaceae. In both cases the eggs are deposited in strictly parallel rows. The majority of miners, however, deposit their eggs singly, with usually one egg in or on each leaf, thus ensuring under normal conditions sufficient food for the young larvae. This fact alone makes it possible for the mine to maintain a quite definite shape and formation; if mines always occurred in numbers on each mined leaf, it would be impossible to distinguish one from the other, since the entire leaf substance between upper and lower epidermis would be eaten away. No individual characteristics could be ascribed to different species — there would be no minology. Reaumur defined the characteristic feature of mining insects: "they live in great isolation". The fact that in one leaf only a single mine is found makes it possible to work out distinguishing characteristics of mines and miners.

If we consider this isolation as an essential characteristic of the mining habit, we must pay some special attention to cases where exceptions to this principle are the normal rule or which can be considered as anomalies. We must therefore discuss the cases in which several mines are found in one leaf. Two groups can be immediately distinguished: in one case members of the same species are concerned, leading a joint life in a common mine or for other reasons meeting in the same leaf; in the other case different species form mines which meet in the same place. In both cases it will be of interest to discover the mutual behaviour of the two parties in such an encounter.

1. Encounters between members of the same species.

Under normal circumstances this will apply to species, whose larvae are gregarious. Solitary miners of the same species will only rarely be found with more than one in the same leaf. The ovipositing female seems to be able to detect with great accuracy whether any given leaf has already been laid on by another female; if this is so, no further eggs are deposited. However, if several or even numerous mines of the same species do occur in a single leaf, there are two possible explanations. Either a number of eggs were deposited by a single female — which might happen where there was an urgent need to lay, when the female had been perhaps for a long time fruitlessly seeking a suitable food-plant, or different females had each successively laid one egg on the leaf in question. This latter possibility seems to be the more frequent. We have already referred to the larvae of the Anthomyidae and the *Chylizosoma*, which usually live gregariously; the leaves of Polygonatum, of Liliaceae and Orchidaceae which they mine are usually large enough to provide all the larvae with sufficient food. In addition, the Anthomyidae possess the ability to change their mines; thus in a case where the first leaf does not prove adequate for all, some of the larvae can move and find another leaf. A number of grass-mining Agromyzidae also normally live gregariously, although they are not able to change to other mines or leaves. In such a case the ovipositing female must judge in advance whether the selected leaf will suffice for the complete development of the entire batch of larvae and on our present information errors of judgement in this respect are not made. The number of larvae feeding in such a communal mine vary considerably. It largely depends on the width and length of the blade of grass concerned. A number of species of *Dizygomyza* invariably feed gregariously in this way and they also pupate together in the mine, each puparium being attached with a weak strand of silk. However, even among species of *Agromyza* which pupate outside the mine, several species live gregariously in the leaves of Gramineae. It is interesting to observe how all the larvae move forward in a row feeding as they go, usually working down from the apex of the leaf towards the broader leaf-base, so that as they increase in size, even though packed close together, they nevertheless find sufficient space in the broadest part of the leaf. If one falls behind as a result of being parasitised or diseased in some other way, once behind the others it is unable to continue feeding much longer and soon perishes. *Rhamphus* species do not actually live gregariously; each individual constructs its own mine but apparently a single female often deposits numerous eggs on one leaf. In this case the mines are sufficiently small to ensure that the larvae do not get in each other's way, and even in *Rh. oxyacanthae* Mrsh. in leaves of Crataegus it is rare to find mines coalescing, even though they may be present in numbers; if this did occur it would of course have a harmful effect on the larvae's future development.

In all these cases the gregarious habit is in no way optional, as it is a direct result of oviposition. This is not, however, always the case; there are species where a distinct propensity for gregarious mining can be observed, For instance, the larvae of *Atemelia torquatella* Z. in the leaves of Betula and Ulmus are invariably present in some numbers in each mine, although they are able to change to another mine, and space for mining sometimes becomes very limited. This is seen even more distinctly with the young larvae of *Hyponomeuta malinella* Z. and *H. padella* Hbn. The eggs are deposited in heaps on the cortex of the twigs in autumn, where they overwinter. When the young larvae hatch in spring they immediately burrow into the leaves which are just commencing to unfold and produce small blotch-mines, before joining forces to construct a communal nest or web. However, this inclination to gregariousness which later develops so distinctly can already be detected in the original mines, where in many cases several larvae enter a leaf at the same place and live together in a communal mine.

Many cases arise in which several females appear to have oviposited successively on a single leaf without any obvious explanation as to what particular attraction of the leaf induced such abnormal behaviour. It is difficult to assume that their powers of detecting previously deposited eggs were so weakened by some unknown factor that they failed to notice the activities of their predecessors. It must rather be assumed that the leaf in question has such an outstanding attraction for the females that they act against their normal instinct of avoiding previously infested leaves. Just what these attractions are cannot be determined. In many cases the females act in a way detrimental to their offspring, since full development of the larvae is frequently no longer assured. However, even when present in great profusion, a large number of individuals may be able to complete their metamorphosis successfully. For instance, 14 normally developed mine channels of *Nepticula microtheriella* Stt. were found in one leaf of Carpinus. Also with *Leucoptera laburnella* Stt. in the leaves of Laburnum a very large number of larvae are not infrequently found on a single leaf; the individual mines run into one another and coalesce without any harm ensuing for the larvae; these thus lead a secondary gregarious existence and it is only possible to recognise the borders of the original individual mines from the primary thicker frass deposits. *Phytomyza senecionis* Kltb. was once found in a leaf of Senecio nemorensis L. in such numbers that individual channels could no longer be recognised at all; the whole upper surface of the substantially-sized leaf had assumed a uniform whitish-green colour, and the number of larvae which had been present could only be determined from the exit slits, through which the mature mature larvae had left the leaf to pupate. Similarly, a profusion of mines of *Nepticula splendidissimella* HS. was found in a leaf of Rubus; this species usually forms a linear-mine which follows the leaf veins and occurs only as a single specimen in each leaf. In the most extreme case 123 linear mines of *Nepticula quinquella* Bed. were found in one leaf of Quercus.

A profusion of mines of the same species in one leaf is found most frequently in years and in localities where an abnormally large number of ovipositing females have congregated together. This occurs again and again with the same few species. The most striking example is provided by *Lyonetia clerkella* L. which from time to time constantly re-appears in great numbers, when its mines are so common that on some trees not a single leaf remains unaffected and every leaf contains a large number of its ornate channels. Its mines are then found in equal abundance on various Rosaceae, such as Malus and Prunus, and also on Betula. That this is not merely a result of our modern habit of growing trees in pure stands is suggested by the fact that the first little book dealing with leaf-mines refers to the abundance in this species as early as 1680. In this case the numerous larvae in a single leaf seem to in no way prejudice, each other's development; the channels cross and re-cross each other without the larvae suffering any ill-effects. Among the genus *Lithocolletis*, occurrence in profusion is found in the same way on Populus as, for instance, with *L. populifoliella* Tr. on P. nigra L. and *L. petrazzaniella* Cst. on P. alba L., without the numerous larvae in one leaf in any way disturbing each other; in these cases the outbreaks of abundance are not of such frequent and regular occurrence. Populus has a sufficient number of large leaves to ensure the development of numerous larvae in a single leaf. On the other hand when *Nepticula hübnerella* Hbn. occurs in profusion, with the numerous mines being constructed in each of the much smaller leaves of Crataegus, considerable difficulties stand

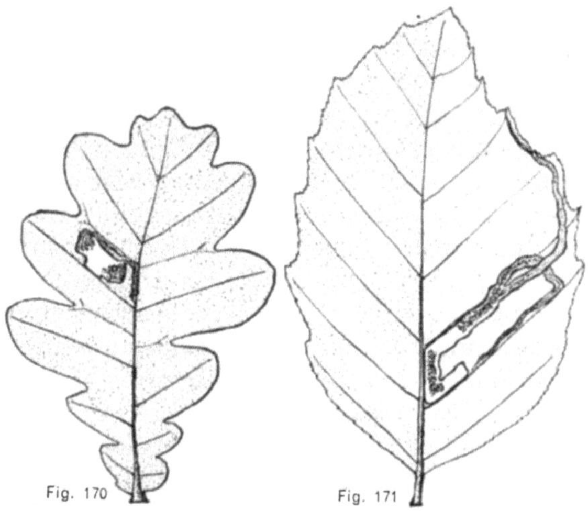

Fig. 170. Quercus: *Nepticula albifasciella* Hein., two mines merge to form a joint blotch.

Fig. 171. Carpinus: *Nepticula floslactella* Haw. Two linear mines run together. The second larva has passed through the first channel without feeding, which therefore has a stretch without frass.

in the way of normal development of the larvae. One finds in such leaves that a number of the young larvae commence mining on the lower surface when the others have eaten all the upper surface parenchyma. Many dead larvae however, can be found in the leaf, a sign that this unusual food did not suit them. *Agromyza alni-betulae* Hend. shows a preference for the young off-shoots from the base of the stem where large numbers can often be found on a single leaf. In this case the larvae do not appear to cause each other any harm and they usually develop normally. On Alnus the same species is invariably found with only a single larva to a leaf.

However, it is not so unusual to find two mines of the same species together in one leaf, with the eggs often having been deposited in close proximity. It is reasonably safe to assume that both eggs were deposited by the same female; this is also suggested by the fact that in such cases, if they are bred, the two turn out to be a male and a female. On a number of occasions the mines of the future "couple" coalesce but they live in such complete harmony that they in no way impede each other's development in this communal mine. Fig. 170 shows two mines of *Nepticula albifasciella* Hein. in a leaf of Quercus. The initial mine of the one adjoined the mid-rib, of the other a lateral vein; the blotches which developed from these early mines have coalesced and reveal their origin only by the two distinct deposits of frass. Even stranger results arise when two linear mines run into each other, as is shown in Fig. 171 with *Nepticula floslactella* Hw. on Carpinus. The first of the two larvae commenced its mine on the margin of the leaf and completed its development before reaching the mid-rib. The second began near a lateral vein and its channel entered that of the first. There it found an area it had to cross without being able to feed, as a complete channel had been eaten away by its predecessor; therefore in this double mine, there is a length which is free of frass. Only after reaching the mid-rib was it again able to feed and from now on mine and frass-line were formed in the normal way. Such encounters of two mines of the same species may be met with some frequency; the reactions of each larva to the presence of the other are always in accordance with the existing circumstances and consequently vary.

2. Encounters by separate species in the same leaf.

When oviposition has already taken place on a leaf and is followed by a second female of a different species, the powers of discrimination of the second female invariably fail; mines of the most varied species are therefore frequently found together in one leaf at the same time. An additional relevant factor is that the time of oviposition varies far more among different species than among individuals of the same species. If the mines of two separate species meet in one leaf, the reactions of the larvae will again vary. The least difficulties are caused when the first species to commence mining constructs a shallower mine than its successor. The new arrival will in most cases completely engulf the original mine in its own; it consumes the entire parenchyma inside the leaf but it leaves quite untouched the frass left by the first miner; in all such cases where mines cross, it is therefore not difficult to detect the course of the old mine within the new from the fully preserved frass-line. For instance, we were able to trace the entire course of the channel of *Phytomyza conyzae* Hend. through a full-depth blotch-mine of *Coleophora conyzae* Hend. on Inula; in the same way we can frequently discover the original channel of *Agromyza alni-betulae* Hend. in the full-depth blotch-mines of *Scolioneura betulae* Klg. In such cases the frass is never eaten as well, not even in species still more closely related to the saprophytic forms, such as the species of *Chylizosoma*. In practically each individual case miners which meet each other in a leaf behave differently, so that we can only mention here some of the most important possibilities. Frequently, the action taken is determined by the species which was present in the leaf first. For example, when *Nepticula microtheriella* Stt. and *Lithocolletis coryli*

Fig. 172

Fig. 173

Fig. 172. Betula: *Scolioneura betuleti* Kl. with an isolated patch of mine, to which the larva passed through a lower surface mine of *Lithocolletis ulmifoliella* Hb.

Fig. 173. Betula: *Nepticula lapponica* Wck. Left, two normal linear mines. Right, encounter of a linear mine with the blotch-mine of *Scolioneura betulae* Zadd.

Nic. encounter each other in a leaf of Corylus, *Lithocolletis* may have been the first to make its mine. When the channel of the *Nepticula* runs into such a tentiform mine (Plate 2, Fig. 4), the larva is unable to feed in the mine in the normal way, presumably because pressure from the surrounding tissue during feeding is absent; it therefore passes through the mine without feeding, and continues its own mine only on emerging on the other side. When, on the other hand, the *Nepticula* mine was the first to be made, the "formidable depths" of this mine apparently cannot be crossed by the first instar *Lithocolletis* which is busy removing the outer wall of the epidermis; we therefore find that the *Nepticula* channels are delimited by the *Lithocolletis* blotches. However, once the first vulnerable stage as sap-feeder has been passed, the upper-surface mining *Lithocolletis* larva was able to continue its mine right through the lower-surface mine of *Parornix avellanella* Stt.

Scolioneura betuleti Klg., which produces full-depth blotch-mines on Betula, is not able, like the Anthomyidae, to leave the leaf at one point and burrow in at another. In Fig. 172, however, we see a small area of mine more in the basal part of the leaf isolated from the rest of the blotch-mine of this species, which seems to suggest such a possibility. But this is only apparently so; this isolated area of mine was caused by *Lithocolletis ulmifoliella* Hbn. mining on the lower surface of the leaf at this spot, which the Tenthredinid larva had passed through without feeding and only re-commenced feeding on reaching its boundary. This is possible with species with such a slight degree of specialisation in their feeding habits as saw-flies. On the other hand a larva of *Phylloporia bistrigella* Hw. perished when its channel entered the upper-surface tentiform mine of *Lithocolletis betulae* Z. The so much more highly specialised *Phylloporia* larva was apparently unable to adapt itself to the changed environment in the *Lithocolletis* mine even for the short time, which would have been necessary for crossing the mine. On occasions, by a change of instinct, the larva is able to avoid in time the dangers which might arise from an encounter with another. This is shown by the behaviour of a larva of *Nepticula lapponica* Wck. in a Betula leaf on encountering a mine of *Scolioneura betulae* Zadd. (see Fig. 173). The *lapponica* larva in its last instar invariably directs its channel towards the margin of the leaf, there crosses a vein and then proceeds towards the mid-rib, as can be seen in the normal mines on the left-hand side of the leaf. One of these larvae found its way blocked by the blotch-mine of the Scolioneura. It did not accept the uncertain risk of crossing the mine but immediately went backwards, as we also noticed previously with *N. microtheriella* Stt. and mined out the area between the two lateral veins, which did not, however, provide it with sufficient food. It was therefore forced to cross one of the veins at a point close to the mid-rib and where the vein was thus of substantial strength; this doubtless caused it difficulties but it succeeded and reached maturity.

Such encounters do not always have such a happy ending. In other cases they lead to the death of one of the two larvae involved. Two closely related species, which are markedly different in the method of oviposition, live in the leaves of Crataegus. In *N. regiella* HS. the egg is always deposited on the edge of the leaf. On hatching the young larva forms a channel which closely follows the margin of the leaf; later this is expanded to a large blotch-mine which likewise begins on the leaf margin. By contrast in *N. corvimontana* Hering the egg is always deposited some distance from the edge of the leaf; on hatching the larva usually mines a number of coils around the point of oviposition and then proceeds straight towards the edge of the leaf; like the previous species its course follows the margin and then also expands to a large blotch-mine. For both species it appears to be quite essential to form the first part of their mine in immediate juxtaposition to the margin of the leaf. Presumably their mouth-parts are only able to cross the ends of the vascular bundles, where they contain the smallest possible quantities of hard elements. The consistent occurrence of the course of the mine along the margin of the leaf at least suggests that the practice has some biological significance. Fig. 174 shows a leaf-tip of Crataegus, in which both species

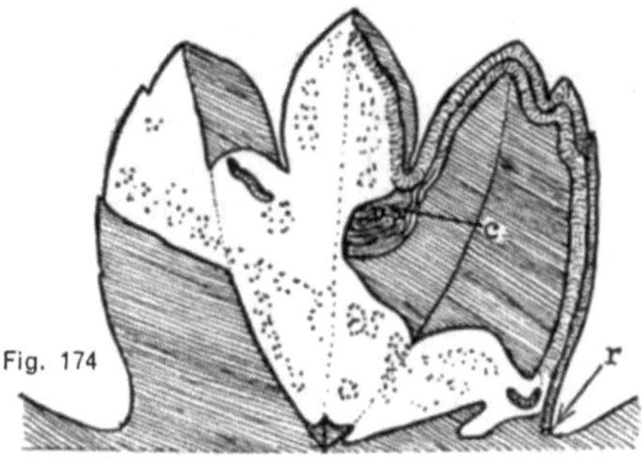

Fig. 174. Crataegus (Leaf-tip): Encounter of *Nepticula corvimontana* Hering (beginning at c) and *N. regiella* H.S. (beginning at r). *N. corvimontana* Her., which was unable to gain the margin of the leaf, died before maturity.

have laid their eggs. The first to appear, *N. regiella* HS. (r), has been able to form its mine undisturbed; it constructed its channel first along the margin of the leaf and then expanded it to a blotch spreading right across the tip of the leaf. This had all taken place before the larva of *N. corvimontana* Her. (c) hatched from the egg. Following its instinct, it attempted to gain the margin of the leaf but the way was already blocked by the *regiella* mine. Its indecision is expressed in the numerous bends, which in addition do not show the frass deposit in the customary transverse rows of grains, but there is only a narrow central line of frass, an indication that feeding was not being pursued normally. Finally, the larva decided that the course of its channel should follow not the margin of the leaf but the track of the other species; in this way it successfully continued its development, as can be seen from the normal frass deposits. It then reached a projecting part of the blotch of the first species and was unable to make progress with its own mine; it was perhaps already weakened by not being able to mine along the edge of the leaf and at this point it died. Such "mine tragedies" resulting from a shortage of space probably occur with some frequency.

An analogy with this importance of being beside the edge of the leaf can be seen in *Phyllocnistis suffusella* Z. which bends a portion of the leaf-edge over the end of the mine when pupating. Sorhagen (1886) gives the following description of this process: "If two larvae are mining on the same leaf, each seeks to be the first to gain the edge of the leaf without crossing the track of the other; the larva is helpless outside its mine, which it must attempt to leave in the immediate proximity of the edge of the leaf, as this is essential for its pupation .... I have observed on a number of occasions that a second mine ran beside the normal marginal mine; however, as it was unable to reach the margin before the first, it followed a lateral vein to the mid-rib and then returned to the margin, but, having arrived too late, once again returned beside a second lateral vein to the mid-rib and after a third fruitless attempt to get in front of its rival, crossed over the mid-rib to the other side of the leaf, where it reached its desired goal and produced an irregular course along the margin of the leaf until pupating." In the same way as the larva finds it necessary to reach the margin of the leaf for pupation, for some species feeding on the margin of the leaf represents a compelling necessity.

A true gregarious life of several species in a communal mine is not to be found among our European species. If several species do occur in a single mine, this is usually not at the same time and, as far as possible, the larvae avoid each other. A mine of *Phytomyza nigra* Mg. was on one occasion found in a leaf of Holcus with the puparium already in the mine. *Hydrellia griseola* Fall. was mining on the same plant. The larvae of this species frequently change their mines and two of them had entered the mine of the *Phytomyza* and there had continued feeding, without in any way disturbing the puparium of the originator of the mine. Of these two, one pupated in the mine and produced an imago, while the second died. However, such cases represent rare exceptions. According to Maulick (1937), gregarious living of several species does occur with the mines of certain Hispinae in Central America. Various species of the genera *Arescus* and *Chelobasis* were frequently found in communal mines on Heliconia, a genus of Musaceae. Maulick presumes that this communal living in a single mine must frequently give rise to hybridization between the closely related species and he believes that the very variable wing pattern found in the species concerned can be ascribed to such hybridization. This seems all the more plausible, since the Hispinae normally pupate inside the mine and the emerging imagines, which in Coleoptera always remain for some time in the mine to harden up before eating their way out, might thus very easily find a suitable mate. If the emergence of different species took place in a single mine simultaneously, it is very reasonable to assume that in all probability hydribization would ensue. It is particularly interesting to note the possibility of a crossing of species as a result of the mining habit and it would be most desirable for detailed observations to be made on this point to confirm it by immediate examination.

Even though, as we have seen, Reaumur's remark about the great solitude of the life of the miner is not accurate in every instance, nevertheless, encounters of two mining insects in a leaf usually lead to disaster for one of them, unless it manages to escape this fate by a suitable change in its feeding instincts.

19.

PARASITES AND ENEMIES OF MINERS; INQUILINES AND SYMBIONTS

We mentioned earlier how the mine can be considered as a sort of "hot-house"; greater humidity and higher temperature create particularly favourable conditions for the larva feeding within, which is as a result in many cases able to reach maturity more rapidly than is possible with external feeders. If conditions in the mine are thus in every way favourable for the development of the miner, this applies also simultaneously to organisms which are working for its destruction and annihilation, among which parasites figure predominantly. It is true that the mining habit affords the larva greater protection against certain enemies which normally attack free-living larvae but not against all. Parasites, which have adapted themselves to leaf-miners and share living conditions in the mine with the producer, also experience favourable conditions which accelerate their development. Everyone who has spent time breeding miners knows that they are not free from parasites. Some groups of parasites, however, are completely excluded from this field. As far as our experience goes, parasitic Tachinidae (Larvaevoridae) never occur on larvae which pass their whole life in their mine. The female of these flies is not able to insert its egg inside the mine channel, since it does not possess an ovipositor capable of penetrating into the mine cavity. All known cases of Tachinidae parasitising miners apply to temporary miners; as long as the larva lives inside its mine, it remains unmolested and it is attacked by the parasite only after having left the mine and is feeding on the surface of the leaf; here it is exposed to the ovipositing fly or may become infected by an egg on the leaf surface, as for instance with species of *Phytomyzoptera* on Gracilariidae, which after leaving their mine live under leaf rolls.

The remaining parasites are able to penetrate inside the mine with their ovipositor and to lay eggs on the larva inside; the parasitic larva develops either inside or on the mining larva and finally causes its death. It is not even always necessary for the parasite to puncture the leaf tissue; it may insert its ovipositor through natural openings in the plant, such as stomata and hydathodes and so introduce the death-dealing guest into the mine. In some cases even the eggs of the miners are "stung"; this occurs to the eggs of Anthomyidae and many Lepidoptera which are glued to the surface of the leaf, as well as those of Agromyzidae and Trypetidae which are inserted into the leaf tissue. Finally, there are also parasites of pupae and these can be expected in large numbers in species whose pupa is particularly exposed, especially in the puparia of many species of Phytomyza and Ophiomyia which transform in a "pupal blister" inside the leaf. At all events the number of parasites of miners belonging to insect Orders is extremely large and they are largely responsible for the fact that so many species known from their mine for many years have never been secured as adults, since too high a proportion of the larvae are parasitised. The breeder who might despair at the fruitlessness of many of his attempts to breed these miners, can draw little comfort from the fact that some of the parasites are for their part again attacked by other parasites — hyperparasites; certain Chalcididae and Proctotrupidae are the best-known examples of these. This explains how, when breeding a single species of miner, a mixed bag of the most varied specimens may be obtained, while the mine-producer itself is not represented.

Among the parasitic Hymenoptera, the *Ichneumonidae* with their numerous species are found only very rarely on mining larvae and then only on the larger species, usually on temporary miners. Attacks by *Braconidae* and *Chalcididae* are extremley common, with the latter frequently as hyperparasites of the former. *Proctotrupidae* are found less frequently in mines, while the *Cynipidae*, parasitic gall-wasps, are found most rarely of all and have been bred almost exclusively from species of *Phytomyza*.

One might expect the hymenopterous parasites of leaf-miners to be specially adapted species, possessing a specially modified ovipositor, enabling them to pierce the leaf tissue. On the contrary, they tend to be extremely polyphagous parasites and parasitic species are found, which are known not only from miners but also from free-living larvae. It is no rare occurrence to breed species of Braconidae from *Lithocolletis* species, which are already known from larvae of Noctuidae. This occurs frequently and at first sight appears very surprising, since one would expect, with so highly specialised hosts as miners, to obtain equally highly specialised parasites. The reason why this is not so can perhaps be explained by the fact that the systematic classification of these parasites has not been worked out in sufficient detail and that classification of these Hymenoptera based solely on morphological characteristics is not adequate for distinguishing different species. Subsequent investigations may well lead to differentiation between species, which has so far not been apparent. Parasites of mining insects thus possibly represent a comprehensive new field for future investigations.

Less frequently larvae of gall-midges (Cecidomyidae) are also found in mines but without their parasitic nature having in every case been proved. Vimmer (1908) found them in Lonicera mines of *Phytomyza (Napomyza) xylostei* Kltb. but was unable to prove their parasitic nature; Trägårdh, however, was able to make a direct observation of a *Rhynchaenus* larva being sucked out by a Cecidomyid larva in a mine. A large number of parasitic Cecidomyidae are known and it is to be expected that future investigations will bring to light many further examples of their parasitism on mining larvae.

The parasitic larva frequently leaves the miner only after pupation but in some cases earlier. In rare cases it is possible to find that the mining larva, after having been abandoned by the parasite, continues feeding, pupates and produces a well-formed imago; we observed this on one occasion in *Phytomyza senecionis* Kltb. In most cases, however, the mining larva sooner or later dies. An experienced minologist can often detect the presence of a parasite from the nature of the mine, which shows considerable deviations from the normal type, when the larva has been parasitised. Numerous instinct changes develop in the parasitised larva which cause it to proceed quite differently in the construction of its mine from normal, healthy members of the species. For instance, species which commence their mine as a spiral may change to a straight course, while species which mine in a straight line may produce a spiral. Abnormalities may occur in the choice of the side of the leaf or of the tissue which is mined, the selection of the pupal site may deviate from normal, mines which usually avoid the veins pass straight over them, normal development from linear to blotch-mine may fail to occur and other changes of this type may be found. De Meijere (1911) describes the changes in the behaviour of a parasitised *Pycnoglossa cinerosa* Zett. which mines in the fronds of Pteridium. The egg of this species is deposited towards the centre of a pinnule and from here the larva proceeds to the apex of the leaf, where, on arrival it commences mining downwards on the opposite margin. In this way, while in its first instar, the larva is able to feed on the softest tissue of the apex of the frond. The parasitised larva showed its change of instinct by mining downwards right from the outset, whereupon it soon reached the hardest tissue of the leaf and died.

Parasitic fungi also play a part in the lives of mining insects, which has not so far in most cases been investigated in detail. Their presence should cause no surprise in view of the microclimate in the mine, which is highly favourable to the development of fungi. The higher the temperature and the relatively greater humidity inside the mine channel also favour the growth of fungi in the mine. Parasitic fungi must therefore be considered as responsible for the destruction of mining insects on a far wider scale than has so far been realised. Botrytis (Beauveria) species, in particular, cause the death of far more mining larvae than has generally been accepted. These species of fungi cause "muscardine" in larvae. Starý (1938) pointed out that the larvae of *Lyonetia clerkella* L., which are only rarely attacked by parasitic Hymenoptera, frequently succumb to muscardine. The fungal mycelium, beginning with the internal organs, spreads over the larva's entire body, which finally assumes a wax-like appearance. Without exception the larva dies when infected by this fungus. The Botrytis mycelium thrives on after the death of the host and finally penetrates the leaf epidermis. It then forms its spores on the surface of the leaf, which may also infect larvae which have already left their mine. It is well-known that this species of *Lyonetia* appears in great profusion in some years, so that on some trees not a single leaf can be found without a mine; in the following year it may be so scarce that practically no mines can be found, although parasitic Hymenoptera have undergone no corresponding increase. Starý presumes that this striking decrease in numbers can be ascribed to infection with this parasitic fungus.

Enemies of miners play only a subsidiary role by comparison with true parasites. Ants figure largely among these enemies. In temperate latitudes they attack miners far less than in tropical regions, where they are very active in pursuit of insect larvae. Larvae in mines enjoy a certain degree of protection and the quite common adoption of the mining habit, which can be observed in more southerly latitudes among certain families, such as the Gelechidae, is ascribed to the more frequent attacks by ants to which they are subject. It is also much more common in the tropics for ants to seize miners from within their mines. The main enemies of mining insects in Europe are birds, which peck out the larvae or pupae from the mines. In the first larval generation of miners this is not of frequent occurrence. At this period birds can find an abundance of food and they are not constrained to supplement their sustenance by painstakingly searching out larvae in their mines. Parallel with this is the fact that the first generation of miners is far less numerous than that appearing in autumn. Also, the first generation develops to maturity far more rapidly than the second, as a result of the more nutritious plant tissues which are at their disposal. Experienced breeders of miners find it difficult with many species to find the first generation of mines still inhabited by the larvae; this applies, for instance, to *Nepticula*. Many species of this genus have only developed one generation, which lives in autumn. In winter, especially during periods of frost and snow, birds suffer a far greater shortage of food and at such times all mines which can be found are carefully examined and the larvae or pupae, which have remained in the mine, are systematically pecked out. This happens with great regularity with all the species mining leaves, which remain for the greatest part of the winter on the tree or the bush, such as with *Lithocolletis* on Quercus, Carpinus, etc. In autumn large numbers of these mines can be found which are inhabited, while an examination in spring reveals that the inhabitants have been pecked out from most of the mines. Some birds do not even take the trouble to open the mine but eat the entire mined area or the whole of a small leaf, which contains a mine. According to Baer (1906) investigations of the stomachs of *Parus ater* L. and *Regulus cristatus* Vieill. showed numerous spruce needles containing the young larvae of *Epinotia nanana* Tr.

By comparison with these main enemies — birds and ants — other enemies from the insect kingdom are of relatively little importance. The *Chrysopidae* should be mentioned which suck out the larvae. The mine offers no protection against them, since they bite open the epidermis and are thus able to reach the larvae. Heinrich & De Gryse even observed that a *Chrysopa* had bitten through the mid-rib and sucked out the Gracilariidae larva hiding below. Attacks of this kind have not often been noted; the frequent occurrence of empty larval skins in the mine suggests, however, that these preying marauders attack our miners much more often than is realised.

The defence measures adopted by miners against the attacks of enemies and parasites are in general quite inadequate. While the mining habit itself represents a considerable degree of protection against persecution of this type, it fails whenever the assailants have become specialised in the selection of their prey. We already saw that mining does not prevent the parasites from laying their eggs, and only gives protection against the Tachinidae. The relative immobility of the larva in the mine makes it difficult for it to take any evading action against attacks by a parasite which has already selected its victim. External feeders attempt by vigorous body movements or by moving away to escape from the attacker. Any kind of flight is, of course, automatically excluded in the case of mining insects. Their best means of protection is afforded by making themselves as inconspicuous as possible. The shape and situation of the mine have a direct connection with this need for increased protection. There can be some doubt whether a linear-mine or a blotch-mine is the more conspicuous; both the one and the other may stand out more clearly according to varying circumstances. The exact position in the leaf, direction of light, and many other factors can influence this. It is, however, certain that a larva which is resting below the mid-rib or one of the larger lateral veins may frequently be protected from discovery. This habit is accordingly extremely widespread among mining larvae of the different Orders. Among Diptera we may mention as the best-known examples of this: *Liriomyza strigata* Mg. (Fig. 12), *Phytomyza ramosa* Hend. in the leaf of Knautia and various species of *Melanagromyza;* among *Lepidoptera*, *Heliozela*, some *Gnorimoschema* and *Cosmopteryx zieglerella* Hb. (*eximia* Hw). In all these species the larva withdraws to the protection of the vein when it is not actually feeding and here it is very difficult to detect. The case mentioned above where a *Chrysopa* actually bit through the hard vein in order to reach the larva shows that the protection offered is not infallible. When *Dizygomyza morionella* Zett. is infesting Phlomis, it was found that the sizeable linear-blotch mines occurred almost exclusively on the bracts, where they were completely hidden by the whorls of the flowers. Against enemies seeking their prey solely by sight, there is little doubt that they enjoyed really effective protection. The mine is also relatively inconspicuous if the surroundings are of a dark colour, such as we find with the mines of *Phytomyza succisae* Hering in the leaf of Succisa (Pl. 1, Fig. 4). The mine channels can be found only with difficulty among the dark parts of the leaf. These areas, once the habits of the insects are sufficiently known, provide a means for hunting birds and insects (and also for the entomologist!) of finding the mines more easily. Such habits offer no protection against parasitic Hymenoptera, since they seek out their prey by means of their acute sense of smell.

The mining larva probably enjoys really effective protection in cases where it envelopes the mine in a foam. Fiebrig (1908) reported a case of this kind. He found a Buprestid larva of the genus *Pachyschelus* mining in the leaves of Sapium, a South-American species of Euphorbiaceae. Fiebrig discovered that the mine was enveloped throughout its length in a foam arising from the bubbles of rubber sap, which in this plant is particularly abundant and watery. This foam which covers the mine occurs on both the upper and lower surface of the leaf. It is not produced by the spiracles or the digestive apparatus, as with the other foam producing Diptera, Cicades, etc. In this beetle larva the forepart of the thorax possesses on each side a bulbous swelling. When a lactiferous canal is bitten into, the watery sap flows in quantities into the mine cavity and to the particular spot where the larva has just been feeding. It now withdraws its head somewhat into its thorax, which thereby becomes thickened and presses firmly against the walls of the mine. Simultaneously, a somewhat concave space arises in front of its mouth. It now rapidly advances its fore-body; the sap which has entered is unable to escape laterally and is squeezed out through the epidermis onto the surface of the leaf. Fiebrig did not investigate whether the sap passes through natural openings in the plant, such as stomata and hydathodes. He reports, however, that the bubbles of foam appear polygonal and not round and therefore one might perhaps assume that the passage takes place through the inter-cellular spaces of the epidermis cells, the structure of which is mirrored in the bubbles of foam. On reaching the surface of the leaf the sap readily hardens and provides a covering of foam for the mine channel, representing an effective protection for the miner. The thoracic callosities lie on the sides of the thorax, because the largest space between the body and the leaf tissue is in this area, while the dorsal and ventral areas are in any case pressed closely against the leaf tissue. If these callosities were not present, the plant sap would escape past the sides of the larva's body. Once the sap has been expelled from the mine channel in this way, the larva continues feeding, until a further quantity of sap has flowed into the mine, whereupon the pumping action is renewed. Another species of *Pachyschelus* living on another genus of Euphorbiaceae (Croton) does not possess these thoracic callosities. In this plant the sap is far more viscous and less fluid and consequently does not flow into the mine cavity in such abundance; pumping action by the larva is therefore in this case not necessary. In this species the leaf above the mine is thus not covered in foam.

After this description it is apparent that the foam-producing mining larva is acting primarily in order to remove the undesirable milky sap of the Euphorbia plant from its mine channel and that the protection for the mine from the hardening foam on the leaf surface is only achieved incidentally. Production of foam by insects frequently however, has a more direct connection with the living habits of the species concerned; it is not merely a reaction to the flow of sap from the plant. This is clearly indicated by the fact that various Trypetid larvae live in the inflorescences of the European Senecio without any production of foam, while this is very distinct in the Muscid *Egeria senecionis* Meade living on the same species, and, after hardening, it represents an effective protection for the larva.

Minology is a science of too recent origin for every detail to have been investigated, as is perhaps the case with its sister science, cecidology, which is much older. While we possess a great deal of information on the harmless co-inhabitants of plant-galls, which are not parasites on the gall-producer and are known as inquilines, similar information is lacking for miners or is limited to a few isolated cases. Even in these it is not generally known whether they are inquilines in the strict sense of the term, where a different species merely seeks shelter in the living quarters of the miner; or whether they represent examples of true communal living, in which the visiting species also finds food, whether this be plant tissue or crumbs from the master's table in the form of the mining larva's frass which, as we know, contains a considerable quantity of nutrient substances, even though the protein constituents are missing. In contrast to the parasites which can also be found in the mine, inquilines cause no damage to the actual mine-producer. Possibly the Cecidomyidae, in as far as they are not true parasites, may occupy the role of inquilines; as also saprophytic larvae of *Drosophila* and *Lycoriidae*, which are encountered in mine cavities. The part played by *Thysanoptera* is still very much open to doubt. That they are not merely inquilines but also at least co-feeders is shown by the discovery of their frass in mines (or natural cavities of the plant) which they inhabit. At least in some cases they consume parts of the leaf substance in the mine channel. The construction of their mouth-parts would preclude the possibility of their producing a genuine mine but they certainly produce openings in the mine by perforating the epidermis and so maintaining their connection with the outer world. On the other hand they frequently visit mines merely for the protection they offer, particularly when climatic conditions are unfavourable. It is reported that *Diarthrothrips coffeae* Will. is very commonly found in the mines of *Leucoptera coffeella* Guér.-Mén. during the rainy season, where it is protected against the inclemencies of the wheather. It is claimed that this species of *Leucoptera* damages the coffee leaves not so much by its own mining as by its indirect attraction of this species of Thysanoptera. The significance of the Phoridae, which are also sometimes found in mines, is equally controversial. Members of the genus *Megaselia* are especially addicted to this habit. *Phytomyza sönderupi* Hering lives at the end of the petiole of Caltha, together with which *Megaselia limburgensis* Schmitz has also been bred. From the petiole strange mine off-shoots were constructed into the leaf-blade and it has so far not been possible to show beyond doubt whether they are to be ascribed to the *Phytomyza* or the *Megaselia* which lives with it. As early as 1909 and 1911 Malloch and Wood believed that Phoridae, which they had bred and which today are included in the genus *Megaselia*, were mine-producers; however, it subsequently became clear that these species require further investigation to show whether they really produce the mines themselves or merely feed on waste products which they find in the mines. Pater H. Schmitz, the well-known expert on Phoridae, showed that *Megaselia* species are in fact able to consume green plant tissue, when he obtained the same species of *Megaselia* from the seeds of Pinus, where they had been eating the embryos. As long as no more exact observations are available, it will be better to consider Phoridae found in mines as co-feeders.

Finally, mites (Acaroidea) are also frequently found in mines but these can certainly never be considered themselves as miners. They likewise eat waste products and also the mining larvae themselves, if they have for any reason died. It is also not uncommon for them to attack mining larvae which are still living but are weakened through sickness. There is little doubt, however, that they cannot be considered as true parasites.

A field which has only recently been explored is that of intra-cellular symbiosis. Minute organisms have been found inside the body cells of numerous species of insects belonging to the most varied Orders; these organisms are usually bacteria or yeasts which have adopted a particularly intimate relationship with their host's body and are passed on to its offspring by means of a highly-differentiated mechanism. These cases of symbiosis represent a joint life of two species of organism and both derive advantage from this intimate connection. The insect species permits the yeast or bacterium to live in its own body cells and promotes its reproduction by specialised means; it simultaneously derives the advantage either of the guests breaking down its food and making it more easily digestible, or of itself consuming a proportion of the symbionts. The presence of the symbionts facilitates a greatly increased consumption of carbohydrates, cellulose or other substances by the insect larva. These reciprocal relations, often involving great refinements of adaptation, occur in numerous phytophagous insects.

It was natural to seek cases of intra-cellular symbiosis among miners, with their highly specialised habits. Portier, in particular, has devoted himself to discovering such symbionts among miners. He did not succeed in detecting organisms of this sort in the body-cells of true miners but he did discover them floating free in their digestive tract. He assumed that here also they have a rule to play and render the foodstoffs more suitable for digestion. Portier (1918), however, also considers that the larva in a mine is in a completely sterile medium, so that, as long as the larva is feeding in the mine, no microbes can reach it. He was thus unable to discover any organisms either in the intestines or the frass deposits of *Nepticula malella* Stt. He reached entirely different conclusions with *Gracilaria syringella* Fbr. While these larvae, which are common on Syringa and Ligustrum, mined inside the leaf, he found no microbes in their intestine but discovered them in large quantities after the larvae had left their mine and had constructed a leaf-roll, in other words were in contact with the outside world. The author assumed that all miners are free of microbes in their intestine, as long as they are inside their mine but that microbes are present in all species living outside their mine, owing to the constant possibility of suffering infection. Temporary miners occupy an intermediate position, since the larvae are free of micro-organisms as long as they are in their mine but contain them after becoming external feeders.

The investigations of Portier have not been checked in every respect but theoretical considerations alone show that they cannot be accurate. Ilse Schwarz (1929) proved that the so-called symbionts observed by Portier in *Nonagria* larvae are nothing but Microsporidae, in fact genuine Protozoa, which can be commonly found in the body-cells of larvae, without this necessarily leading to any serious disease of a larva so infected. The occurrence of such *Nosema* species, which these in fact are, has long been known in connection with the breeding of silk-worms. Under no circumstances can they be considered as symbionts, as was thought by Portier. As far as miners are concerned, Portier's conclusions are not even in accord with the facts. He completely accepted the idea that the larva in the mine exists in a sterile milieu. We know, however, that the mining larva has three possible ways of penetrating into the tissue of the leaf. In many cases the egg is affixed to the leaf externally; on hatching, the larva consumes part of the egg-shell, before burrowing into the leaf. In this case it is obvious that the larva will already have become infected by microbes from the outside world. With other species the egg is also deposited on the surface of the leaf but the young larva enters the leaf direct through the base of the egg, as in species of *Nepticula*. It must be doubted, however, whether the underside of the egg and the surface of the leaf at the point of oviposition are completely sterile; the imago which laid the eggs certainly took no precautions in this respect. Thus in this case also it would seem that the possibility can be excluded of the miner which has entered the leaf from its egg having completely sterile intestines. There remain only species in which the female inserts the egg deep into the leaf tissue, as occurs with the majority of mining Diptera, with all Tenthredinidae and the *Rhynchaenus* species. One might assume that here some degree of sterility would be achieved inside the mine channel; however, this is not so, as we saw when discussing the phenomena of regeneration of such tissue in mines. We noted the presence of micro-organisms in considerable quantities on callus tissue, which is particularly sensitive to infection, and this we found precisely with species which insert their ova inside the leaf. We know also that airtight isolation of the mine channel from outside cannot be attained, even when the point of entry of the larva is closed by a completely intact egg-shell. A number of natural openings exist in the plant, such as the stomata, consisting of the stoma with the respiratory cavity beneath, and the hydathodes, the water pores, through which the leaves of the plant are capable of discharging excess water. Micro-organisms are able to enter at any time through these openings and during our discussion of callus tissue, we saw that this actually does occur (Fig. 154). Portier's view that the larva in a mine exists in a completely sterile milieu is not accurate and the conclusions he draws cannot be accepted. He claimed that the great variation in the mouth-parts which can often be noted between the mining and free-living instars of the same species, to which the investigations of Trägårdh had already drawn attention, can be explained by the fact that while the larva remains in the mine it is able to break up the leaf substance more extensively. After leaving the mine its mouth-parts are only capable of a much coarser break-up of its food and parallel with this begins the activity of the "symbionts" which achieve a further break down of the food. Portier sees this as the essential feature in the symbiotic relationship between miners and their guests.

We are thus faced with the rather surprising fact that an ecological group as highly specialised as miners possess no true symbionts, although they are so widely distributed in the insect kingdom. This is possibly connected with the fact that in general the mining larva receives its food in an easily digestible form; it contains indigestible elements only in small quantities. Thus the development of symbionts by miners was not found to be essential for their successful survival.

20.

DAMAGE CAUSED BY MINERS, AND THEIR PRACTICAL VALUE. GEOGRAPHICAL DISTRIBUTION

In general, each mining insect consumes only a relatively small quantity of leaf substance during the course of its life; when comparing miners with free-living species one might assume that miners could never cause real damage to trees and plants cultivated by man. There are nevertheless a few species which must be considered as distinct pests. They occur as such either when they appear in excessive numbers, as in some years with *Lyonetia clerkella* L., when on some trees no single leaf is spared, or when they appear so early that substantial damage is caused to the development of the very young plants. This applies primarily to the spring generation. Damage caused by excessive numbers is larger than might immediately be expected. One must consider not only the parts of the plant destroyed by feeding, which are no longer available for the synthesis and translocation of foodstuffs; far larger areas of the leaves are "ringed" by the activity of the miners; the vessels for conducting sap are severed, translocation is interrupted and nutrient materials from these areas can no longer serve the plant as a whole. Furthermore, large-scale damage is caused, even before desiccation processes commence. In years in which miners occur in large numbers, the results of their activity can be seen in a reduction in the growth ot wood in the plant. The numerous small injuries tot up and represent a larger hindrance to the development of the infested tree than might at first sight appear. The development of herbaceous plants also is appreciably impaired, when they are attacked in an early stage in their growth. *Ceuthorrhynchus contractus* Mrsh. and *Phyllotreta nemorum* L., which mine in the first days of spring on young Cruciferae plants, impede their development to a very marked degree and cause damage which the plant is not able subsequently to make good. *Coleophora serratella* L. causes less damage by its mines than by its attacks on the buds of fruit-trees, which commence immediately after hibernation. The worst pest in our latitudes is undoubtedly the beet-fly *Pegomyia betae* Curt. whose activity considerably reduces the yield of sugar from beet; in its first generation it also seriously affects the development of the immature young plants. A few species also inflict appreciable damage in gardens and parks, among which *Gracilaria syringella* Fbr. especially makes such widespread attacks on the leaves of Syringa and Ligustrum that the infested bushes look completely dried up. According to Bail (1908) indirect damage occurs in this case, owing to the asymmetrical growth disturbances resulting from the act of oviposition of the first generation, which might be termed lateral campylotropy (see Fig. 151). In addition, a tendency to split occurs in the affected leaves, leading to a laciniate form. The constantly recurring attempts of this species to transfer to Caprifoliaceae indicate the probability of a further threat to these ornamental plants in the future. There are also a few species which by their excessive numbers may become pests in forest areas. Species of *Rhynchaenus*, such as *Rh. quercus* L. and *Rh. fagi* L., can have a devastating effect on forest trees and may considerably reduce the yield of wood. According to Boas (1921) a premature fall of leaves as early as June may be caused when these species occur in excessive numbers. In some areas *Coleophora laricella* Hb. recurs annually in great numbers on Larix; as this species mines at a time when the tree is beginning to form its shoots, the damage is all the more extensive; it can frequently lead to the complete death of trees. In years of profusion of *Lyonetia clerkella* L., Betula is frequently attacked and damaged as much as the normal fruit-trees. Many ornamental flowers are also attacked by miners, particularly by the widespread and polyphagous *Phytomyza atricornis* Mg. It attacks plants in great numbers especially in green houses. Our Azaleas are damaged by *Caloptilia azaleella* Brants which, originally indigenous to Japan, has spread as a pest throughout most of Europe and North America and can appreciably impair the growth of these plants. A beetle larva, *Mordellistena cattleyana* Champ., similarly ravages Orchids. In all these cases where a species has been imported, the damage is even more extensive, since the natural enemies are lacking which might limit the increase of these pests. In some cases, by importing the parasite, it will be possible to introduce effective control but a detailed knowledge of the ecology is a necessary pre-supposition for this. Control of mining pests is an extremely difficult problem, since they are so well protected inside the leaf, that they are impervious to any form of spray. This difficulty applies particularly to grain pests, among which figure in northern Europe some species of *Agromyza* and in the Mediterranean countries *Syringopais temperatella* Led.; with this latter species very deep ploughing proved effective, since the larvae which had entered the ground for pupation were destroyed. *Hydrellia griseola* Fall. occurs as a grain pest only on very damp ground in the neighbourhood of water; crop rotation will be the best means of combating this pest. The prevention of agricultural and forestry pests developing in excessive numbers will be best achieved by mixed cultures, while mono-cultures encourage their development.

Apart from direct damage by miners, indirect damaging effects may occur, as a result of the construction of the mine facilitating the work of other pests. J. G. Leach (1927) observed that miners *(Scaptomyzella, Elachiptera)* are primarily responsible for the occurrence of Bacillus carotovorus which causes heart rot in Celery (Apium). This bacillus is only injurious in conjunction with damp and dies off in dry conditions. In practice one sees, however, that outbreaks of this disease usually occur during dry periods. This results from the fact that the mining insects which in damp weather live in the tips of the leaves, proceed to the hearts of the plants when it becomes dry and in this way cause infection by the bacillus.

In this connection we must also discuss which plants are particularly subject to attacks by pests. According to some observers, retarded and poorly-developed leaves and plants prove a special attraction to miners. Other observers by contrast note that plants with an abundant supply of sap are attacked by preference. These two observations do not necessarily contradict each other. Individual species behave differently in this respect; some species are particularly fond of attacking weaker plants, while others visit precisely the most flourishing. Miners are especially suitable for investigations of this type. On the one hand the damage they cause to the leaf is frequently not so large that the development of the leaf is greatly impaired; on the other hand they possess such limited powers of movement that any retardation of development can always be ascribed to the respective insect. Poorly developed plants or plant parts invariably have far less sap than those which are normal or of luxurious growth. While this represents a disadvantage for some free-living insects, it need not be so for miners. We know that the inside of the mine cavity possesses abundant moisture, so that even in stunted leaves miners are ensured of an adequate supply of moisture. Food which is too damp is not suitable for most insects; every breeder of insects has experienced the outbreak of various diseases which decimated his larvae, after the food-plants had been freshened up by damping with water.

In this respect, similar habits to those of miners have been observed in parasitic fungi, just as previously we were able to note parallels between these two ecological groups. Among these fungi, certain species attack by preference retarded host-plants and are known by mycologists as "weakness parasites". Other species by contrast seek out food-plants of distinctly luxurious growth, as for example, the rust fungi. Here also we thus find two groups of parasites, whose behaviour as regards selection of their food-plants is as conflicting as that of miners. This is a problem affecting not only parasitic fungi but also phytophagous insects in general and calling for comprehensive investigations; the limited powers of movement of miners makes them a particularly suitable field for future research in this direction.

The effect of the damage caused by these "weakness parasites" among miners is substantially reduced by their habits; one might even consider them to have certain value, since in this way particularly weakly plants, if not eliminated, are at least so prejudiced in their future development that miners distinctly assist natural selection. It is desirable that analogous investigations to those undertaken on parasitic fungi should also be carried out on miners and gall-producers in order to clarify this question.

Specifically useful insects are not found among miners to the same extent as among gall-producers, where the structures some of them cause are of commercial value. One case, however, deserves mention here, where insect mines are used as food, as reported in detail by Dampf (1924); the species concerned is one of the Hesperiidae, *Aegiale hesperiaris* Wlkr. which mines the succulent leaves of Agaves. Usually several of these larvae occur in the same mine in a leaf and in which considerable quantities of sap then accumulate; this, presumably under the influence of the excretory products of the larvae, produces a certain type of ferment which is particularly tasty. In Mexico these mined leaves with the larvae inside are collected by the Indians and sold at markets as "gusano del maguey", a special delicacy. In addition to this one particular case of material value, the study of leaf-miners has a much greater value in a quite different direction. As we have seen on a number of occasions, miners offer a specially suitable field for studying scientific problems involving both zoology and botany, and they are a better source of study than any other group of phytophagous insects for drawing conclusions on the relations between plants and animals. Investigations can be made which illuminate our views on the ontogeny and phylogeny of insects and animals, and which suitably illustrate the extent of the connection between anatomy and living habits and how the one is influenced by the other. On the other hand our consideration of monophagy and oligophagy showed that under certain circumstances plant relationships may be reacted to far more sensitively by mining insects than the classification adopted by systematic botany would appear to warrant. Even when serum diagnosis fails, the oligophagy of miners can detect such relationships. Finally, we also find here a link with plant biochemistry, since food selection by miners may indicate the presence of similar chemical substances in different groups of plants, as we find with the *Antispila* species in connection with plants containing the "wine-red" pigment. As miners represent a unified ecological group but do not belong together systematically — in many instances they are of polyphetic origin — one can follow up particularly well the development of analogous structures as opposed to true homologies and thus add to our knowledge of the possibilities of development of different organisms.

In general the geographical distribution of phytophagous insects, and thus of miners also, is dependent on the distribution of their food-plants. The two do not, however, coincide and it is probably true to say that the distribution of phytophagous insects is less extensive than that of their food-plants. Many species of miners are not found in all places where their food-plant occurs. It is sufficient to mention as an example *Antispilina ludwigi* Hering, which has so far only been discovered in a very small, confined locality, while its food-plant, Polygonum bistorta L. possesses a widespread distribution. Many species of miners do not follow their host-plant to higher altitudes. The leaves of the plants then usually become smaller and no longer provide the larva with sufficient food to reach maturity. Other species, however, occur which are specially adapted to small-leaved plants; the adaptation usually takes the form of being able to change from one leaf and mine to another. One finds here as elsewhere that the species with the widest distribution are those which are at least highly oligophagous, if not actually polyphagous, so that they can everywhere find suitable food. We also find real cosmopolitans among the miners, such as *Phytomyza atricornis* Mg., which occurs practically throughout the world, wherever flowering plants are found. However, even species with a less extensive range of food-plants may possess a world-wide distribution, such as *Plutella maculipennis* Curt., which lives on all possible Cruciferae, or even *Bedellia somnulentella* Z. which is confined to members of the Convolvulaceae.

As one proceeds southwards from Europe or North America, many changes occur in the groups of species which have adopted the mining habit. For example, there are less and less species from among the Tenthredinidae and in the Mediterranean area one finds already a considerable reduction in the number of species. Among Lepidoptera also different species occur as miners in warmer countries, even though they are not reduced in numbers. Among the Agromyzidae, *Phytomyza* species, which throughout the Holarctic region are the most numerous group, are considerably reduced in numbers; *Melanagromyza* and *Ophiomyia* species are more common as miners in the tropics. The *Hispinae* are represented in Europe by only three species; in the tropics numerous species are found and they attack the most varied families of plants, while in Europe they are limited to Gramineae and Cistaceae. In the tropics they live on a large number of plant families. Buprestidae are also found there as miners very commonly. Among Lepidoptera in tropical regions *Gracilariidae* dominate all other genera. This family together with *Phyllocnistis* represents the majority of miners which are found there. Tropical mining insects as a whole have been investigated only incompletely and unsystematically; information so far collected refers largely to chance discoveries, unless an entomologist living in the area was able to undertake systematic breeding. Breeding of tropical miners is far more difficult than in our latitudes, owing to the great humidity and the extensive formation of mould which this causes. This represents a field where much work remains to be done; a wealth of special and interesting adaptations can certainly be expected, as individual observations have already shown. We may mention a Javanese *Phyllocnistis nepenthae* Hering mining exclusively in the leaf blade of the insect-eating plant Nepenthes, which is modified to a "pitcher" (Plate 2, Fig. 8). A special habit of this kind demands extensive adaptations, about which we so far have no or only inadequate information. It is to be expected that a systematic investigation of mining insects in tropical regions will reveal a wealth of such specialisations in taste. Unfortunately, investigations of mining insects outside Europe are only in their very beginning; Needham, Frost and Tothill have produced a comprehensive work for North America. The investigation of the miners of New Zealand has been carried out on a large scale by M. N. Watt. All other work is prejudiced by the fact that when mining insects have been bred at all and have been described together with their mines, the worker concerned has invariably studied only insects of a single Order and has not related them to the mines of other Orders in the same locality. Frequently also the descriptions of mines are not made in sufficient detail to be able to distinguish mines of one insect Order from mines of other Orders. Systematic investigations bearing particularly in mind the habits as miners must necessarily embrace all Orders, in order to be able to work out the characteristic differences in the mines and to give the right emphasis to problems common to them all.

We have even less knowledge of the distribution of mines in time than of their distribution geographically. Very few fossilised mines are known and what has sometimes been described as a mine, in fact was not one at all. The higher insects which alone come into consideration as miners developed relatively late in the history of the world; their development occurred parallel with that of flowering plants only from about the middle of the Cretaceous onwards. We are therefore familiar with mines in a fossilised condition only from the tertiary, and such a mine is illustrated on Plate 2, Fig. 7. All pre-tertiary specimens described as mines cannot be considered as such; they are in many cases other types of feeding patterns, mostly furrows in the leaves caused by worms in leaves lying at the bottom of lakes. In other cases they are fungal infections, which even to-day frequently resemble insect mines. It is to be desired, however, that palaeo-botanists should pay special attention to such specimens and send any they may find for the opinion of a minologist. We cannot, however, expect from this source any new evidence on the origin of the mining habit.

PLATE 2.

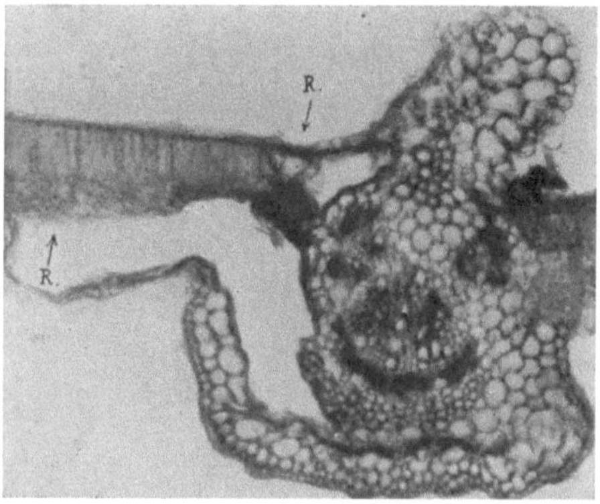

1

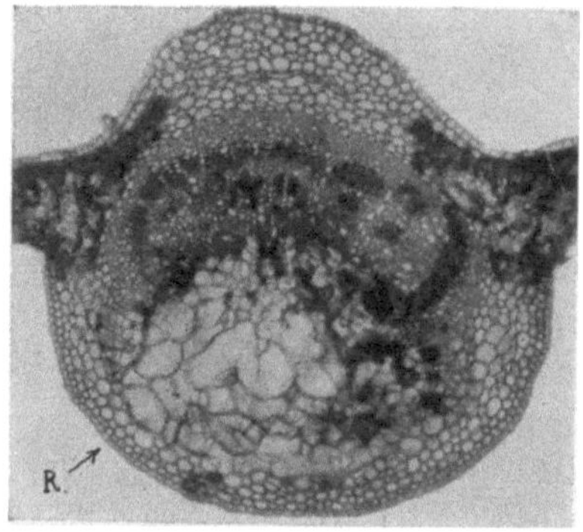

2

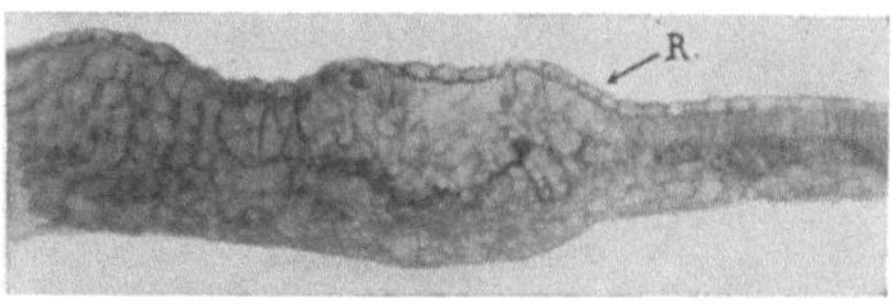

3

4

5

6

7

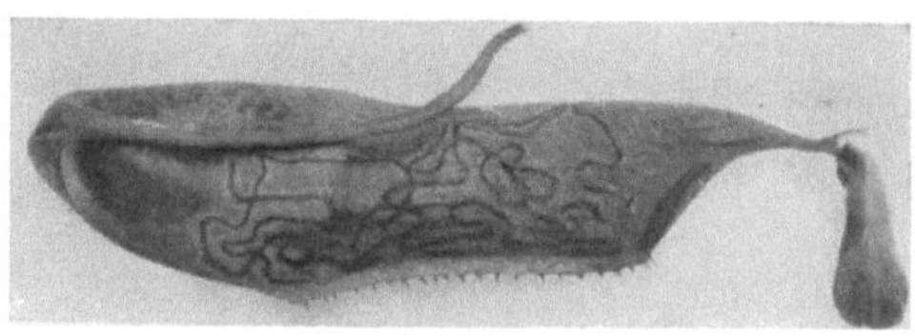

8

1. Cross-section of leaf mid-rib of Dahlia with linear mine of *Liriomyza strigata* Mg. Beginning of callus formation with hyperhydric cells of regenerated parenchyma (R). **2.** Cross-section of leaf mid-rib of Quercus with mine channel of *Rhynchaenus quercus* L. The mine cavity is completely filled up by regenerated parenchyma (R). **3.** Cross-section of leaf of Dahlia with mine of *Liriomyza strigata* Mg. The abundant development of regenerated parenchyma has distended the upper and lower epidermis. **4.** Leaf of Corylus with a blotch-mine of *Lithocolletis coryli* Nic. encountering the linear mine of *Nepticula microtheriella* Stt. The *Nepticula* larva has crossed the *Lithocolletis* mine without feeding. **5.** Linear mine of *Liriomyza strigata* Mg. on the gall caused by *Gnorimoschema cauliginella* Schmid. **6.** "Natural" mine in leaf mid-rib of Tragopogon of *Cnephasia chrysantheana* Dp. and *Argyroploce lacunana* Dp. in the upper (lighter) portion of the vein. **7.** Fossilised mine of *Phytomyza lethe* Hering from the Miocene. **8.** Linear mine of *Phyllocnistis nepenthae* Hering in the wall of the "Pitcher" of Nepenthes (Java).

21.

THE STUDY OF MINERS

The investigations of the past 100 years can be considered to have completed the study of mining insects in its broad outlines; the intensive work of the last three decades in particular has clarified the systematic position of the various leaf-miners to such an extent that, on finding a mine, its characteristics alone in most cases suffice to reveal the name of the producer without having to breed it out. This is certainly true as far as the genus is concerned but frequently applies to the individual species as well. Much detailed work remains to be done, however, and in the last section we will recapitulate the main gaps in our knowledge, in the clarification of which every entomologist can play his part. The principal requirement is for numerous individual mines to be bred out in the species where the producer cannot with sufficient certainty be recognised from the characteristics of its mine; in this way it can be hoped to discover features of the mines which are characteristic of individual species. A wealth of new information will then certainly come to light, especially in areas where breeding has not been undertaken as intensively as in Central and Northern Europe. In this respect Western and Southern Europe have not been worked on anything like an adequate scale and many surprises and new discoveries can be expected; this applies also to the mine fauna of the higher mountains. It may therefore be of value to record here the experience gained in breeding miners over a period of more than 30 years; even though each entomologist employs his own methods, this chapter may save newcomers to this subject making many otherwise unavoidable mistakes.

1. Collecting mines and miners

There is no season of the year during wich the minologist is forced to be idle. Each month offers some opportunity for collecting mines for breeding. Even in winter there is much work to be done, even though the period of active growth of most plants has come to a stand-still. The best plan at this time is to search on plants which have retained green, assimilating leaves and here mines will be found still containing larvae or pupae. The blotch-mines of *Phytomyza ilicis* Curt. can be found on Ilex, while the larvae of *Phytomyza ranunculi* Schrk. may be found feeding beneath the snow on Ranunculus; this is the best time of the year for looking for the various mines on Luzula; *Mompha* larvae can be found in Epilobium and mines with the puparia of *Phytomyza crassiseta* Zett. on Veronica. Thus in this apparently unfavourable season the minologist can find and breed quite a number of species. In addition, dead leaves still on the trees, as on Quercus and Carpinus, should be examined during these winter months; mines can also be found in fallen leaves and if one recalls that it has not so far been possible to determine *Lithocolletis* species on Quercus solely by their mines, a fruitful task can be tackled in these winter days by collecting and breeding such species in individual isolation and assisting in the clarification of these open questions by making a detailed examination of the characteristics of each mine.

Most mining species occur in two generations, one appearing in spring and early summer and the second again in autumn. The number of specimens in the first generation is usually very small and development takes place very rapidly; the minologist can thus to begin with very frequently arrive too late and find that the mine is empty. He can then try his luck again in autumn or otherwise rather more early in the following year. From the freshness of the empty mine, one can soon recognise how many days or weeks earlier the search must begin in the coming year. In the time between these two generations when there is much work of various kinds to be done, the species which only occur as a single generation can usually be found as larvae.

Only very little indispensible equipment is necessary on mine expeditions. What is quite essential is in the first place a thorough botanical knowledge. Usually it is sufficient to know the genera of plants or to be able to identify them without delay. However, for many groups a knowledge of the species of food plant is important for determining the miner and this applies especially to Ranunculaceae, Umbelliferae and Compositae. The minologist should get used to identifying the food plant of every mine that he finds. If he cannot manage this himself, he should collect a piece of the unknown plant with a leaf, a flower and if possible a fruit, which he can then have identified by someone with the requisite botanical knowledge. The best idea is to carry at all times a small pocket botanical guide, such as are available in most countries and with the aid of which plants on which mines have been found be identified on the spot; this is always easier when the plant is fresh than when it has dried up and with some practice it soon becomes possible to work out the names of the food plants. One or two magnifying glasses are necessary, both for investigating plants and important characteristics of the mines; it is sufficient to have one low-power with a magnification of 2 and one of higher power with a magnification of 20; with these two all essential details can be observed. All the mines which have been collected must be kept in such a way that as far as possible the parts of the plant they are in lose no moisture; it is important that they should not lose their turgidity by evaporation, as when this happens the majority of larvae die. The leaves or parts of the leaf must therefore when ever possible be kept in air-tight containers. Cylindrical jars which can be tightly closed with a cork are the most suitable. The larger the jar, the more successful the breeding is likely to be, but less of these large jars can of course be carried in the field. Jars 8 cm high with a diameter of 3 cm have proved to be the most convenient size; they provide adequate space for the development of the mine and a sufficient number can be carried on excursions. The bottom of the container should be lined with a layer of short-stemmed moss, which should be pressed in firmly to prevent it working loose. The moss serves as a humidity regulator; it absorbs surplus moisture given off by the leaves and if there is any danger of these drying up they can in turn absorb moisture from the moss. It is often a good idea to include beneath the moss a layer of damp but not wet earth, in which the larvae of Tenthredinidae and some Coleoptera and Lepidoptera can pupate. This is not necessary for mining Diptera. A number of mined leaves are put into each jar, after first ascertaining beyond doubt that they belong to the same species. If the leaves are small they should be kept as they are but with larger leaves a suitable area should be cut out around the mine. Experience will soon show how much of the leaf the larva requires to reach maturity. With species which change to other leaves and mines, additional leaves should be kept in a tin in order to give the larva further food, should this later be necessary; this precaution need not, of course, be taken if the food plant is readily available.

If a locality has been found with numerous mines of a species which one wishes to breed, as many as possible should be collected, since one must always reckon with some degree of loss through disease, parasitisation, etc. Other leaves should be kept for the mine herbarium but for this empty mines can be used, although a few with the larva inside should be retained. They can also be put in a tin or between the pages of a notebook. A single mine may have features not normal for the species and therefore as many leaves as possible should be selected for the herbarium.

Many characteristics of the mine can most easily be recognised in the fresh leaf which has not yet dried, and it is therefore a good plan to make an immediate examination of the most important features for identification purposes. Special attention should be paid to the method of oviposition, the shape of the mine, its size and depth, its colour and also that of the surrounding parts of the leaf; one should investigate any existing openings, the points of entry of *Coleophora*, the position of exit slits on the upper or lower surface of the leaf in empty Agromyzidae mines, the inside of the mine, any possible silk, pupal chambers, frass deposits and their shape and colour in the various parts of the mine, the extent to which the mined area can be stretched (representing a difference between *Lithocolletis* and *Parornix*) and any prepared emergence holes. A search should then be made of the surroundings of the mine; this is particularly important in mines of *Coleophora*, where it is necessary to discover the larval case and without which identification is frequently impossible. Many species of this genus construct only a small mine and then proceed to another leaf; it may often be necessary to examine many leaves, before finding the first case. It is desirable to collect as many cases as possible, in order to establish a certain identification; when pupating, many species attach their case to quite other plants than those on which they had been feeding, as with *C. serratella* L. and *S. fuscedinella* Z. A pupal site of this kind can easily suggest that a species has been feeding on a plant, where in fact it had not been mining at all. When dealing with older mines, the cases must often be looked for at the site of pupation, which is frequently on the stem or on the twigs of the plant, where the cases are much more difficult to see than on the leaves.

Many collectors use only tin boxes for breeding out the mines; the maintenance of a constant humidity is not, however, so easy as with jars provided with a layer of moss and closed with a cork. However, tins are useful for keeping the mines in temporarily, before transferring them to the final breeding jars after arriving home. If this practice is adopted, it is to be recommended to include in each tin with the few mined leaves additional leaves to prevent the atmosphere becoming too dry.

2. Breeding miners and their preparation for the cabinet

It is of great importance to keep a breeding diary. Every mine collected should be numbered serially and entered up in the diary. Under each number the following information should be noted: place and date of discovery, the most important characteristics of the mine which have been detected, the botanical name of the plant, details of the environment which may also be of importance, date of leaving the mine, of pupation and finally of the emergence of the imago. Information as to whether the mine is common or rare and which plants or parts of plants are favoured should also be noted. The breeding jar receives the same number; it must also be included on the data label of every insect bred, in other words of the imagines of both miners and their parasites and it is also given to the corresponding species in the mine herbarium. By this means it is possible to say at a moment's notice and without fear of mistake which insect belongs to which mine and vice versa. For ease of reference it is a good idea to keep the diary on one side of the page only and for each number to include a page with the mine in question affixed. Attention to great accuracy in this numbering ensures that no mistakes occur and is of the greatest importance.

The breeding jars now require constant and careful observation by the entomologist. If moisture accumulates on the jar, this shows that there is excessive humidity and the moisture must be removed. If the leaves are seen to be drying up, the inside of the cork should be damped with water. The jars should not be kept at too high a temperature and they should under no circumstances be left exposed to the sun, as they then become too warm. If one is forced to travel, it is best to pack them in a box which can be carried by hand, in order to prevent any change in their normal position; the jars must at all times remain upright, so that the moss does not work loose and roll around. If the larvae have left the mine for pupation and have gone into the moss or the earth, the empty pieces of leaf should immediately be removed from the jar to prevent any unnecessary development of mould. If they have pupated inside the mine, they can be carefully removed and laid on the covering of moss in the jar. In some cases, as with *Lithocolletis*, it will be better to place the leaves containing pupae in another jar without any moss, merely ensuring the necessary humidity by regularly damping the cork. In such cases it is best to cut away as much as possible of the leaf around the pupation site, so that only the piece of leaf containing the cocoon or the pupal chamber remains. The jars should be carefully examined daily, if possible both mornings and evenings. The humidity can be checked at the same time. If the larvae were found while very young, after a certain time the food should be renewed. The new leaves are placed in the jar, and as soon as the larvae have transferred to them, the old leaves are removed. There are a few species in which one finds that, after leaving the mine, the larvae wander around in the jar for some considerable time, apparently unable to find a suitable place to pupate, such as the Elachistidae (except *Elachista*) and some *Cosmopteryx*. Such species, when they have left their mine, should be transferred to a jar with fine saw-dust, in which they soon construct their pupal cocoon.

In the first generation, emergence takes place after a very short time, the pupal quiescence lasting usually only a few weeks. The second generation, as also the species having only a single generation, normally produce the imago only after overwintering. Overwintering species are always much more difficult to breed than those which emerge the same year and many specimens perish during the winter. One must make every effort during this period to provide them with conditions in the jars corresponding as closely as possible with the wild state, and this is not always easy; it is frequently especially difficult to ensure just the correct degree of humidity. The best plan is to sleeve out such species on the twigs on which they live but this may be impracticable for many reasons and is quite impossible if one has to travel. The one essential is to keep all species in a cold place during the winter; under no circumstances should they be put in a warm room. The most suitable places are lofts, balconies, etc., where the jars can be kept cold but protected from the worst ravages of the weather. With some species development can be accelerated by "forcing". After leaving the breeding-jars for a short period (say two weeks) exposed to extreme cold, they should be brought first for a few days into an unheated, and then into a warm room. The imagines will then emerge prematurely. The adult insects can in this way be obtained from the middle of January onwards and an immediate start can be made preparing them for the cabinet, thus saving time when one is later busy with collecting and further breeding. Forcing cannot, however, be adopted with all species and, for instance, with many *Nepticula* such a produces absolutely no results. On the other hand it can be satisfactorily employed with *Lithocolletis* and a large part of the *Agromyzidae.* If only a few mines of a certain species have been found, it is better not to attempt to force them, as at best this produces less perfect insects than if they are allowed to develop normally.

On examination of the breeding-jars, any insects which are found to have emerged should be immediately removed and put in a dry, empty jar or glass-topped box. This is achieved by placing the two jars one above the other and inducing the insect to crawl up towards the light into the new jar or pill-box. This is then closed (with the lid or cork downwards) and the insect kept alive until the following day. If a jar stands with the cork uppermost, the insects frequently squeeze between the glass and the cork and get damaged or crushed. Most insects require at least a day to harden up and to develop their full colouration. With very delicate species such as *Nepticula,* they may safely be killed after a few hour's; with insects where the development of their colour takes longer, such as Coleoptera, it is to be recommended to keep them alive for several days or even a week and during this time to feed them on leaves of their food plant.

Killing is undertaken most efficiently with ethyl acetate. A drop should be poured onto blotting-paper with a suitable backing and when it has been absorbed and the paper is no longer wet, the uncorked jar containing the insect to be killed should be placed upside down and exposed to the fumes of the ethyl acetate. Diptera and Coleoptera should be left for half an hour or more; with the former one must wait at least until their wings are folded above their body (Fig. 31). If they do not die in this position but fold their wings around their body, it is possible to induce the desired position by pressing gently with the head of a pin on one side of the thorax. One soon learns by experience just where this pressure should be applied to the thorax to cause the wings to reverse their position. Mounting of Diptera is simpler than with the other Orders: they are merely pinned on the tip of stainless steel micro pins or points. A mounting strip with a hole prepared for the bearer pin is laid upside down and a micro pin is stuck through from the underside, projecting sufficiently to impale the fly ventrally but not to protrude above the surface of its thorax. The setal pattern on the upper surface of the thorax, which is so important for purposes of identification, must remain clearly visible. A suitably long pin is then inserted through the mounting strip, and, after labelling, the fly is ready for the collection. With Coleoptera, a small drop of mounting gum is placed on the ventral side of the thorax and with this the beetle is glued to the mounting strip. As soon as it is firmly fixed by the hardening of the gum, the legs should be set in a position which shows all the features important for identification and is at the same time attractive to look at (Fig. 46). The best plan is to learn the exact technique of mounting from a coleopterist. Subsequent treatment is the same as for Diptera.

Mounting Lepidoptera is more difficult, since it is important to spread out the wings in such a way that no scales are rubbed off; the colour and pattern of the wings are particularly important for identification. Tenthredinidae are mounted in the same way and require no special discussion (Fig. 44). Specimens of both Orders need only be exposed to the fumes of the ethyl acetate until they are completely unconscious and show no movement. The jar should then be removed and they should be pinned with a micro pin in the exact centre of the thorax but, in contrast to Diptera, from above. It is most important that the pin should enter the thorax exactly in its centre. The more accurately the insect has been pinned, the easier is its subsequent mounting. Before pinning species of *Lithocolletis*, one should note if there is a longitudinal white line on the dorsal side of the thorax; this is an important characteristic for identification which, after pinning, is often difficult to detect.

Mounting takes place on special micro setting-boards. A good selection is available at any entomological store. A number of the boards should have a groove 1 mm wide, which is suitable for *Nepticula* and other moths of this size. Other boards should have a groove of 2 mm (*Lithocolletis* and others), while finally one needs a groove of 3 mm for the larger Gelechiidae and Tortricidae. The moth is pinned on the setting-board so that the base of the wings lie flush with the surface of the board. Braces of smooth, transparent setting paper, wide enough to cover the wings, are now pinned on each side of the groove and first one, then the other, wing is moved up under these braces, until they have reached approximately the position illustrated in Fig. 25. The wings should be moved with the setting-needle only by pressure on the inner margin; in order to prevent the removal of any scales, the surface of the wings should not be touched. Sometimes the correct position can be obtained merely by blowing the wings forward. Once the wings are as desired, the braces are fixed by pins at each end. A small pad of cotton-wool should be inserted beneath the abdomen to support it in a horizontal position. The setting-board should then be put in a tightly closed tin, containing several drops of ethyl acetate to complete the killing of the insects. After two to three weeks the braces should be carefully removed. Mounting strips should be prepared by making a hole at one end for the large bearer pin, and at the other end a smaller hole with a finely pointed needle for the micro pin. The micro pin carrying the insect which has been set and is now ready for the collection is inserted into this hole and the bearer pin into the other; after labelling, the insect can be placed in the collection. It is a good idea to learn this method of mounting from a microlepidopterist. If one feels that it is not possible to carry out this mounting without causing considerable damage to the moth or if, as when travelling, there is insufficient time to set the insects properly, it is adequate merely to pin them correctly and to spread out the wings quite provisionally. They should be kept in an entomological store-box and pinned so that the bodies are resting on the cork; the fore- and hind-wings can then be eased forward either with a pin or by blowing and they can be allowed to harden in this position. They can subsequently be set properly without great difficulty. For mounting insects which have already dried in an unset condition, they should be placed in a relaxing tin with relaxing fluid or damp sand, and after a short time (at the most several hours) they can be set as if fresh. Beetles need relaxing for rather longer, and Diptera should be placed higher up the pin to prevent any possibility of the insect coming loose after drying.

It is of the utmost importance to ensure that each insect is accurately labelled with as much detailed information as possible. This is best done immediately after setting. The label cannot contain too much information. It is essential to include: locality, date of emergence, name of food-plant (genus and species), number in breeding diary, if possible the date of the discovery of the mine; also the name of the collector and, if applicable, of the identifier should not be left out. It is best to attack several such labels to each pin, as a single label would have to be too large. The more comprehensive the data on the label, the greater is the insect's scientific value. When breeding, not only the mine-producer but invariably also its parasites will be obtained. These must not be thrown away; they are mounted in the same way as Diptera and should be labelled with the same amount of detail, with in addition the name of the mine-producer which was their host. While at the moment it is not possible to determine the species of Braconidae and Chalcididae obtained in this way, they nevertheless represent valuable material for future detailed work on these groups, for which the discovery of the host is a step of great importance.

3. Arrangement of the collection

Every minologist's collection consists in reality of two collections — one of mine-producers (the insects obtained by breeding), and of a mine herbarium. The minologist does not collect, like other entomologists, on a systematic basis but on an ecological basis. This is also reflected in the arrangement of the collection. The insects are grouped not according to their systematic classification but the dominating principle in their grouping is their occurrence on the various plants, which they mine. Accordingly, each species obtained by breeding is labelled in two ways: first, there is a coloured label (say green) bearing the botanical name of the plant species; below this is the name of the first mine-producer found on this plant and then follow the bred specimens from as many localities as possible. The next label gives the name of the following species of miner found on the same plant, again followed by the actual specimens and so on. It is best to arrange the plants according to their natural family groupings, so that one drawer contains the plants of a given family with the miners found on them. Anyone having sufficient room can put after each miner any parasites which may have been obtained from it, whose names should be recorded on differently coloured paper;

the majority of such labels will probably have to remain blank for the time being or will contain only the generic designation of the parasite, if this can be obtained from a specialist. The arrangement according to the food-plants, which are themselves grouped according to the plant families, facilitates ease of reference, especially as many identical species occur on closely related plants. When wishing to discover the extent to which the same or different species have been obtained from related plants, the insects one wishes to examine will in this way be found close together. The arrangement of the mine herbarium follows the same system as any other botanical collection. The leaves containing the mines must first be dried. This is done by placing them between sheets of blotting-paper or newspaper and slightly weighting them, so that they are subjected to slight pressure. When necessary the individual leaves must be laid out in such a way that they show up all characteristics of the mines. It is important to speed up as far as possible the process of drying the leaves. This is best achieved by frequently changing the sheets of paper between which the mined leaves are drying, in other words keeping the paper at all times dry, and drying the damp sheets before using them a second time. The more often the paper is changed, the more rapidly does drying take place and the better preserved is the natural colour of the leaves. Greater pressure does not accelerate this process. With leaves which are particularly difficult to dry, such as those of Crassulaceae, the process of drying can be speeded up by "ironing" them with an iron which is not too hot, and thus rapidly expelling a large part of their water content, before placing them between paper. This can also be recommended for all parts of the plant, which experience shows turn black on drying, such as certain species of Populus and Salix, Lathyrus niger L. and others. When one finds on changing the paper that the leaves no longer feel cool to the lips, this shows that the mined leaves are sufficiently dry and can be transferred to the mine herbarium.

The preparation of the mine herbarium depends to some extent on the taste of the minologist, but also on the available space. In principle, it is desirable to use for each species as large a sheet of paper as possible, to which a large number of leaf-mines of the same species can be attached. Lack of space and the requirement of easy reference, however, make it necessary to set a limit on the size of paper used. We therefore do not use the large folio sheets commonly found in herbaria preserving whole plants, but use sheets only half this size. We have found most suitable a sheet of 33 × 21 cm folded in two. The necessary data are noted on the outside of this folded sheet: first, the name of the plant (to which can be added the name of the family), then the name of the producer of the mine, below this the place and date of finding the mine, date of emergence of the imagines, any desirable information on the local environment and the name of the collector, with if necessary the name of the person identifying the species. Inside this folded sheet as many mines are affixed, as there is room for. The mines should be attached in such a way that they can at all times be easily removed, since they will often be required for detailed examination. The leaves should under no circumstances be covered with any sort of glue or paste and stuck down, which would make out of the question any subsequent examination of the characteristics of the underside. It suffices to attach them very loosely by sticking a narrow strip of thin adhesive paper over the petiole, of the kind used by philatelists for sticking in stamps. Leaf-mines stored in folded sheets in this way are unlikely to fall out of place and they can easily be examined whenever desired, by removing with forceps the strip of paper by which they are attached. It is better not to attach the mines to plain sheets of paper, as they are much more likely to become damaged when frequently looking through the collection. The type of paper used should have some capacity for absorbing the traces of moisture, which still remain in the leaves, but it is also necessary to be able to write on it with ink without this "running". Certain types of duplicating paper best incorporate these two qualities.

Each folded sheet should bear the name of the plant and will contain each species of mine found on it; these are best arranged alphabetically according to the name of the mine producer. The plant species are arranged according to their genera, which are again in alphabetical order within each family. Each plant family is best stored in a separate folder. These folders should be constructed in the form of port-folios, as is often done in herbaria: two strong cardboard covers are fastened together on one side by adjustable strips of tape and on the side where they open they can be closed by clips or additional tapes. In this way a uniformly equipped herbarium can be formed, which can be kept in any book-case and which does not take up too much room.

It is advisable to keep each drawer of the cabinet well supplied with paradichlorbenzole, which effectively kills the beetle larvae and Psocidae feeding on the insects in the collection. Mine herbaria are less exposed to such dangers, especially when they are frequently examined, as will be inevitable with any thorough study of mining. It is nevertheless desirable to keep the herbaria in an air-tight box or case, which should be thoroughly disinfected at some suitable time when the collection is not in constant use. It can be left to the taste of the minologist whether the files are arranged according to the natural relationship of the plants or alphabetically; for rapid reference we found the arrangement according to plant families the most useful.

Experimental Minology

A quite recent development in the study of miners is represented by experiments undertaken during the course of breeding. The insects are not kept under conditions most likely to guarantee successful development but on the contrary the environment is subjected to extensive modifications in order to test the ability of the mining larvae to react to changed conditions and thus to gain some idea of their ecological potentialities. This is a most fascinating field of study and a few words about its technique may well prove useful. Possibilities are offered to every enthousiast to carry out experiments by simple methods, the results of which may prove of real interest and value in many directions. We have at various times referred to the fact that no satisfactory explanation of the origin of certain peculiarities of the mine or of its surroundings in the leaf have yet been discovered. When breeding mining insects, various substances which are not normally present can be introduced into the leaves containing the larvae; the twigs bearing the mined leaves should be placed in solutions of these substances, deposits of which are then absorbed by the leaves. Changes of temperature, humidity and other factors can easily be arranged without the necessity of any elaborate preparations. If it is suspected that a particular species reproduces parthenogenetically (as with many Tenthredinidae, some species of *Phytomyza*, *Lithocolletis emberizaepennella* Bché. etc.), it is easy to establish proof of this. This is desirable for all species in which in captivity the males only occur infrequently. In such cases the growing food-plant or, failing this, whole shoots or twigs should be placed in water in a breeding cage, and all the freshly emerged females should be segregated here, as long as no males have been obtained. If oviposition takes place and mines develop, breeding should be continued until the imagines appear, in order to obtain the proof desired. We managed in this way without difficulty to establish the parthenogenetic development of *Phytomyza crassiseta* Zett. on Veronica for several generations.

It is particularly interesting to check the various host-plants found acceptable by a species which is not strictly monophagous. A male and female obtained by breeding should be put in a breeding-cage with, as food, some plant related to their normal food-plant and one should observe whether oviposition and the construction of a mine takes place. This transfer to abnormal food-plants may be undertaken with the larva itself, which has already begun its development in another plant. Buhr (1937) for the first time carried out systematic transfers of this type and obtained interesting results which we already discussed in the chapter on the miners' choice of food. It is very simple to carry this out with the mining larvae possessing greater freedom of movement, in other words which are able to change from one mine or leaf to another. The larvae are removed from their mine and placed in a container with only leaves of the new plant. The more closely this is related to the original food-plant, the more easy will it be for the larva to eat its way into the new leaf. An artificial transfer must be made with species which never change their mine and are unable to penetrate into another leaf. A slit under the epidermis is made in the new leaf, into which the larva, which has been removed from its original leaf, is inserted. If the transfer is successful the larva will soon construct a mine channel in the new food-plant, which in many cases, however, differs from that in its normal food-plant. It is not easy in all plants to make a convenient hole below the epidermis; in such cases the larva can be introduced into an empty mine of some other species. When this is not possible, Buhr's example can be followed and the whole portion of leaf containing the larva can be transplanted. The area of the leaf around the larva is cut out and this, with the larva inside, is inserted into the new leaf; this is best done on the mid-rib near the leaf-base. The larva continues feeding first in the tissue of the leaf to which it is accustomed and when there is none of this left, it is compelled to penetrate into the tissue of the new leaf and feed on this, or it must die. The transfer will be effected easily, with difficulty or will fail, according to how closely the two plants were related and to the ecological potentiality of the species concerned. Experimental transfers were carried out, for example, with the larva of *Trypeta zoë* Mg., which is so common on Senecio. It was found that normal development continued on the genera Lactuca and Galinsoga which also belong to the Compositae and on Artemisia vulgaris L. Solidago, however, was completely refused (on which *Vidalia spinifrons* Schröd. lives in completely similar mines). Transfer to Taraxacum was in many cases unsuccessful (on which *Trypeta immaculata* Macq. lives in similar mines); when a mine was produced it differed considerably from normal. Normal development took place with a transfer to Cucurbita pepo L. (belonging to the closely related family Cucurbitaceae), but not on Cucumis sativa L. *Philophylla heraclei* L., which has developed oligophagy (first degree) within the Umbelliferae family, will not, however, accept indiscriminately all genera of the family, as the experiment showed. Larvae originating on Pastinaca could not be reared to maturity on Laserpitium siler L. or on Siler trilobum L.; during our experiments only small mines were formed on Aegopodium, although in the wild state fully developed mines have been found on this genus. The more monophagous species will attempt to mine on very closely related plants, but even here the larva is unable to reach maturity. We transposed larvae of *Fenusa pumila* Klg. from birch leaves into empty linear mines of *Agromyza alni-betulae* Hd. on Alnus; very minute gnawings on both sides of the channel of the fly larva showed that the larvae had eaten some of the unaccustomed leaf substance before they died. On the other hand when the closely related *Fenusa dohrni* Tischb. from Alnus leaves was transferred to such channels, it continued its normal development. It succeeded in this only if it was placed with its ventral side uppermost. If it was introduced with its dorsum uppermost, narrow, full-depth outlines of the channel showed that it had been able to obtain a slight amount of food; however, it soon turned over and was then able to continue its normal feeding. The larva of *F. dohrni* Tischb. (from Alnus) introduced into the mine of *Fenusa pumila* Klg. (Betula) did continue feeding but was not able to reach maturity. *Phyllotoma vagans* Fall. (from Alnus) showed itself to be less discriminating; the larva introduced into the mine of *Scolioneura betuleti* Klg. on Betula continued feeding normally and pupated as in an Alder leaf, while the reverse exchange of food-plants was not successful: the *Scolioneura* was unable to continue its development in the mine of the *Phyllotoma*. Even *Tischeria ekebladella* Bjerk. from Quercus was able to continue its development inside the mine of *Fenusa dohrni* Tischb. on Alnus; in this joint mine the shallower and less transparent *Tischeria* portion was easily distinguishable. *Nepticula albifasciella* Hein., introduced from Quercus into *Agromyza* mines on Alnus was unable to adapt itself to the change. A full-depth, transparent outline of the channel showed that the larva had commenced feeding but development did not continue. Buhr (1937) quotes a number of interesting transfers of this type, among which the successful transfer of *Scaptomyzella flava* Fall. to Phlox is particularly striking. Experiments of this kind can be arranged without difficulty and provide interesting insight into the possibilities of the mining larva adapting itself to different plants of varying degrees of relationship to its accustomed food-plant. They also represent a means of graduating the ecological potentialities of the mining species concerned. Work of this kind, however, does not enable conclusions to be drawn about the value of food selection by miners as a guide to plant relationships. In such cases there is no selection of food either by the larva or by the ovipositing female but compulsion exercised on the larva to adapt itself. To this extent we cannot agree with the conclusions of Buhr who decided, from the extensive potentialities of mining larvae of adapting themselves to abnormal food-plants, that the choice of food by oligophagous mining insects cannot be utilised to indicate relationships between the food-plants.

Experimental minology is still in its first beginnings. Future work in this field using more subtle methods will enable it to be developed into a particularly interesting and fruitful specialised branch of minological research.

5. Difficulties of Determination

We suggested in our introductory chapter that the mined leaf represents a "visiting Card" left by the larva of the insect producing the mine; the insect's name is revealed provided that the lettering of the visiting card can be deciphered. In general it is true to say that with adequate experience and knowing the species of food-plant, it is possible to determine the name of the mine-producer from the characteristics of the mine. There are, however, a number of exceptions which fortunately are not numerous, where the name cannot so easily be decided and which must be briefly mentioned here.

a. Oligophagy and Polyphagy

A very large proportion of miners are monophagous, in other words confined to one definite plant genus or species, and this greatly facilitates determination. This immediately becomes more difficult with species which are less restricted in their choice of food, that is, with oligophagous and polyphagous species. In such cases it can be anticipated that mines of these species will constantly be found on plants, on which they have never previously been observed; we may breed out a mine which is strange to us, only to find that the producer is already familiar from other plants. Frequently the systematic position of the newly infested plant will give some indication of the producer. Thus, even before results have been achieved from breeding, a knowledge of plant relationships can indicate which species of miner will emerge. This applies particularly to all cases of oligophagy (first degree), in which the mine is found on a different genus of the same family. Thus, *Dizygomyza lamii* Kltb. and *D. labiatarum* Hend., which are markedly oligophagous within the Labiatae, will certainly be found on many additional genera, on which they have so far not been noted. Similarly, the characteristic mines of *Liriomyza congesta* Beck. will be found on many Leguminosae, on which they have so far not been discovered; in such cases the experienced minologist can make the correct identification before the adult insects have been bred. The mines of *Ceuthorrhynchus contractus* Mrsh. and *Phyllotreta nemorum* L. will be found on many new genera of Cruciferae, which from the point of view of the miner are so very similar.

At least among certain groups of Gramineae similar conditions apply. This family of plants causes the minologist particular difficulties, in as far as he is not a specialist on Gramineae. At the time when mines are found in the leaves of Gramineae, the flowers have usually not yet appeared, so that the exact identification of the species of grass will frequently cause difficulties. It is therefore advisable to identify as many as possible of the grasses when they are in flower and to prepare cross-sections of the leaves of all species which have been identified with certainty (when held between split Sambucus pith they can easily be cut with a razor blade). After staining the sections with safranin and light-green, the vascular bundles and the sclerenchyma stand out clearly from the rest of the leaf; the arrangement of these organs and the structure of the epidermis cells, of the stomata and the hairs are so characteristic for each species of Gramineae that leaves of grasses which are not in flower can subsequently easily be identified by comparison. If mines are found in later years in leaves of grasses of which the species are not known, a rapidly prepared cross-section of the mined leaf will enable the species of grass to be easily identified by comparison with the preparations.

There are a number of especially polyphagous species, which presumably may still be found on a large number of new food-plants, and which we will briefly mention here. They are:

Liriomyza strigata Mg. (Fig. 12). The mine is easily recognisable by its juxtaposition to the leaf veins and the branching and forking, which it undergoes as a result. The initial part of the channel begins anywhere in the leaf but then soon turns towards the mid-rib, along which the main length of the channel is constructed and from which lateral channels branch off. The frass is deposited in threads or like strings of pearls; there is frequently no trace of frass in the side channels, as the larva constantly returns to the main portion of the mine on a principal vein, where the main frass deposits are found. This species may be found on all dicotyledons.

Phytomyza atricornis Mg. (Fig. 120). The larva lives in lower or upper surface linear mines, the grains of frass are usually separate and distinct and lie alternately on the left and right sides of the mine channel. Pupation takes place in a "pupal blister", usually on the lower surface of the leaf and often some distance from the end of the mine. Sometimes (as when on Phlox) the frass may be deposited in threads. This species may be found on all dicotyledons.

Coleophora paripennella Z. lives in autumn in a case with lobate appendages and from within this case produces the well-known *Coleophora* blotch-mines. If the case can no longer be found, the occurrence of this species can be suspected when one notices that in a number of blotch-mines holes have also been eaten out of the upper epidermis or that in addition to the normal small openings in the lower epidermis, others with a much larger diameter occur. In both instances the larva has cut out additional material for the enlargement of its case. This species lives on a wide variety of flowering trees and shrubs but may also be found on bushy plants.

Orthochaetes insignis Aubé is found in irregular, full-depth mines, in which the frass forms a central line. The mines sometimes tend to be linear, at other times more blotch-like and they usually have short lateral offshoots. They are not so invariably associated with the mid-rib as those of its near relative, *Orth. setiger* Beck., which is confined to Compositae. This species represents a clear example of polyphagy, since it occurs not only on dicotyledons but also on monocotyledons.

There are a few other groups among the oligophagous and polyphagous species which demand special attention. Among them there are a number of very closely related species, which produce very similar mines, and which it has not yet been possible to identify from their mines alone. They will be discussed individually below.

b. The *Cnephasia* species of the "*wahlbomiana*" group

Among the few mining Tortricidae there are several species which are extremely similar and which have only recently been separated. Even the imagines resemble each other so closely that it is only possible to distinguish them by examining the genitalia. These species are in the genus *Cnephasia;* they were previously grouped under the name *wahlbomiana* L. Recent investigations have shown that these are a group of very similar species, which can only be effectively identified from their genitalia. Existing information shows that among these species there are only three which produce mines in their early instars, subsequently living among leaves which are spun together or entwined in some other way. We thus have to distinguish three species whose mines also are so similar that it has not so far been possible to identify them by this means. They are also so polyphagous that there is no possibility of distinguishing them from their food-plants. Frequently all three species are found mining on the same plant. The mines are very inconstant in shape, sometimes tending to be linear, at other times more blotch-like.

They are full-depth mines; some or all of the frass is ejected and is often found in little piles at the beginning of the mine. Inside, these mines are somewhat drawn together by silk; they therefore do not appear clearly transparent but are always to some extent opaque. They invariably appear in early spring but according to Buhr (i. l.) a second generation may occur in more southerly latitudes. It is not difficult to breed them. The larvae soon leave their mines and will accept practically any food that may be offered them. They mature very rapidly, pupate between spun-together leaf-parts and in a short time the moth emerges. When the minologist finds mines of members of this genus in April or May, he should arrange to breed them out — which is very simple, and will then have three means for identifying the species. In contrast to the superficially very similar imagines, the early stages are much more easy to distinguish. The species can be identified from the larvae, this can later be checked by the characteristics of the pupae, and finally there is the possibility of examining the genitalia of the male, in order to obtain absolute certainty.

We shall now briefly outline the method of examining genitalia, in order to provide a few practical hints for anyone who has not carried out this work. The chitinised apparatus should be removed from the end of abdomen of the freshly killed male, before the moth is set. Pressure exerted on the tip of the abdomen with a scalpel or a pin is sufficient to expose the genital organ, which can then be completely removed with forceps. This causes so little damage to the abdomen that it is not noticeable in the set insect. If the moths have already been set and are thoroughly dry, the tip of the abdomen must be cut off with a sharp dissecting knife. The genitalia obtained in this way should be soaked in a tube of 4% caustic potash, preferably after a previous immersion in spirit, which facilitates penetration of the caustic solution. The genital organ should be left for 24 hours and then placed in clean water, where the chitinous apparatus can be freed from all other remnants of tissue; this can be done without difficulty with the help of two needles or fine-pointed pins. When it is clean, it should be washed several times in water in order to remove the caustic potash; a few drops of citric acid can be added to the last wash to neutralise any remnants of the caustic solution. It should now be extended in such a way that the essential parts can be easily examined. This is done by opening out to the sides the two shell-shaped valves on a slide in a small quantity of water, producing a position as shown in Fig. 175. When this has been achieved a cover slip or a portion of a second slide should be placed above them to hold them in position. If necessary the cover slip can be slightly weighted. Spirit, preferably pure alcohol, should now be added from the side to expel the last remains of water and in which the organ hardens. Additional alcohol must be added from the side of the cover slip, if evaporation makes this necessary. About a quarter of an hour suffices to fix the specimen which can then be prepared as a permanent slide. It should be put into clove oil to remove the last traces of water and is best left there for several days until no moisture remains. Suitable stains may also be used (magenta with gradations to orange, or light green) in order to obtain a more attractive slide. This is done before putting the specimen into clove oil, and alcohol-soluble substances must be used, as water-soluble stains will affect the hardening and fixed position of the specimen. From the clove oil the specimen is placed on a slide, in the normal way for preparing all microscope slides, and is embedded in Canada-Balsam or some similar medium and covered with a cover-slip. It is important to label the slide with as much detail as possible. However, it is more convenient if the specimen can be mounted on the same pin as the moth itself. This is a rather more complicated process but has the advantage of the genital organ always being with the insect it belongs to. For this purpose the specimen is mounted between two cover slips. One cover slip is affixed to a slide — it will adhere firmly to the condensation produced by breathing — and on this a small drop of Balsam is placed to receive the specimen, which is now covered with a second cover slip the same size as or (better) a little smaller than the first. One should not use too much Balsam, so that the upper cover slip can be adjusted with-

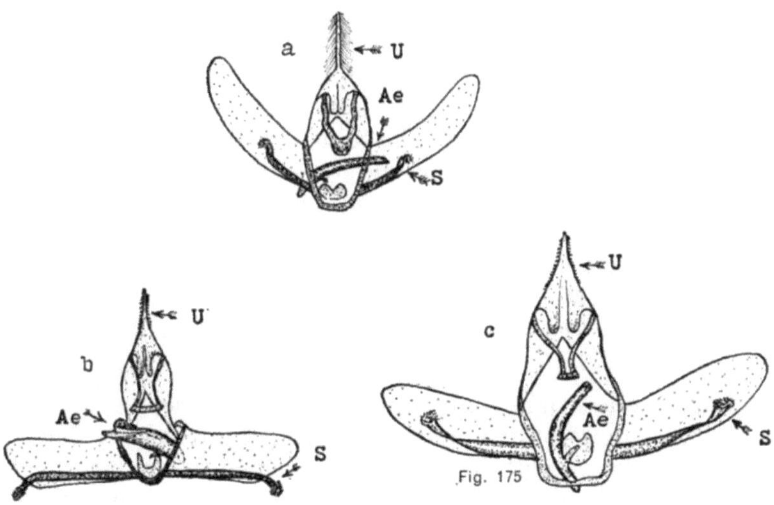

Fig. 175. Male genitalia of *Cnephasia*: a. *incertana* Tr., b. *virgaureana* Tr., c. *chrysantheana* Dup.

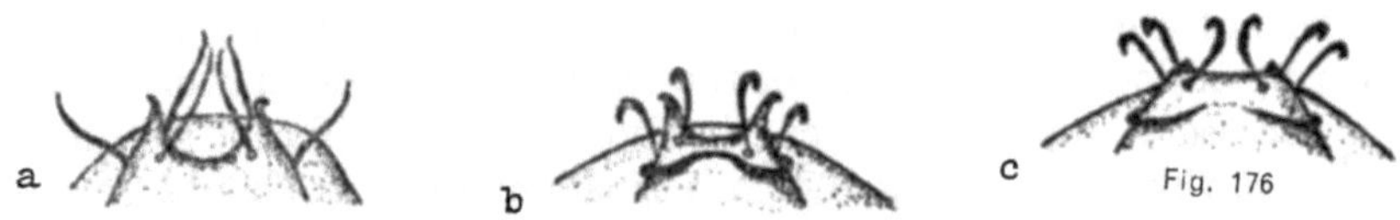

Fig. 176. Terminal area of pupa of *Cnephasia:* a. *chrysantheana* Dup., b. *virgaureana* Tr., c. *incertana* Tr.

out being itself smeared by Balsam. When in position the one cover slip should exactly cover the other and the slide should now be left to dry for some considerable time. When completely dry the slide should be edged with adhesive white paper, sufficiently wide to cover all but the actual specimen. The paper should be allowed to project slightly on one side, in other words is not flush with the edge of the slide, and the pin can be inserted through this area so that the moth and the prepared specimen of its genital organ are together. The paper edging of the slide should bear the same data as the label of the moth itself.

The shape of the following three organs is especially important for identifying the three species concerned: the unpaired uncus (U), the sacculi which are on the inner side of the paired valves (S), and the aedeagus (Ae) which glides between them. These parts are indicated by small arrows in Fig. 175.

The larvae can be distinguished by the bristle-bearing warts. One species possesses below the anal orifice a peculiar chitinous structure, which is characterised by its darker colour resulting from its more marked chitinisation. The pupae also can be identified down to species, if the anal end is freed from the threads of silk in which in normally hangs and if the structures on the terminal segment are examined under a good lens. The spine-covered extension known as the cremaster and the bristles in the area of this structure are characteristically constructed for each species (Fig. 176) and enable the species to be identified. It is advisable to apply all three checks and thus make quite certain of the identification.

Taking all three stages into account one can thus identify miners of the genus *Cnephasia* according to the following key:

1. Larva: warts light, only the base of the setae dark.
   Pupa: spines on the cremaster not bent into hooks, short; their length less than the distance to a ridge lying ventrad. Bristles on the end of the body noticeably thickened away from their point of origin, suddenly bending over distally to a hook (Fig. 176c).

Male genitalia: uncus very long, covered with long hairs, sacculus very short, aedeagus slender (Fig. 175a) *Cnephasiella incertana* Tr.

- Larva: all warts completely black.
  Pupa: spines on the cremaster as long or longer than their distance from the ventral chitinous ridge, if such exists (Fig. 176a, b).
  Male genitalia: uncus short, covered with short spinules, sacculus longer............................ 2

2. Larva: a dark, comb-like structure lies below the anal orifice.
   Pupa: terminal bristles noticeably thickened away from their point or origin, bent into a hook at the end. A distinct transverse ridge ventrad of the cremaster spines (Fig. 176b).
   Male genitalia: aedeagus short and thick, with a toothed lobe before its end, sacculus as long as the lateral valve (Fig. 175b) .... *Cnephasia virgaureana* Tr.

- Larva: Anal comb lacking.
  Pupa: bristles on the end of the body thin, gradually tapering to a point. No distinct transverse ridge ventrad of the cremaster spines. Empty pupal case always darker than in preceding species (Fig. 176a).
  Male genitalia: aedeagus slender, with no distal spine-covered lobe, sacculus shorter than the valve (Fig. 175c) .... *Cnephasia chrysantheana* Dup.

It cannot be too strongly recommended to examine all *Cnephasia* mines from these three points of view. Quite possibly other species of this group mine in their early instars, although this has not yet been discovered. A routine three-fold check of all these mines that may be found will either bring to light other points of difference indicating one or more mining species in this group or it will confirm that no others exist.

c. *Lithocolletis* species on Rosaceae

The species of the genus *Lithocolletis*, which are so easily recognisable from their tentiform mines and from the pupation which takes place inside the mine, cause the minologist considerable difficulties, if he studies the species of this group which live in lower surface mines on Rosaceae. There are a number of very similar species, which

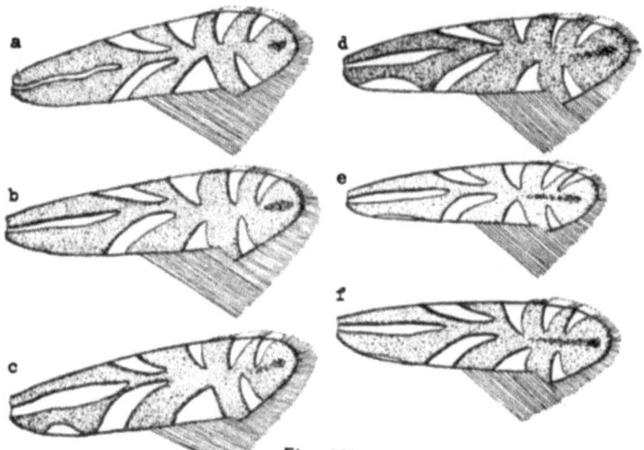

Fig. 177

Fig. 177. Forewing of *Lithocolletis:* a. *cerasicolella* HS., b. *sorbi* Frey., c. *pomifoliella* Z., d. *oxyacanthae* Frey., e. *cydoniella* Fb., f. *mespilella* Frey.

are extremely difficult to distinguish and whose association with specific plants is not completely certain. Each of these species has a food-plant which it prefers, so that identification from the host-plant is possible with some degree of accuracy; however, a small percentage of the species also occur away from their normal food-plant on other genera of the family. If one wishes to be certain of the identification, it is therefore necessary to breed the imago. This is extremely easy with this genus, since the larva pupates inside its mine and the mined areas containing the pupae need only to be cut out and they will almost certainly in due course yield the perfect insects. Their characteristics, however, are somewhat difficult and variable, that it is also advisable to prepare a specimen of the genitalia for certain identification. This is done as explained in the previous section, only in this case the genitalia should not be spread out, as with this genus this is far more difficult; Fig 178 therefore shows a lateral view of the genitalia. The most important feature are the sacculi (indicated by arrows and coloured darker), which in many cases are asymmetrical in shape. The following key sets out the differences between the species both in the fore-wings (Fig. 177) and in the genitalia; if both sets of characteristics are used in identification, it will be possible accurately to determine the species:

Key of the Lithocolletis of the Rosaceae

1. Basal streak of the forewings sinuate, notably long and slender (Fig. 177a). Valva tapered, with spines near the apex (Fig. 178a, b) (on Prunus) 2

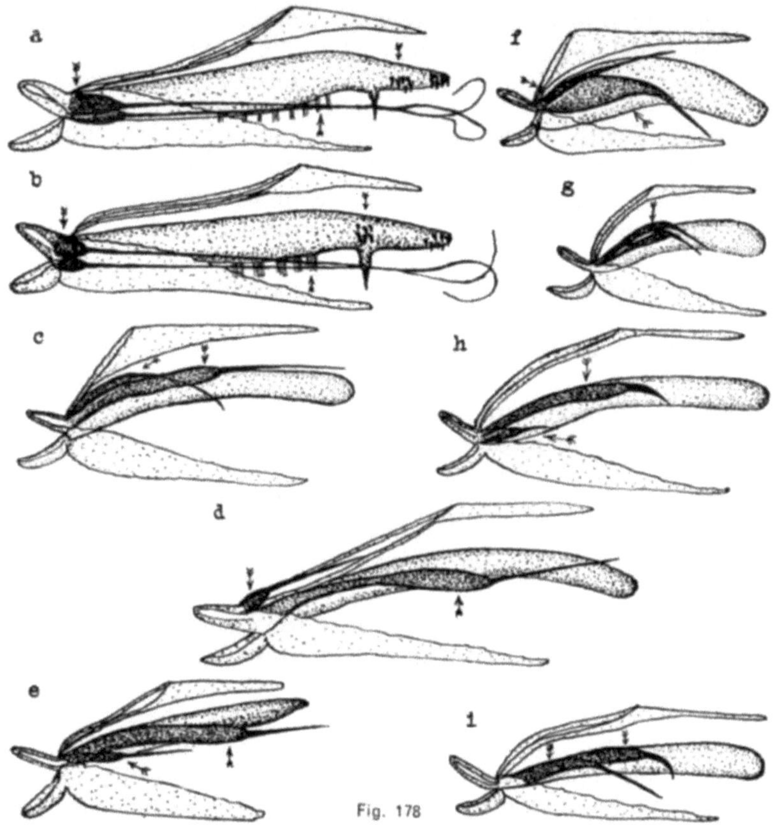

Fig. 178. Lateral view of male genitalia of *Lithocolletis:* a. *spinicolella* Z. b. *cerasicolella* HS., c. *mespilella* Frey., d. *pomifoliella* Z., e. *oxyacanthae* Frey, f. *cydoniella* Fbr., g. *sorbi* Frey., h. *blancardella* Fbr., i. *domesticella* Sorh.

Basal streak straight, not markedly long or slender (Fig. 177b—f) Valva not spined, filaments of the sacculi shorter (Fig. 178c—i) 3

2. Forewings golden-ochreous, basal streak dark-margined above (P. spinosa). Valvae: both groups of spines near the apex (Fig. 178a, 180) .... *L. spinicolella* Z.
- Forewings golden-orange, basal streak unmargined (Fig. 177a) (P. avium, cerasus a.o.). Valvae: a group of spines above the strong spine, the other at the apex (Fig. 178b, 179) .... *L. cerasicolella* H.-S.

3. Posterior tarsi unspotted. Forewings see Fig. 177b (Sorbus) Sacculi equal in length, filaments short (Fig. 178g) *L. sorbi* Frey
- Posterior tarsi normally dark-spotted, sacculi asymmetrical 4

4. Forewings with well-marked white spot or dorsal line towards the base (Fig. 177c, d). Valva rounded at end 5
- Forewings without well-marked white dorsal line or spot towards the base. Valva tapered or rounded at end 7

5. Basal streak broader, often connected with first dorsal spot (Fig. 177c). Valva very long, ribbon-shaped (Fig. 178d) (On Malus, preferring cultivated forms) *L. pomifoliella* Z. (*concomitella* Bankes)
- Basal streak narrower, never connected, Valva shorter, filaments of the sacculi shorter.......... 6

6. Expanse 8—9 mm, forewings golden-orange. Valva rounded at end. Sacculi with agglutinated, short filaments (Fig. 178h) (on Malus, preferring subsp. acerba, Amelanchier a.o.) .... *L. blancardella* Fbr.
- Expanse 7—8 mm, forewings orange-fuscous or brownish-orange (Fig. 177d). Valva tapered at end (Fig. 178e) (On Crataegus) .... *L. oxyacanthae* Frey

7. Forewings clear red-orange, the white spots broad (Fig. 177e). Valva shortened, right sacculus broad (Fig. 178f) (On Pyrus and Cydonia) .... *L. cydoniella* Fbr.
- Forewings darker. Valva longer 8

8. Expanse 7—8 mm, forewings deep ochreous orange (Fig. 177f) Sacculi with long filaments (Fig. 178c) (On Mespilus, Sorbus) .... *L. mespilella* Hbn.
- Expanse 8—9 mm, forewings brownish-orange or orange-fuscous. Sacculi as before. (On Malus and Pyrus) *L. pyrivorella* Bankes

As many specimens as possible bred from genera of Rosaceae should be examined utilising this key, in order to obtain accurate information on the distribution of species on the various genera of Rosaceae. In particular, the classification of *L. pyrivorella* Bankes as a separate species is not yet fully justified. Simultaneously, attention should be paid to any minute differences in the mines (type of folds, mottling on the upper surface of the leaf, position and colour of the cocoon, time of emergence), so that it may be possible in future to work out differences between the species on the basis of their mines.

d. Mines on Aquatic plants

When discussing species living on aquatic plants we mentioned that identification of the Tendipedidae (Chironomidae) is only possible if the larvae or pupae are also examined. The parts of these early stages which are of importance in identification were illustrated in Figs. 32 and 33. These mines are only found on submerged parts of the plants; they are all similar and take the form of a linear channel with at least two openings of the same size. As pupation takes place in the mine, it is not difficult to collect either larva or pupa in the mines and from these to establish the identification. This is best done from the empty pupal cases and it is therefore advisable to wait until the imagines have emerged, when the characteristics of the pupa are more easily recognisable.

Our knowledge of the distribution of the individual species of *Hydrellia* on their various food plants is also very incomplete. In this case examination of the very similar larvae gives no assistance. It is therefore necessary to await the emergence of the flies before the species can be identified. We indicated previously that breeding of these insects is very easy; pupation takes place inside the leaf and the imagines emerge after a very short time and can then be examined.

With Tendipedidae, as with *Hydrellia*, we lack detailed information on the association of individual species with definite plants. Apparently such association is lacking or only exists to a very loose degree. Any minologist with access to lakes or ponds has the opportunity of carrying out specialised investigations into this problem, which is outside the scope of most entomologists. One point which should perhaps be followed up is the varying times of emergence of the individual species.

e. Mines in fallen leaves.

A number of species do not live in fresh, green leaves but mine the tissue of leaves which have already fallen and are dead. They thus represent a transitional stage between true miners and inhabitants of rotting matter (saprophytes) and need not concern us here, since the feeding pattern they produce no longer accords with our definition of a mine. Among them we do nevertheless find transitions to true miners. A particularly interesting case is *Coenorrhinus (Rhynchites) pauxillus* Germ., which mines predominantly, but not always, in dying leaves, which have fallen as a result of the oviposition puncture. Accordingly the mine invariably commences in the leaf-base. Examination of mines of this kind, which are found especially on Rosaceae

(Malus, Cotoneaster, Crataegus, Mespilus, etc.) enables this species to be identified without difficulty from the distinct head capsule of the larva. No other species with similar habits have as yet been observed, apart from those mentioned below; however, their very different larval morphology excludes any possibility of confusion.

In the *Lauxaniidae* (Sapromyzidae), a family of Diptera, there are a number of species which feed between the upper and lower epidermis of fallen and already rotting leaves. These species can be immediately recognised from the larva, which is entirely without head capsule or legs. The other members of this family live on all possible kinds of rotting matter, and this type of mining represents a case of specialised saprophytic habits, not of mining. They accordingly produce no mines of characteristic shape, and the minologist wishing to identify them must breed the larvae and have the imagines determined by a specialist on Diptera. The larvae are found in autumn and winter on fallen leaves and they can be bred without great difficulty.

The species with these habits, which have so far been discovered, are *Halidayella aenea* Fall., *Peplomyza litura* Mg., *Sapromyza obsoleta* Fall. *notata* Fall. and *subfasciata* Zett., *S. (Minettia) plumicornis* Fall. and *Tricholauxania praeusta* Fall. It is fairly certain that further species of this family with similar habits will be encountered, and it seems desirable that minologists with time and opportunity for such investigations, should carry out some detailed study of these transitional forms between miners and saprophytes, which have been so neglected in the past. Particular attention should be paid to whether specialisation occurs on certain species of plants, although, from the origin of such feeding habits, this appears unlikely.

True miners in falles leaves *(Nepticula)* are invariably accompanied by "green islands" around the mine.

f. Pseudo-mines

In an earlier chapter we clearly defined a mine as the living and feeding quarters in the green tissues of a plant; anyone commencing the study of these structures can nevertheless easily confuse other very similar feeding forms as mines, but which do not, however, accord with our definition. We will briefly outline below a number of such possibilities.

Frequently fungal patches are found on leaves, causing a discolouration of the area of leaf concerned and often being mistaken for mines, unless a careful examination is made. The commonest example of this is the overwintering form of an ascomycete, Rhytisma acerinum Pers., which occurs in autumn on Acer. These patches are even more likely to be considered as blotch-mines of some insect species, since "green islands" are also found in association with them. Fungal patches are quite commonly found on Populus very similar to the blackish, roundish blotch-mines of *Leucoptera sinuella* Rtti. Beautiful red fungal patches were found on Rumex lunaria L. in the Canary Islands, which most deceptively resembled the serpentine mines of *Xystophora rebeli* Hering, occurring on the same plant.

The egg slits made by various species of insects form another group of mine-like structures. The female makes a slit in the leaf tissue with its ovipositor (frequently also in the stem), into which the egg is inserted. This slit may still remain for some time after the larva has hatched from the egg and then be mistaken for a mine. The Tenthredinidae, for example, construct egg slits of this kind. During the development of the embryo the egg absorbs moisture from the tissues of the plant and undergoes considerable expansion, thus enlarging the slit. For some time after the larva has left its egg shell, the slit may be mistaken for a mine. It can always be recognised by the absence of frass. Beetles of the genus *Dytiscus* and ovipositing Odonata also produce egg slits of this type, in both cases on plants which are near or in water so that on hatching the larva has immediate access to the water. Any structures of this kind which may possibly be encountered can therefore be ascribed to one of these two groups.

At the beginning of this book when discussing mines vis-à-vis other feeding forms, we referred to "window-feeding", which can often closely resemble mining but which is distinguishable by the fact that only one, usually the upper, epidermis remains intact. Confusion with mining can often arise when the window-feeding occurs in long strips, thus producing a superficial similarity with linear mines. Such linear window-feeding is frequently found with the larvae of *Phytodecta* species which feed on the lower surface of Populus leaves. A very similar feeding pattern may also be found on Comarum with *Phytobius comari* Herbst, a weevil. In all these cases a closer examination will show that only epidermis remains and that therefore these feeding forms cannot be considered as mines.

It is more difficult to decide this when the leaves concerned are very hairy, as it is in any case not easy to discover whether the epidermis remains beneath the surface of hairs. A number of species feed in this way, especially among Lepidoptera. They are frequently considered as miners, although their feeding patterns are not mines. The Noctuid *Eublemma paula* Hb. even consumes both upper and lower epidermis but leaves the hairy covering intact and the resulting effect is a striking similarity with a mine; the extremely hairy leaves of Helichrysum are particularly conducive to such feeding habits.

In some cases cavities arise in the tissue of the leaves, as also of the stem or fruits, into which air penetrates, thereby giving them the appearance of mines. In a similar way chlorotic areas arise (having a scarcity of chlorophyll), likewise producing the effect of a mine. Such cavities and areas lacking in chlorophyll are found in the stems of Umbelliferae, especially in the region of the inflorescence (for instance in Anethum), but a close examination suffices to show that no feeding has been taking place and no frass is present. Very similar structures, particularly resembling linear mines, are found in the pods of Leguminosae; in these too, especially in damp weather, a fungal infection develops and the resemblance to mines is then even more striking. Pale patches known as chlorotic areas, which can often be found in the leaves of a wide variety of plants and particularly in cultivated garden plants as a result of one-sided manuring, resemble mines but can easily be distinguished from them by the invariable absence of frass.

Frequently very numerous, transparently-white, short channels can be seen in the leaves of Nicotiana as a result of a certain virus disease, and these might also be mistaken for mines. However, there is no suggestion that oviposition has taken place, frass deposits are lacking, and in particular their extreme profusion in the leaf makes any confusion with mines unlikely.

A number of Tenthredinidae larvae occur on Salix, especially of the genus *Pamphilius*, and their feeding on the margin of the leaf results in one-sided growth and the leaf curls downwards. The *Pamphilius* larvae live beneath the folded-over leaf-edge, in the same way as *Parornix* and some *Caloptilia*. The minologist may easily be inclined to consider these as mines of *Lithocolletis*, in which the leaf-edge often folds over the mine. However, if part of the leaf-edge is folded back, the complete absence of silk will soon show that this is neither a mine, nor a lepidopterous leaf-fold, as may often be found with the later instars of miners, but that it is a genuine gall formation, a cecidium.

g. Aids for the study of mines

As the bibliography which follows shows, the literature on mines and miners is extremely scattered. There are very few comprehensive works and these are either incomplete or fell victim to the war and are no longer obtainable. Furthermore, systematic investigation has so far been largely limited to Europe. The book by Needham, Frost and Tothill certainly represents a detailed study of North American miners but it can be expected that a systematic investigation of North

America by specialists on minology will result in many more species being discovered; everything which has been described so far is based primarily on investigations by specialists in individual Orders but not in the whole field of mining insects. The group of species around *Liriomyza pusilla* Mg., as also the *albiceps* and *obscurella* groups of *Phytomyza*, will produce a vast increase of species. The investigations of Morris N. Watt in New Zealand represent a complete study but in this case also it can be anticipated that future work will result in a greatly increased number of species.

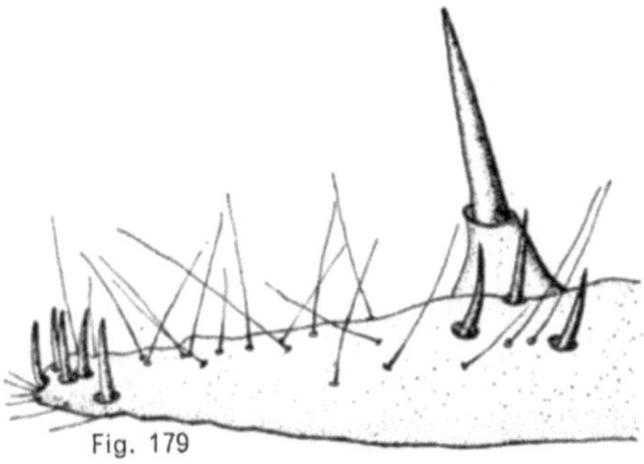
Fig. 179

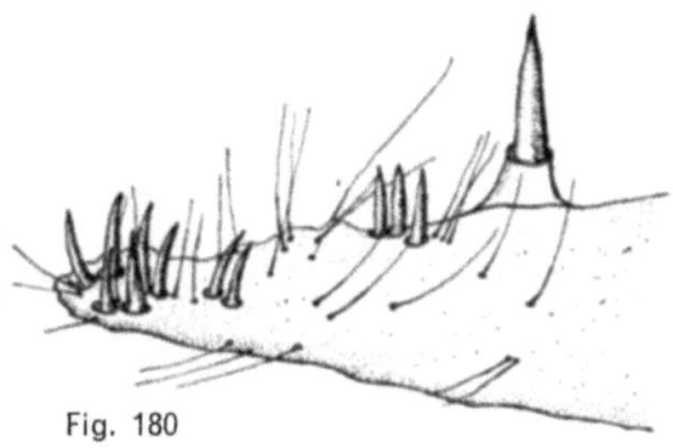
Fig. 180

Fig. 179. Apical part of valva of *Lithocolletis cerasicolella* H.S.
Fig. 180. Apical part of valva of *Lithocolletis spinicolella* Z.

Until the appearance of the book (in preparation) for identifying the mines of the whole of Europe, which is provided with a key for each plant genus by which the mines can be determined without having to breed out the imago, the only way of identifying the producer of any leaf-mines which may be found is to breed the imagines and, after careful mounting, to send them for determination to a specialist on the appropriate group. It is desirable that as many minologists as possible should work in contact with each other and exchange information on their breeding and identification problems, in order to avoid too much extra work for the specialists who are in any case fully occupied. During winter months the author of this book will willingly determine mines or the perfect insects bred from them in any cases of doubt. When breeding, as many specimens as possible of each species should be obtained, so that several examples of each species can be made available to the specialist; this is necessary when studying variable characteristics and the specialist will certainly appreciate a few spares in reward for his often painstaking work. The greater the care and attention given to mounting and labelling the specimens, the more easily can the identification be established.

Once again it should be emphasised that as many minologists as possible should get together and where convenient work in groups; it then becomes possible to exchange experience, to show each other species of mines which have been identified, so that the experience of one becomes available to the others and in this way the unnecessary and frequently difficult breeding of known species can be avoided. Reciprocal exchange of information on breeding experiences of every kind will facilitate everyone's work and the amount of new information discovered will in this way be greatly extended. Assistance with each other's determination problems will eliminate the necessity for duplicating breeding of the same insect, and the pleasure attendant on every fresh discovery will in its turn help to increase interest in minology.

FINAL REMARKS: PROBLEMS AND AIMS OF FUTURE RESEARCH ON MINES

Although the work of the last hundred years has resulted in a great deal of knowledge about mines and miners, very much still remains to be done and future generations of minologists can rest assured that their investigations will constantly produce new information, whether it be of a systematic or of a general biological nature. Throughout this book reference has continually been made to problems which can be taken up by future research and to points still requiring investigation and explanation.

From the systematic point of view the fauna of Central and Northern Europe has in general been fully investigated; however, this does not exclude the possibility of many new discoveries of so-far unknown mining insects. Anyone who collects intensively in the same area for a number of years will discover specialities which have not been found elsewhere, and thus add to our store of knowledge. Even with our commonest species, existing information can still be enlarged in many directions, and in this the work and breeding done by everyone can be of assistance. For example, until to-day it is still not possible to identify the *Lithocolletis* on Quercus and on Salix solely from their mines. Here is a problem which can be tackled by anyone, by working out the essential differences between the mines of the individual species from the study of numerous individual mines. In many species the first instar mines are still not known — for example, in Gracilariidae and Incurvariidae. Much new information remains to be obtained from observations of oviposition or the breeding of unknown first instar mines which may be found. The females can frequently be caught and if they can be induced to lay, the formation of the first instar mine can be observed. The available descriptions of larvae of mining Lepidoptera are in many cases very inadequate; they usually deal with the slight variations in the larva's colour. A morphological description of the mouth-parts and the setal pattern will provide valuable additional information and a more certain guide to identifiaction. For the identification of mines it is necessary in many cases to rely on an examination of the larvae or pupae. In such cases — for instance, with the species living on Gramineae — there is little doubt that in due course differences in the mines will be discovered, which will enable the mines to be recognised and identified. Particularly with the Gramineae it will be of value to discover to what extent individual species confine themselves to certain genera. Anyone making such investigations must of course possess detailed knowledge of the Gramineae genera. When investigating miners on aquatic plants, study of the Tendipedidae will result in the discovery of a number of new species, and in this case also it will be of value to discover to what extent species have specialised on individual aquatic plants.

The more miners that are bred from outside Central and Northern Europe, the more new discoveries will be made, and this applies especially to the Eastern parts of the Mediterranean area where many new finds will come to light. Only some new species can be expected from Eastern Europe; but the further south one goes from there, the more new discoveries can be anticipated. The more extreme is the environment in our latitudes, the more certain it is that fresh await discovery. Areas of raised bogland, xerothermic localities, and especially the medium and higher mountains ranges will doubtless provide many an interesting new discovery. The smaller and more insignificant are the plants examined, the greater is the likelihood of finding on them species which have so far been overlooked.

Much remains to be explained in the general biology of miners. Any observations likely to throw any light on the origin of the mining habit would be of particular importance. It seems certain that the mining habit arose polyphyletically but any observations suggesting that it arose in any other way would be of the greatest interest. With *Coleophora* there are already various indications as to how mining may haven arisen and there is no doubt that the same will apply to other groups in specialised ecological field. The development of mining from xylophagous forms is of particular interest, such as we find combined in the ontogeny of *Heliozela betulae* Stt., which for the greater part of its larval existence is a wood-feeder and only in the last instar becomes a leaf-miner. Stem-borers represent a border region in this respect, in the systematic elucidation of which Buhr did such valuable work. Experimental minology can be of assistance in such cases and by using suitable host-plants it may be possible to produce mines in species, where so far no mines have been known.

More attention should be paid to the morphology of mining larvae as related to the peculiarities of the food-plants of the various species and there is little doubt that numerous adaptations will be discovered in this connection. As the brilliant investigations of Trägårdh showed, the variability of the head-parts in miners is very considerable. Trägårdh was naturally only able to give a small selection of examples and it will be left to future investigations to go into greater detail and to illustrate further cases of specialisation. In this respect detailed morphological investigations of mining Coleoptera and Hymenoptera will reveal analogous developments; it will be possible to show here also that the shape of the larva's body is dependent on the particular nature of the leaf substance they favour.

Further information on the special conditions inside the mine cavity may be obtained by physiological and bio-chemical investigations. This will lead to a greater understanding of the numerous features of mines and miners, which at the moment are without explanation. Detailed investigations of the influence exerted on the plant tissues by the frass deposits of the mining larva will prove particularly interesting and will provide the key for understanding many hitherto obscure points. On the one hand interesting interactions between insect and plant inside the tissues of the latter will come to light, while on the other hand it will be seen to what extent the mine and its surroundings represent a self-contained unity within the leaf.

An immediate task is to make a more exact list of the plants selected by mining insects, which should then be correlated with the plant relationships as worked out by botanists. All possible information on oligophagy, polyphagy, xenophagy and organoxeny should be collected, and in this way interesting side-lights on pure botany will be produced as a by-product of the information we have obtained from insects.

Such investigations will show more and more clearly that the study of mines represents a border region covering both the insect and plant kingdoms. If minology is studied systematically from this point of view, it can lead to new discoveries in both botany and zoology and will provide the link uniting these two sciences; they will both then in their turn be interested in its further development and expansion.

May these remarks therefore attract new supporters for our science, who will find both emotional and intellectual satisfaction in the study of these attractive structures and who will be prepared to assist in the further construction of the great edifice of minology.

REFERENCES

Anonym, Curiosities of Entomology. London, 64 pp. (p. 25—32: The Microlepidoptera or Leaf-miners. p. 33—38: The Coleophora, or Tentmakers).

Abderhalden, Handbuch der biologischen Arbeitsmethoden. IX. Teil I/II: Fischer, E., Züchtung der Lepidopteren, p. 277—356. Heikertinger, F., Züchtung von Dipteren. p. 357—390 & Züchtung von Coleopteren. p. 399—458.

Adamczewski, St., 1947, Notes on the Lepidoptera of Poland. — The Entomolog. 80. p. 101—106, 133—136.

Aldrich, J. M., 1918, Notes on Diptera. — Psyche (Boston) 25, p. 30—35.

———1918, Seasonal and climatic variation in Cerodonta. — Ann. Soc. ent. America 11. p. 63—66.

———1922, A new Sugar-cane Miner. — Bull. Brookl. ent. Soc. 18. p. 20—23.

Alfieri, A., 1924, Notes sur Anister raffrayi Grouv. et sa larve (Coleopt.). — Bull Soc. R. ent. Egypt. 1924. p. 82—83.

Altum, B. 1874/75, Forstzoologie III. Insekten. 1. 335 pp., 2. 365 pp. Berlin.

———1876, Der Buchen-Springrüsselkäfer. — Ztschr. f. Forst- u. Jagdwesen 1876. p. 283. f.

———1894, Das Rabenfederchen (Coleophora coracipennella Hb.) ibid. 1894, 639.

Amsel, H. G., 1930/31, Die Microlepidopterenfauna der Mark Brandenburg nach dem heutigen Stande unserer Kenntnisse. — Dtsch. ent. Zeit. Iris 44, p. 83—132. 45, p. 147—201.

———1932, Die Microlepidopterenfauna der Stilfserjochstrasse und des Ortlergebietes. — Dtsch. ent. Ztschr. 1932, p. 1—17.

———1932, Neue mitteleuropäische Kleinschmetterlinge und Bemerkungen über Melasina lugubris und M. ciliaris. — Dtsch. ent. Ztschr. 1932. p. 18—24.

———1933, Die Lepidopteren Palästinas. — Zoogeographica 2. p. 1—146.

———1935, Neue palästinensische Lepidopteren. — Mitt. Zool. Mus. Berlin 20. p. 271—319.

AMSEL, H. G., 1936, Zur Kenntnis der Kleinschmetterlingsfauna Sardiniens. Eine neue Schützeia-Art aus Palästina. — Veröff. Dtsch. Kolon. & Uebersee-Mus. 1. p. 344—368.

AMSEL, H. G. & HERING, M., 1931, Beitrag zur Kenntnis der Minenfauna Palästinas. Dtsch. ent. Ztschr. 1931. p. 113—152.

———1933, Beitrag zur Kenntnis der Minenfauna Sardiniens. Boll. Lab. Entom. Bologna. 6. p. 79—92.

AMYOT, M., 1863, Histoire de la Teigne Syringella (Tinea syringella Fabr.) — Ann. Soc. entom. Fr. (4) 4. p. 5—12.

ANDRES, A., 1920, Ein Schädling an Azaleen in Gewächshäuser. — Ztschr. ang. Ent. 6. p. 404—405.

ANKEL, W., 1938, Frass-Spuren von Schnecken. — Natur & Volk 68, p. 333—337. Frankfurt/M.

AURIVILLIUS, CHR., 1882, Orchestes populi L. såsom skadedjur. — Ent. Tidskr. 3. p. 30.

BAER, W., 1906, Ein Frass von Steganoptycha nanana Tr. nebst Bemerkungen über ähnlich lebende Kleinfalter. — Naturw. Ztschr. f. Land- u. Forstwirtsch. 4. 429.

———1909, Gracilaria simploniella F. R. und die Eichenrindenminen. — ibid. 7. p. 45.

BAIL, TH., 1908, Über Pflanzenmissbildungen und ihre Ursachen, vornehmlich über mannigfaltige Entwicklung der Fliederblätter unter dem Einfluss der Raupen der Fliedermotte, Gracilaria syringella. — Ber. Westpr. zool. bot. Ver. 30. p. 239—256.

BANKS, CH. S., 1919, Two Philippine Leaf-mining Buprestids, one being new. — Phil. Journ. Sci. 15. p. 287 ff.

BARASCH, A., 1934, Natürliche Gruppierung der mitteleuropäischen Coleophoriden. Dtsch. ent. Ztschr. 1934. p. 1—116.

BARCA, E., 1932, Drei verkannte Gelechiiden, die aus Norwegen beschrieben sind. — Ent. Tidskr. 53. p. 31—35.

BAUDYŠ, E., 1929, Muškovitost hrachu. — Českosl. Zemědělce 11. č. 17. p. 64.

———1936, Hospodářská Fytopathologie. II. Hubení Škůdců Živočišných. — Brno. 624 pp.

BAUER, E., 1917, Beiträge zur Mikrolepidopteren-Fauna von Naumburg a.S. — Mitt. Ent. Ges. Halle 11. p. 1—71.

BECKER, L., 1861, Observations sur quelques chenilles de Tinéides. — Ann. Soc. ent. Belg. 5. p. 52—57.

BECKER, TH., 1894, Dipterologische Studien I. Scatomyciden. — Berlin. ent. Zeit. 39. p. 77—196.

BECKER, TH., 1894, Revision der Gattung Chilosia Meigen. — Nov. Act. Acad. Leopold. Halle 62. Nr. 3. p. 194—512.

———1894, Dipterologische Studien IV. Ephydriden. — Berlin. ent. Zeit. 41. p. 91—276.

———1902/03, Ägyptische Dipteren. — Mitt. Zool. Mus. Berlin 2. pt. 2/3. p. 1—195.

BECKER, TH., BEZZI M., KERTESZ, K. & STEIN, P., 1905/07, Katalog der palaearktischen Dipteren vol. 3. 4.

BECKER, TH., 1908, Dipteren der Kanarischen Inseln. Mitt. Zool. Mus. Berlin 4. p. 1—180. Dipteren der Insel Madeira. — ibid. p. 181—206.

———1926, Ephydridae. in Lindner, Fliegen d. paläarkt. Reg. 56. 115 pp.

J. C. B. (BECKMANN, JO. CHRIST.), 1680, Bericht von denen auff den Blättern der Bäume in diesem 1680ten Jahre häufig gefundenen Schlangen-Gestalten. Frankfurt/O.

BEIRNE, B. P., 1945, The Male Genitalia of the British Stigmellidae. — Proc. Roy. Irish Acad. 50. B. Nr. 9. p. 191—218.

———1940, Stenoptilia saxifragae Fletcher, in Ireland. — Ent. Rec. & Journ. Var. 52. p. 61—62.

———1941, The Life-History of Stenoptilia saxifragae, Fletcher. — ibid. 53, p. 85—86.

BELING, TH., 1888, Beitrag zur Metamorphose einiger zweiflügeliger Insekten. — Verh. zool. bot. Ges. Wien 1888. p. (3).

BENANDER, P., 1928, Gelechiidae. in: Svensk Insektfauna 10. Tineina 1. p. 1—97.

———1929, Über die Raupen von vier Arten der Cnephasia wahlbomiana-Gruppe. — Ztschr. wiss. Ins. Biol. 24. p. 164—167.

———1931, Zur Biologie einiger Kleinschmetterlinge IV. — ibid. 26. p. 48—54.

———1936, Fjärilar från Kivik och Kinnekulle. — Ent. Tidskr. 1936. p. 262—269.

———1936, Förteckning över Sveriges småfjärilar. Catalogus Insectorum Sueciae. VI. Microlepidoptera. — ibid. 1946. p. 1—82.

———1937, Notiser om svenska microlepidoptera, med beskrivning av Lita arenariella n. sp. — Opuscula entom. 1937. p. 29—32.

———1937, Die Gelechiiden-Raupen. Eine vergleichend-morphologische Untersuchung. — ibid. p. 49—109.

BENANDER, P., 1938/39, Die Coleophoriden Schwedens. — ibid. 1938, p. 107—124 & 1939, p. 30—110.

———1944, Sveriges Lithocolletider. — ibid. 1944. p. 79—137.

BENSON, R. B., 1934, Some new or little known Sawflies. — Entom. Monthl. Mag. 70. p. 201—203.

———1935, The Alien element in the British Sawfly Fauna. — Ann. Appl. Biol. 22. p. 754—768.

———1936, Two new European Sawfly Genera of the Subfamily Fenusinae. Ann. Mag. Nat. Hist. (10) 18. p. 620—626.

———1936, Some more new or little known British Sawflies. — Ent. Monthl. Mag. 72. p. 203—207.

BERGER, K., 1936, Über Blattminen und Pflanzengallen der westlichen Oberlausitz. — Isis Budissima 13. p. 89—191.

BEUTENMÜLLER, WM., 1890, On the habits of the N. A. Rhynchophora. — Canad. Ent. 22. p. 200—203, 258, 261.

———1890, The food habits of the Chrysomelidae. — Entomol. Americ. 6. p. 175.

BEZZI, M., 1928, Diptera and Atherica of the Fiji Islands. — London, p. 1—220.

BISCHOFF, H., 1927, Biologie der Hymenopteren. Berlin, 598 pp.

BLANCHARD, E. E., 1926, A dipterous leaf-miner on Cineraria, new to science. — Revist. Soc. Entom. Argent. 1. p. 10—11.

BLATCHLEY, W. S., 1910, Coleoptera or Beetles known to occur in Indiana. 1386 pp.

BLATCHLEY, W. S. & LANG, C. W., 1916. Rhyncophora or weevels of North America. — 682 pp.

BLUNCK, H., 1920, Die niederen tierischen Feinde unserer Gespinstpflanzen.— Ill. Landw. Zeit. 40. p. 259—260.

———1925, Der Stand der Rübenfliegenfrage. — Mitt. dtsch. Landwirtsch. Ges. 25 p. (1—4).

———1927, Der Stand der Rübenfliegenfrage im Jahre 1926. — Die deutsch. Zuckerindustr. 1927, Nr. 1. p. 1—4.

———1932, Zur Kenntnis der Lebensgewohnheiten und der Metamorphose getreidebewohnender Halticinen. — Ztschr. angew. Ent. 19. p. 357—394.

——— & MEYER, 1932. Erdflöhe. — Flugbl. Biol. Reichsanst. 121.

BLUNCK, H., BREMER, H. & KAUFMANN, O., 1933, Untersuchungen zur Lebensgeschichte und Bekämpfung der Rübenfliege Pegomyia hyoscyami Pz. 11. Die Lebensgeschichte der Rübenfliege. — Arb. Biol. Reichsanst. 20. p. 517—585.

BOAS, I. E. V., 1896/98, Dansk Forstzoologie. — Kjöbenhavn.

———1921, Bladfald fra Bøg i Juni. — Fra Skoven og Traemarkedet III. Nr. 14. p. 105—107.

BOEKER, 1908, Die Bekämpfung der Runkelfliege. — Deutsch. landw. Presse 35. p. 388.

BOIE, F., 1847, Zur Entwicklungsgeschichte mehrerer Trypeta-Arten. — Stettin. ent. Zeit. 8. p. 326—331.

———1848, Entomologisch-biologische Notizen. 3. Fenusa rubi B. — ibid. 9. p. 340.

BOISDUVAL, J. A., 1867, Essai sur l'entomologie horticole. — 648 pp.

BONDAR, G., 1925, Lagarta minadora das folhas do algodoeiro, Acrocercops helicometra Meyr. n. sp. — Corr. Agricult. 3. Nr. 2. p. 44—46. Bahia.

BONNET, CH., 1745, Traité d'insectologie; ou observations sur les pucerons. — 228 pp.

BÖRNER, C., 1906, Zwei neue Möhrenschädlinge aus den Gattungen Ceutorhynchidius und Phytomyza. — Arb. Biol. Reichsanst. 5. p. 289—292.

BONSDORFF, E. J., 1866, Finnlands tråvingade insecter (Diptera) förtecknade och i korthet geskrivne II. — Bidr. F. Vet. Soc. Helsingfors 7. 306 pp.

BOSELLI, F., 1928, Elenco delle specie d'insetti dannosi e loro parassiti ricordati in Italia dal 1911 al 1925. Portici, 264 pp.

BOUCHÉ, P. Fr., 1834, Naturgeschichte der Insekten. Berlin, 216 pp.

———1847, Beiträge zur Kenntnis der Insektenlarven. — Stettin. ent. Zeit. 8. p. 142—143, 162.

BOX, H. E., 1922, The White Coffee Leaf-miner (Leucoptera coffeella Stn.). — Farmers' Journ. Nairobi 4. p. 9—10, 19—21.

———1923, The bionomics of the White Coffee Leaf-miner, Leucoptera coffeella Guér., in Kenya Colony. — Bull. Ent. Res. 14. p. 133—145.

BRASHNIKOW, W. C., 1897, Zur Biologie und Systematik einiger Arten minierender Dipteren. — Nachr. Moskau. Landw. Inst. 3. p. 1—22.

BRAUN, A. F., 1908, Revision of the North American species of the Genus Lithocolletis Hb. — Trans. American. Ent. Soc. 34. p. 269—357.

———1910, New species of Tineina from California. — Ent. News 21. p. 171—179.

Braun, A. F., 1912, Notes on North American species of Nepticula, with descriptions of new species. — Journ. Cincinn. Soc. Nat. Hist. 21. Nr. 3. p. 84—102.

———1912, Notes on some North American Tineina. — Canad. Ent. 44. p. 159—161.

———1914, Notes on North American Tineina, with descriptions of new species. — Ent. News 25. p. 113—117.

———1914, Notes on North American species of Nepticula with descriptions of new species. — Canad. Ent. 46. p. 17—24.

———1914, Notes on Coleophora, with descriptions of two new species. — Journ. Cincinn. Soc. Nat. Hist. 21. p. 157—167.

———1915, Notes on some species of Tischeria, with descriptions of new species. — Entom. News. 26. 271—273.

———1916, Notes on Lithocolletis with descriptions of new species. — Ent. News 27. p. 82—84.

———1917, Nepticulidae of North America. — Trans. Amer. Ent. Soc. 43. p. 155—209.

———1918, New genera and species of Lyonetiidae. — Ent. News 29. p. 245—251.

———1919, Notes on Cosmopterygidae, with descriptions of new genera and species. — Ent. News 30. p. 260—264.

———1919, Descriptions of new species of Coleophora. — Ent. News 30. p. 108—112, 127—131.

———1920, Notes on Elachista with descriptions of new species. — Ohio Journ. Sci. 20. p. 167—172.

———1921, Notes on Elachista. II. — ibid. 21. p. 206—210.

———1923, Microlepidoptera: Notes and new species. — Transact. Amer. Ent. Soc. 49. p. 115—127.

———1924, The Chambers specimens of Tineina in the collection of the American Entom. Soc. 49. p. 347—358.

———1925, Some undescribed Microlepidoptera and notes on Life-histories. Trans. American. Entom. Soc. 51. p. 13—17.

———1925, Microlepidoptera of Northern Utah. — ibid. 51. p. 183—226.

———1930, Notes on new species of Microlepidoptera from the mineral springs region of Adams County, Ohio. — ibid. 56. p. 1—17.

———1933, Pupal tracheation and Imaginal venation in Microlepidoptera. ibid. 59. p. 229—268.

Braun, A. F., 1935, Notes and new species of Microlepidoptera. — ibid. 61, p. 45—52.

Bremer, H., 1927, Die Zuckerrübe und ihre Schädlinge. — Die Dtsch. Zuckerindustrie 1927. Nr. 19. 2 pp.

Bremer H., & Kaufmann, O., 1931, Die Rübenfliege Pegomyia hyoscyami Pz.-Monogr. Pflanzenschutz 7. Berlin, 110 pp.

Brèthes, J., 1923, Sur un Diptère mineur des feuilles de Salvia splendens et deux Hyménoptères, ses parasites. — Rev. Zool. Agr. & appl. 22. p. 153—158.

Brischke, C. G. A., 1868, Kleinere Beobachtungen über Insekten.-Schrift. naturf. Ges. Danzig N.F. 2. Heft 1. Nr. 6.

———1880, Biologische Bemerkungen. — Entom. Nachr. 6. p. 56.

———1880, Die Blattminierer in Danzigs Umgebung. Danzig, p. 1—58. 1880. & Schrift. naturf. Ges. Danzig N. F. 5. p. 233—290, 1881.

———1888/90, Insekten auf Farnkräutern. — Schrift. naturf. Ges. Danzig N. F. 7. Heft 1. p. 107 & 3. p. 10.

——— & Zaddach, 1883, Beobachtungen über die Arten der Blatt- und Holzwespen. Abt. 2. — Schrift. naturf. Ges. Danzig N. F. 5.

Britton, W. E., 1895, Notes on some leaf-miners. — Connecticut Agr. Exp. Sta. Ann. Rept. 18. p. 143—146.

———1896, A leaf-miner of the cauliflowers (Drosophila flaveola Mg.) ibid. 19. p. 204—205.

———1912, Chrysanthemum Leaf-miner or Marguerite Fly. — 11. Ann. Rep. St. Ent. Connect. (Connect. Agr. Exp. Sta. Ann. Rept. 35), p. 342.

———1916, Hymenoptera of Connecticut. — Bull. Connect. St. Geol. & Nat. Hist. Surv. 22. p. 1—824.

Bruand d'Uzelle, M., 1859, Essai monographique sur le Genre Coleophora. I. — Ann. Soc. ent. Fr. (3) 7. p. 375—414.

Bryk, F., 1929, Nachschrift zur Klärung des Teratologiebegriffes. — Ent. Zeitschr. 42 p. 287—289.

Buchner, P., 1921, Tier und Pflanze in intrazellularer Symbiose. Berlin. 462 pp.

Buhr, H., 1930, Einige Blattminen und Gallen von der Insel Lesina (Hvar) in Dalmatien. S. B. & Abh. naturf. Ges. Rostock (3) 2. p. 125—148.

———1932/37, Mecklenburgische Minen. — Stett. ent. Zeit. 93. p. 57—115 (1932); 94. p. 47—96 (1933); 96 p. 131—159, 262—292 (1935); 97. p. 304—317 (1936); 98, p. 109—124 (1937).

BUHR, H., 1937, Parasitenbefall und Pflanzenverwandschaft. — Botan. Jahrb. 68. Heft 2/3, p. 142—198.

———1940, Über Verbreitung und ausländische Nahrungspflanzen von Neptikeln. Ztschr. Wien. Ent. Ver. 25. p. 193—196, 209—211, 226—235.

———1941/42, Einiges über bekannte und unbekannte europäische Schmetterlingsminen. — Ztschr. Wien. ent. Ver. 26. p. 201—208, 235—236. 246—247. 283—287 (1941) vol. 27. p. 32—33, 47—55, 72—78. 110—120, 130—135. (1942).

———1941, Dipteren-, insbesondere Agromyziden-Minen aus Südeuropa. — Stettin. ent. Zeit. 102. p. 73—119.

BURKE, H. E., 1917, Flat-headed borers affecting forest trees in the United States. — Bull. U. S. Dept. Agric. Nr. 437.

BUSCK, A., 1904, Tineid Moths from Brit. Columbia, with descriptions of new species. Proc. U. S. Nat. Mus. 27. p. 745—778.

———1906, A review of the American moths of the genus Cosmopteryx. — ibid. 30. p. 707—710.

———1907, Revision of the American Moths of the Genus Argyresthia. — ibid. 32. p. 5—24.

———1909, Notes on Microlepidoptera, with descriptions of new North American species. — Proc. Ent. Soc. Washingt. 11. p. 87 ff.

———1909, New Microlepidoptera from New Mexico and California and a synoptic table of the N. A. species of Heliodines Stt. — ibid. 11. p. 175. ff.

———1912, Descriptions of new genera and species of Microlepidoptera from Panama. — Smithson. Misc. Coll. 59. Nr. 4.

———1913, Note on a barkmining Lepidopteron of the Genus Marmara Clemens. — Proc. Ent. Soc. Washingt. 15. Nr. 4. p. 150.

———1914, New Microlepidoptera from Hawaii. — Insec. Insc. Menstr. 2. p. 103 ff.

———1914, Two Microlepidoptera on Thurberia thespesioides. — Proc. Ent. Soc. Washingt. 16. p. 30—31.

———1914, Descriptions of new Microlepidoptera of Forest Trees. — Ibid. 16. p. 143—153.

———1914, The chestnut Bastminer. — Insec. Inscit. Menstr. 2. p. 3—4.

———1914, A new Gracilaria on Azalea. — ibid. 2. p. 1—2.

———1914, On Mnemonica auricyanea Wlsh. — Proc. Ent. Soc. Washingt. 16. p. 151—163.

Busck, A., 1915, Descriptions of new N. American Microlepidoptera. — ibid. 17. p. 79—94.

———1919, A new species of Bucculatrix injurious to Hollyhock. — ibid. 21. Nr. 5.

———1920, Description of new Central American Microlepidoptera. — Insec. Insc. Menstr. 8. p. 83—95.

———1920, A new Gracilaria injurious to Avocado. — Canad. Ent. 52. p. 239. ff.

———1931, On the female Genitalia of the Microlepidoptera and their importance in the classification and determination of these moths. — Bull. Brookl. Ent. Soc. 26. p. 199—211.

———1932, Description of a new Cosmopterygid leaf-miner on Helianthus. — Proc. Ent. Soc. Washingt. 34. p. 17—20.

———1934, Microlepidoptera of Cuba. — Entomologia Americana 13. p. 151—203.

Busck, A. & Heinrich, C., 1921, On the male genitalia of the Microlepidoptera and their systematic importance. — Proc. Ent. Soc. Washingt. 23. p. 145—152.

Caesar, L., 1917, Notes on some insects of the season. — Ann. Rept. Ent. Soc. Ontario 47 p. 106—110.

Calmbach, V., 1920, Tischeria complanella Hb. — Ent. Ztschr. 34. Nr. 18. p. 70.

———1921, Lyonetia clerkella L. — ibid. 34. p. 97—98.

———1923, Praktische Anleitung zur Präparation, Fang und Zucht der Microlepidopteren. — Stuttgart, 15 pp.

Campbell, A. S., 1925, Agromyzid Fly in Beans. — Lingnaam Agric. Rev. 3. p. 16—17.

Camerano, Lorenzo, 1879, Gli insetti; introduzione allo studio dell'entomologia. 342 pp.

Cameron, A. E., A contribution to a knowledge of the Belladona leaf-miner Pegomyia hyoscyami. — Ann. Appl. Biol. I, 1. p. 43—76.

———1917, Some experiments on the breeding of the Mangold fly (Pegomyia hyoscyami Pnz.) and the Dock fly (Peg. bicolor Wied.). — Bull. ent. Res. 7. p. 87—99.

Cameron, P., 1882/84, A monograph of the British Phytophagous Hymenoptera. — Ray Soc. vol. I. 340 pp., II. 233 pp.

Carpenter, G. H., 1911, Some Dipterous larvae from the turnip. — Journ. econ. Biol. 6. p. 67—74.

CARPENTER, G. H., 1913, Injurious Insects and other animals observed in Ireland during the year 1911. The Rapsberry Leaf miner Fenusa pumilio Htg. — Econ. Proc. Roy. Dubl. Soc. 2. Nr. 5. p. 72 f.

ČERNIK, L. F., 1925/41, Krankheiten und teratologische Missbildungen an Pflanzen der Olmützer Flora. — I.: Časopis Vlast. Spolk. Mus. Olomouci 36. Nr. 1—4. p. 124 ff. — II. ibid. 38. p. 14 ff (1927). — III. Verh. naturf. Brünn 62. p. 148—164 (1930). — IV. ibid. 63. p. 51—74 (1932) — V. ibid. 64. p. 56—76 (1933). — VI. ibid. 65. (1934. — VII. ibid. 66. p. 29—60 (1935). — VIII. ibid. 67. p. 18—50 (1936). — IX. ibid. 68. p. 49—78 (1937). — X. ibid. 69. p. 91—122 (1938). — XI. ibid. 70. p. 52—88 (1939). — XII. — ibid. 71. p. 56—121 (1940). — XIII. ibid. 72. p. 3—39 (1941).

———1938/41, Fytopathologické správy o rostlinách širšiho okolí olomouckého. I. Čas. Vlast. spolk. mus. Olomouci 51, čis. 191—192. 10 pp. — II. ibid. 52, čis. 195—196. 11 pp. (1939). — III. ibid. 53 čis. 199—200. 20 pp. (1940).

———1941, Fytopathologické správy o rostlinách okolí prostějovského. — Věstn. Přirodov. Klubu Prost. 1938/40. Ročnik 26. 16 pp.

CHAINE, J., 1913, La Cécidomyie du buis (Monarthropalpus buxi Lab.). — Ann. Sci. Nat. (Zool.) (9) 17. p. 269—361.

———1913, Traitement des buis contre le Monarthropalpus buxi Lab. — C. R. Soc. Biol. France 65. p. 156—158.

CHAMBERS, V. T., 1872, On some Leaf-mining Coleoptera. — Canad. Ent. 4. p. 123.

———1872/74, Microlepidoptera. — Canad. Ent. 3, p. 54, 89, 108, 146, 161 ff. 182 ff. 205 ff., 221 ff. — vol. 4. p. 7, 25—29, 41—44, 88. — vol. 5. p. 12 ff, 44—50, 124—128. 185 ff. — vol. 6. p. 8, 72, 96, 128, 149. 197, 217.

———1874, Prof. Frey, of Zürich, and some American Tineina. — Cincinn. Quart. Journ. Sci. 1. p. 193—211.

———1874, Tineina of the United States — ibid. p. 226—259.

———1875, Tineina of the Central United States. — ibid. 2. p. 97—121.

———1877, Notes on American species of Lithocolletis. — Psyche 2.

———1877, Tineina and their Food-plants. — Bull. Survey 3. p. 107—120.

———1877, The Tineina of Colorado. — ibid. p. 121—146.

———1878, Food Plants of Tineidae. — ibid. 4.

———1880, Mouth-parts and legs of Leaf-mining Larvae. — Amer. Entom. (2) 1.

CHAMBERS, V. T., 1880, Habits of Leaf-mining Larvae. — ibid. (2) 1.

———1881, On the early stages of Gracilaria stigmatella F. — Canad. Ent. 13. p. 25—28.

CHAPPELL, A. V., 1929, Biological notes on New Zealand Lepidoptera. — Tr. Pr. New Zeal. Inst. 60. p. 259—264.

CHITTENDEN, F. H., 1898, On Insects that affect Asparagus. — Bull. U. S. Dept. Agric. 10. p. 54—62.

———1902, Notes on Dipterous Leaf-miners on cabbage. — Bull. U. S. Ent. Div. 33. p. 75—77.

———1903, The Beet or Spinach Leaf-miner (Pegomyia vicina Lintn.) — ibid. 43. p. 50—52.

———1904, The leaf-mining locust beetles, with notes on related species. Bull. U. S. Dept. Agr. 38. p. 30—39.

———1907, The Asparagus Miner. — Bull. U. S. Dept. Agr. 66. pt. 1.

———1909, The Parsnip leaf-miner (Acidia fratria Loew.). — Bull. U. S. Dept. Agric. Bur. Ent. 82. p. 9—13.

———1909, Insects injurious to vegetables. 258 pp.

———1911, The Southern Beet webworm. — ibid. 109. p. 1—22.

———1923, Notes on the habits and distribution of North American Phyllotreta. — Proc. ent. Soc. Wash. 25. p. 132—139.

CHRETIEN, P., 1894, Observations suivantes sur les moeurs de trois Microlépidoptères. — Bull. Soc. ent. Fr. 1894, p. CXXVIII—CXXX.

———1896, Description de Microlépidoptères nouveaux. — Bull. Soc. ent. Fr. 65. p. 190—193.

———1899, Description d'une nouvelle espèce de Coleophora. — ibid. 1899, p. 146.

———1899, Description d'une nouvelle espèce de Microlépidoptères. — ibid. p. 112—115.

———1907, Description de nouvelles espèces de Lépidoptères d'Algérie. — Bull. Soc. ent. Fr. 1907. p. 305—308.

———1910, Contribution à la connaissance des Lépidoptères du nord de l'Afrique. — Ann. Soc. ent. Fr. 79. p. 497—531.

———1914, Descriptions de trois nouvelles espèces de Nepticula. — Bull. Soc. ent. Fr. 1914, p. 269—270.

———1922, Microlépidoptères nouveaux pour la faune française. — L'Amateur de Pap. 1. p. 119—121.

CHRETIEN, P., 1923, Les Chenilles des Lavandes. — ibid. 1. p. 229—228.

———1927, Les Chenilles des Astragales. — ibid. 3. p. 193—199, 209—212, 225—229, 241—245, 257—264.

CLAASEN, P. W., 1918, Observations on the life-history and biology of Agromyza laterella Zett. — Ann. ent. Soc. Amer. 11. p. 9—16.

CLAUSEN, C. P., 1927, The Citrus Insects of Japan. — Techn. Bull. U. S. Dept. Agr. 15. 15 pp.

———1931, Two Citrus leaf-miners of the far east. — ibid. 252. 13 pp.

CLEMENS, B., 1872, The Tineina of North America, edit. by Stainton, H. T. — London, 282 pp.

COCKAYNE, E. A. & HAWKINS, C. N., 1932, The Early Stages of Procris globulariae and of P. cognata. H. S. — Ent. Rec. Journ. Var. 44. p. 17—24.

COLLART, A., Contribution à l'étude des Diptères de Belgique. — Bull. Ann. Soc. ent. Belgique 77. p. 306—317.

COLLIN, J. E., 1928, A new species of Hydrellia mining the stems of Water-Cress. — Entom. Month. Mag. 64. p. 128—129.

COMSTOCK, J. H., 1880, The clover Oscinis (Oscinis trifolii Burg.) — Rep. U. S. Commiss. Agric. 1879, p. 200—201.

———1880, The mallow Oscinis (Oscinis malvae Burg.). — ibid. p. 201—202.

———1881, The corn leaf-miner (Diastata? — n. sp.) — ibid. 1880. p. 245—246.

CONDE, O., 1934, Ostbaltische Tenthredinoidea II. Korr. Bl. Naturf. Ver. Riga 61. p. 168—198.

COOK, A. J., 1896, The enemies of grasses and clovers. — In: Beal, W. J., Grasses of North America 2d ed. vol. 1. p. 369—413.

COOLEY, R. A., 1916, The wheat-sheath miner (Cerodonta femoralis Mg.). — 14th Ann. Rept. Entom. Montana Agr. Exp. Sta. Bull. 112. p. 66.

COQUILLETT, D. W., 1895, Two Dipterous leaf-miners on garden vegetables. p. 381—384. Two dipterous insects injurious to cultivated flowers. p. 399—402. — Ins. Life 7.

———1898, On the habits of the Oscinidae and Agromyzidae reared at the U. S. Dept. of Agriculture. — Bull. U. S. Dept. Agr. n. s. 10. p. 70—79.

CORBETT, H. H., 1898, The "pomifoliella-spinicolella" group of Lithocolletidae. Ent. Rec. 10. p. 168—170.

CORY, E. N., 1915, Insect pests of 1914. — Rept. Maryland State Hort. Soc. 17. p. 104—112.

CORY, E. N., 1916, Notes on Pegomyia hyoscyami Pnz. — Journ. Econ. Ent. 9. p. 372—375.

——— 1916, The columbine leaf-miner. — ibid. p. 419—424.

COTES, E. C., 1891, A conspectus of the insects which affect drops in India. — Indian Mus. Not. 2. p. 145—176.

COTTERELL, G. S., 1925, The Hispid Leaf-miner of Oil Palms on the Gold Coast. — Bull. ent. Res. 16. p. 77—83.

CROSBY, C. R. & LEONARD, M. D., 1918, Manual of vegetable garden insects. 391 pp.

CURRIE, G. A., 1937, Galls on Eucalyptus trees. — Proc. Linn. Soc. N. S. W. 62. p. 147—174.

CURTIS, J., 1847, Observations on the natural history and economy of a weevil affecting the pea-crop, and various insects which injure or destroy the mangold and beet. — Journ. Roy. Agr. Soc. Engl. 8. p. 399—416.

———1883, Farm Insects. 2d ed. London. 528 pp.

DA COSTA LIMA, A., 1945. Insetos do Brasil. vol. 5. Lepidópteros I. 379 pp.

DAHLBOM, G., 1837, Kort underrättelse om Skandinaviska Insekters allmännare Skada och Nytta i Hushållningen. Lund. 236 pp.

DAMMERMANN, K. W., 1929. The Agriculture Zoology of the Malay Archipelago. Amsterdam. 473 pp.

DAMPF, A., 1924, Estudio Morfologico del Gusano del Maguey (Acentrocneme hesperiaris Wlk.) — Rev. Mexican. Biol. 4. 4. p. 147—160.

DANIEL, D. M., 1928, Biology and Control of the Blackberry Leaf-miner. — Techn. Bull. N. Y. St. Agr. Exp. Sta. 133. 38 pp.

DEGEER, C., 1752/78, Mémoires pour servir à l'histoire des Insectes. — Stockholm, vol. 1—7.

DE ONG, E. R., 1922, A rice leaf-miner. — Journ. Econ. Ent. 15. p. 432.

DESHUSSES, A. & J., 1929, Parasites nouveuax. Les Mouches de l'Endive (Ophiomyia pinguis Fall. et Phytomyza continua Hend. — C. R. Acad. Agr. Fr. 1. 29. p. 533.

DIAKONOFF, 1936, Bucculatrix maritima. — Tijdschr. Ent. 79. p. LXXXIV—LXXXVI.

DIETZ, W. G., 1907, The North American species of the genus Ornix Tr. — Trans. Amer. ent. Soc. 33. p. 287—297.

———1921, A new species of Coptodisca. — Canad. Ent. 53. p. 44.

DINGLER, M., 1934, Die kleine Spargelfliege (Melanagromyza simplex Löw.). — Anz. f. Schädlingskd. 10. p. 135—139.

DISQUÉ, H., 1901, Verzeichnis der in der Umgegend von Speyer vorkommenden Kleinschmetterlinge. — Iris 14. p. 149—176, 197—250.

———1906, Verzeichnis der in der Pflanz vorkommenden Kleinschmetterlinge. Speyer.

———1908, Versuch einer microlepidopterologischen Botanik. — Iris 21. p. 34—147.

DOANE, R. W., 1899, Notes on Trypetidae with descriptions of new species. — Journ. N. Y. ent. Soc. 7. p. 177—193.

DOBREANU, E., 1936/37, Contribuţiuni la cunoasterea insectelor miniere din România — Bull. Soc. Nat. România 9. p. 11—20.
Noui contribuţiuni la cunoasterea insectelor miniere din România. — ibid. 10. p. 1—12.

———1937, Contribuţiuni la studiul sistematic, morfologic si biologic insectelor miniere din România. — Fac. de Sţiinte, Lab. Zool. Descr. Bucarest. Diss. 128. pp.

DONGE, E. & ESTIOT, P., 1921, Les insectes et leurs dégats. Paris.

DUDA, O., 1921, Kritische Bemerkungen zur Gattung Scaptomyza Hardy. — Jahresh. Ver. Schles. Ins. Kd. 13.

DUFOUR, L., 1857, Mélanges entomologiques (Suite 1). — Ann. Soc. ent. Fr. (3) 5. p. 39—48.

DUFRANE, A., 1930, Les Nepticula des Saules en Belgique. — Bull. Ann. Soc. ent. Belgique 70. p. 29—32.

———1940, Particularités biologiques de quelques Chenilles de Microlépidoptères. — ibid. 80. p. 21—29.

———1942/43, Microlépidoptères de la Faune Belgique. I. Bull. Mus. Roy. Hist. Nat. Belg. 18. Nr. 5. p. 1—12. — II. ibid. Nr. 40. p. 1—12 (1942). III. ibid. 19. Nr. 31. p. 1—8 (1943).

EBERLE, G., 1943, Die Mummelfliege (Hydromyza livens Fall.). — Natur & Volk 73, p. 293—306.

DYAR, H. G., 1900, On the larvae of Atomacera and some other sawflies. — Journ. N. Y. ent. Soc. 8. p. 26—31.

EBELING, CH. W., 1884, Zur Fauna Magdeburgs. — Festschr. 57. Vers. Dtsch. Naturf. u. Ärzte.

ECKSTEIN, K., 1932, Starkes Auftreten der Eichenminiermotte Tischeria complanella Hb. — Anz. f. Schädlingskd. 8. p. 143.

———1934, Die Schmetterlinge Deutschlands 5. Die Kleinschmetterlinge. Stuttgart, 223 pp.

EDMUNDSBURY, W. G. ST., 1937, Unusual Larval Habit of an Ornix. — Entomol. 70. p. 31.

EDWARDS, J., 1914, Celery disease. — Garden. Chron. (3) 55. p. 189.

EIDMANN, H., 1943, Zur Kenntnis des Buchenspringrüsslers Orchestes (Rhynchaenus) fagi L. — Ztschr. f. Pflanzenkrankh. 53. p. 42—61.

ELY, C. R., 1917, A revision of the North American Gracilariidae from the standpoint of venation. — Proc. Ent. Soc. Washingt. 19. p. 29—77.

VAN EMDEN, F., 1924, Bericht über die entomologische Überwachung der Speicher und Kulturen der Firma Caesar & Loretz im Jahre 1923. — Jahresber. Caesar & Loretz 1924. p. 167—232.

———1925, Bericht über die im Jahre 1924 in Speichern und Kulturen der Caesar & Loretz A. G. in Halle aufgetretenen Schädigungen. ibid. 1925, p. 191—274.

ENDERLEIN, G., 1934, Heringiinae, eine neue minierende Chloropiden-Unterfamilie. Zoolog. Anz. 105. p. 191—194.

ENSLIN, E., 1912/17, Die Tenthredinoidea Mitteleuropas. — Beih. zur Dtsch. ent. Ztschr. 1912—1917, 790 pp.

———1919, Beiträge zur Kenntnis der Tenthredinoidea VI. — Entom. Mitt. 8. p. 78—83.

EPPELSHEIM, 1871, Beiträge zur Entwicklungsgeschichte der Kleinschmetterlinge. — Stettin. ent. Zeit. 32. p. 315—318.

———1890, Zur Nepticula gei Wck. — ibid. 1890. p. 229—235.

ESCHERICH, K., 1925, Schäden durch die Eichenrindenminiermotte (Gracilaria simploniella F. R.) in Ungarn. — Anz. f. Schädlingskd. 1. p. 78—79.

FALCONER, W., 1887. The Marguerite Fly (Phytomyza affinis). — The American Florist 2. p. 297.

FALLÉN, C. F., 1814/25, Diptera suecica. — Lund vol. I. II.

FARSKY, F., 1879, Die ersten Ständen zweier Runkelrübenfliegen. II. Die Runkelfliege Anthomyia conformis Nördl. (Fall.). — Verh. zool. bot. Ges. Wien p. 107.

FASSNIDGE, W., 1937, Unusual larval habit of Stenoptilia zophodactyla Dp. — Ent. Rec. & Journ. Var. 49. p. 16—17.

FELT, E. P., 1922, Ocnerostoma piniariella Z., another introduced Insect. — Journ. Econ. Ent. 15. p. 432—433.

———1918, Key to American insect galls. — Bull. N. Y. State Mus. 200. p. 1—310.

FELT, E. P., 1899, List of new species described by J. T. Lintner. — Bull. N. Y. State Mus. 24. p. 306—308.

———1906, Insects affecting Park and Woodland Trees. — Mem. N. Y. State Mus. 8. vol. 1—2.

———1911, Agromyza melampyga Lw. — Bull. N. Y. State Mus. 147. p. 67—68.

———1915, Box leaf midge (Monarthropalpus buxi Lab.) — ibid. 180. p. 42—46.

FERDINANDSEN, C. & ROSTRUP, S., 1921, Oversigt Sygdomme hos Landbrugets og Havebrugets Kulturplanter i 1920. — Tidskr. Planteavl Copenhagen 27. p. 697—759.

FEYTAUD, J., 1914, La mouche du céléri (Acidia heraclei L.) — Soc. Etude Vulg. Zool. Agr. Bull. 13. p. 109—114.

FIEBRIG, K., 1908, Eine Schaum bildende Käferlarve, Pachyschelus spec. (Bupr., Sap.), die Ausscheidung von Kautschuk aus der Nahrung und dessen Verwendung zu Schutzzwecken (auch bei Rhynchoten). — Ztschr. wiss. Ins. Biol. 4. p. 233—239, 358—363.

FIGUIER, L., 1868, The insect world. — 519 pp.

FILIPJEV, N. N., 1926, Lepidopterologische Notizen. IV. — Rev. Russe d'Ent. 20. p. 287.

———1931, id. X. Die Artberechtigung von Gracilaria betulicola Hering. — CR. Ac. Sci. URSS. 1931. p. 236—238.

FIORI, ATT., 1930, Valore sistematico di alcune forme del genere Hyponomeuta Latr. — Boll. Lab. Ent. Bologna 3. p. 246—253.

FISCHER, H., 1936, Die Lebensgemeinschaft des Donauriedes bei Mertingen (Schwaben). — Abh. Naturw. Ver. Schwaben 1936. p. 1—198.

FISHER, W. S., 1922, The leaf and twig mining Buprestid beetles of Mexico and Central America. — Proc. U. S. Nat. Mus. 62. Nr. 2154. p. 1—95.

FINK, D. E., 1913, The asparagus miner and the twelve-spotted asparagus beetle. — Bull. Corn. Univ. Agr. Exp. Sta. 331. p. 409—436.

FITCH, E. A., 1881, The Mangold Fly. — Entomologist 14. p. 8—13, 25—30, 164—165.

FLETCHER, B. T., 1919, Second Hundred Notes on Indian Insects. — Bull. Agr. Res. Inst. Pusa 89.

———1920, Life Histories of Indian Insects, Microlepidoptera. — Mem. Dept. Agr. India, Ent. Ser. 6. Nr. 1—9.

———1921, A List of the Generic Names used for Microlepidoptera. Mem. Dept. Agr. Ind. Ent. Ser. 11.

FLETCHER, B. T., 1930, Report of the Imperial Entomologist 1928/29. — Scient. Rep. Agr. Res. Inst. Pusa p. 67—77.

———1932/33, Life-histories of Indian Microlepidoptera (2d Series). — Scient. Monogr. Imp. Counc. Agr. Res. 2, p. 1—58 & ibid. 4. p. 1—85.

FLETCHER, B. T. & CLUTTERBUCK, C. G., 1939/43, Microlepidoptera of Gloucestershire. — Proc. Cottesw. Nat. F. C. 27, p. 34—53, 93—108, 136—152; vol. 28, p. 1—10, 58—66.

FLETCHER, J., 1905, The Beet-leaf miner (Pegomyia bicolor Wied.). — Rept. Ent. Soc. Ontario 35, p. 52.

FOLOGNE, E. & D'UDEKEM, J., 1860, Notice sur la Coleophora olivacella. — Ann. Soc. ent. Belg. 4. p. 76—85.

FOLOGNE, E., 1860, Observations sur les chenilles de la Gracilaria convolvulella Ann. Soc. ent. Belg. 4. p. 86—87.

———1860, Lépidoptères et Chenilles observés en Belgique. — ibid. p. 108.

———1862, Observations sur quelques Lépidoptères observés en Belgique. — Ann. Soc. ent. Belg. 6. p. 161—169.

———1862, Notes sur quelques lépidoptères observés en Belgique. — ibid. p. 170—175.

FORBES, S. A., 1900, The economic entomology of the sugar beet. — Rept. State Entomol. of the State of Illinois 21, p. 49—57.

———1904, The more important insect injurious to Indian corn. — Bull. Ill. Agr. Exp. Sta. 95. p. 339—340.

FORBES, S. A. & HART, C. A., 1900, The beet leaf-miners (Chortophila floccosa Mcq., Ch. betarum Lintn , Pegomyia vicina Lintn.) — Bull. Illin. Agr. Exp. Sta. 60. p. 407—410.

FOX, W. G., 1925, The Mining Habit of Otiorrhynchus rugifrons Larva. — Ent. Monthl. Mag. 61. p. 273—276.

FOX, H., 1924, Notes from a study of Nepticula pomivorella Pack. — Ann. Rep. Ent. Soc. Ontario 55. p. 67—70.

FRACKER, S. B., 1915, The Classification of Lepidopterous Larvae — Illin. Biol. Monographs 2. Nr. 1., 169 pp.

FRANZ, H., 1944, Biologisches über 4 alpine Curculionidae. — Ent. Blätt. 40. p. 45—47.

FRAUENFELD, G. VON, 1863, Beitrag zur Metamorphosengeschichte aus dem Jahre 1862. Verh. zool. bot. Ges. Wien 1863, I. Die Trypeten. II. Die Rüsselkäfer.

———1863, Entomologische Fragmente. — ibid. p. (66—74).

FRAUENFELD, G. VON, 1864, Zoologische Miscellen .— ibid. 1864, p. 147—158, 379—388.

——— 1865, Bericht über eine Sammelreise durch England, Schottland Irland und die Schweiz. — ibid. 1865, p. 575—594.

——— 1866, Über die Metamorphose von Hydrellia albilabris Mg. ibid. 1866 SB. p. 87.

——— 1867, Ein neuer Gerstenfeind (Hydrellia griseola). — Ann. Landw. Wochenbl. 7. p. 332. 432.

DE FRÉ, C., 1860, Lépidoptères nouvellement découverts en Belgique. — Ann. Soc. ent. Belg. 4. p. 113—115.

FRENCH, C., 1900, The Phytomyza Leaf Tunneler (Ph. affinis). — Handbook of the destr. Ins. of Victoria. 3. p. 71—73.

FREY, H., 1855, Über die in der Schweiz beobachteten Arten des Genus Lithocolletis. — Mitt. naturf. Ges. Zürich 9. p. 600—634.

——— 1856, Die Tineen und Pterophoren der Schweiz. — Zürich. 430 pp.

——— 1857, Revision der Nepticulen. — Linnaea entom. 11. p. 351—446.

——— 1859, Das Tineen-Geschlecht Elachista. — ibid. 13. p. 172—314.

——— 1860, Lithocolletis betulae, Gracilaria pavoniella und Nepticula sericopeza bred. — Ent. Weekl. Intell. 8. p. 36—37.

——— 1861, Ornix pfaffenzelleri and scutulatella bred. — ibid. 10. p. 164.

——— 1863, Das Tineengeschlecht Ornix. — Linnaea Entom. 15. p. 1—41.

——— 1871, Ein Beitrag zur Kenntnis der Microlepidopteren. — Stettin. ent. Zeit. 32. p. 101—130.

FREY, H. & BOLL, J., 1876/78, Einige Tineen aus Texas. — Stettin. ent. Zeit. 1876, p. 209—228 & 1878, p. 249—280.

FREY, R., 1908, Anmärkningsvärda anthomyid-fynd från Finland. — Medd. Soc. Faun. et Flor. Fennica 34. p. 117—122.

——— 1921, Studien über den Bau des Mundes der niederen Diptera Schizophora. Act. Soc. Fauna & Flora Fennica 48. Nr. 3. p. 1—246.

——— 1937, Finlands Agromyzider, bestämda av F. Hendel. — Notulae Ent. 17. p. 82—96.

FRIEDRICHS, K., 1933, Witterung und Insekten in der Gegend von Rostock 1932. — An. Schädlingskd. 9. p. 29—34.

FRITSCH, A., 1882, Tinea araliae Fr. — Beitr. Palaeont. Österr. II. 6. tb. 2. f. 7.

FRITSCH, A., 1901, Tinea araliae. — Studien Kreideformat. 169. f. 9.

———1905, Minierlarven in Blättern. — SB. Böhm. Ges. 1905. 3.

FRITZSCHE, F. W., Beiträge zur Naturgeschichte der Elachista gregsoni Sta. — Korresp. Bl. f. Samml. v. Ins. 1860. p. 31.

FROGGAT, W. W., Australian Insects. 449 pp.

———1911, The French Bean Fly. Agromyza phaseoli, Coqu. — Agric. Gaz. N. S. W. p. 151—154.

———1920, Insects found on Tobacco in N. S. Wales. — ibid. 21. p. 714—716.

FROGGAT, W. M., 1914, Australasian Hispidae of the genera Bronthispa and Promecotheca which destroy Coconut Palm fronds. — Bull. ent. Res. 5. p. 149—152.

FROST, S. W., 1919, The function of the Anal Comb of certain Lepidopterous Larvae. Journ. Econ. Ent. 12. p. 446—447.

———1919, Two species of Pegomyia mining the leaves of dock. — Journ. agr. Res. 16. p. 229—244.

———1921, Lepidoptera injurious to the apple in Pennsylvania. — Bull. Agr. Exp. Sta. Penns. State Coll. 169. p. 1—16.

———1924, Two little known leaf-miners of apple. — Ent. News 35. p. 132—134.

———1924, A Study of the Leaf-Mining Diptera of North America. — Mem. Cornell Univ. Agr. Exp. Sta. 78. p. 1—228.

———1924, The Leaf-Mining Habit in the Coleoptera. — Ann. Ent. Soc. America 17. p. 457—467.

———1925, Convergent Development in Leaf-Mining Insects. — Ent. News 36. p. 299—305.

———1925, The Leaf-Mining Habit in the Hymenoptera. — Ann. Ent. Soc. America 18. p. 399—414.

———1930, Collecting Leaf-Miners in Barro Colorado Island, Panama. — The Scientif. Monthl. 30. p. 443—449.

———1930, The leaf-miners of Aquilegia, with a description of a new species. — Ann. Ent. Soc. America 23, p. 457—460.

———1931, Animal Life on Barro Colorado Island. — Journ. Wash. Ac. Sci. 21. p. 173—175.

———1931, The habits of leaf-mining Coleoptera on Barro Colorado Island, Panama. — Ann. Ent. Soc. America 24. p. 396—404.

FROST, S. W., 1931, New North American Agromyzidae. — Canad. Ent. 63. p. 275—277.

———1931, New species of West Indian Agromyzidae. — Ent. News 42. p. 72—76.

———1932, Cordylura tricincta Loew, a Leaf-Miner on Smilacina racemosa (L.) Desf. — Ent. News 43. p. 75—77.

FUCHS, A., 1897, Microlepidopteren der Loreleygegend. — Stettin, ent. Zeit. 1897, p. 325—340.

FUKAYA, C., 1915, Jitsuyo Engei-shokubutsu Gaichu. — Kujo-ho. 2d Edit. p. 204—207.

FULMEK, L., 1910, Zur Kenntnis schädlicher Schmetterlingsraupen. — Ztschr. landw. Vers. Anst. Österr. 1910.

———1917, Die Lärchenminiermotte. — Natur 1917 Heft 20.

———1920, Zur Kenntnis der Raupe und Puppe von Bucculatrix thoracella. — Ztschr. angew. Entomol. p. 330—337.

———1926, Etliche auf ihrem natürlichen Standpflanzen gezüchtete Falter und Motten in Deli, Sumatras Ostküste. — Misc. Zool. Sumatr. 3. p. 1—4., 5. p. 1—6.

———1930, Sciarinae (Diptera-Mycetophilidae) als Blattminierer. — Zool. Jahrb. Syst. 60. p. 46—48.

FULTON, B. B., 1920, Insect injuries in relation to apple grading. — Bull. N. Y. Agr. Exp. Sta. 475.

FYLES, T. W., 1884, Notes on Pegomyia bicolor (Wied.), a leaf-mining fly new to Canada. — Canad. Ent. 16. p. 69—70.

GARMAN, H., 1892, Some common pestes of the farm and garden. — Bull. Kentucky Agr. Exp. Sta. 40. p. 43—88.

———1904, Insects injurious to cabbage. — ibid. 114, p. 13—47.

GASOW, H., 1928, Die Frühdiagnose des Auftretens der Azaleenmotte. (Gracilaria azaleella Brants). — Arb. Biol. Reichsanst. 15. p. 593—599.

GERASIMOW, A., 1930, Neue und wenig bekannte Microheterocera L. — Iris 44. p. 132—135.

———1931, Zur Lepidopterenfauna von Mittelasien. III. Neue Lithocolletis. Entom. Ztschr. Frankfurt 45. p. 125—132.

———1932, Lepidoptera miners. I. Middle-Asiatic Species of Lithocolletis. — Bull. Leningrad Inst. Controll. Farm & For. Pests 3. p. 197—248.

———1937, Minierende Motten. V. Neue Stigmella- (Nepticula-) und Tischeria-Arten. — Mitt. Zool. Mus. Berlin 22. p. 281—285.

GERASIMOW, A. & GERASIMOW, F., 1932, Notes on the Pests Injurious to Apocynum sibiricum Pall. in the District of Khorezm. — Bull. Leningrad Inst. Contr. Farms & For. Pests 3. p. 279—281.

GERCKE, G., 1878, Über die Metamorphose der Hydromyza livens Fall. — Verh. Ver. naturw. Unterh. Hamburg 5.

———1880, Über die Metamorphose einiger Dipteren. — ibid. 6. p. 1—13.

———1886/89, Dipterologische Miscellaneen. — Wien. ent. Zeit. 5. p. 161—168 (1886) & 8. p. 219—226 (1889).

GERTZ, O., 1918, Kallus hypertrofier och några i samband därmed stående anatomiskt-fysiologiska forhållanden hos minerade blad. — Botaniska Notiser 1918, p. 121—139.

GILL, J. B., 1924, Important Pecan Insects and their control. — U. S. Dept. Agr. Farm. Bull. 1364, p. 1—48.

GILLANDERS, A. T., 1908, Forest Entomology. — 422 pp. Edinburgh & London.

GIRARD, M., 1885, Les insects. — Traité élément. d'Entom. 3. p. 1—1110.

GLITZ, C. T., 1874/77, Verzeichnis der bei Hannover und im Umkreise von etwa einer Meile vorkommenden Schmetterlinge. — Jahrber. Naturhist. Ges. Hannover 24, p. 24—69; 25, p. 23—42; 26, p. 17—52.

GOEZE, J. A. E., 1774, Geschichte der Minir-Würmer in den Blättern. — Naturforscher 1774, Stück 4, p. 1—32. — 1779, Füessly, Mag. Ent. 2., p. 153—160.

———1775, Microscopische Beobachtungen einer vierzehnfüssigen Minir-Raupe in den Apfelblättern. Naturforscher 1775, Stück 5. p. 1—18. 1779, Füessly, Mag. Ent. 2. p. 184—197.

———1775, Beschreibung einer neuen Art Minir-Raupe, die ihre Mine vollkommen zirkelrund und spiralförmig machen. — l. c. Stück 5. p. 62—72. 1779: l. c. 2. p. 197—209.

———1782, Von der Ökonomie besonderer Minirwürmer in den glatten Pappelblättern. — Naturforscher, Stück 14. p. 103—112.

VAN DER GOOT, P., 1930, De Agromyza-Vliegjes der inlandsche Katjang-Gewassen op Java. — Meded. Inst. Plantenz. 78., 97 pp.

GORIATCHOVSKY, VL., 1915, Pests of cultivated plants for 1914. — Warsaw. Hort. Soc. Ann. 1914, p. 64—74 (russian).

GOUREAU, C., 1846, Notes pour servir à l'histoire des Diptères dont les larves minent des feuilles des plantes. — Ann. Soc. ent. Fr. (2) 4. p. 223—234.

———1847, Sur les feuilles de l'Iris pseudacorus minée par la larve d'un Diptère. — ibid. 5. Bull. p. 40—41.

GOUREAU, 1851, Mémoire pour servir à l'histoire des Diptères dont les larves minent les feuilles des plantes et à celle de leurs parasites. ibid. 9. p. 131—176.

———1868, Note. — Bull. Soc. ent. Fr. (4) 8. p. CXIII—CXV.

GRANDI, G., 1929, Contribute alla conoscenza della Tischeria gaunacella Dup. ed appunti Tischeria complanella Hbn. — Boll. Lab. Ent. Bologna 2. p. 122—201.

———1931, Morfologia et Etologia comparata di Insetti a regime specializzato. I. La morfologia comparata die alcuni stati larvali di 20 Microlepidotteri minatori. — ibid. 4. p. 223—337.

———1932, id. II. La morfologi delle larve minatrici di due Coleotteri Crisomelidi della sottofamiglia degli Alticini. — Mem. R. Acad. Ist. Bologna (8) 9. p. 195—202.

———1932. id. III. La morfologia delle larve endofite di due Coleotteri Curculionidi. — Boll. Lab. Ent. Bologna 5. p. 93—103.

———1933, id. IV. La morfologia comparata di vari stati larvali di 30 Microlepidotteri minatori appartenenti a 15 generi ad a 11 famigli. — ibid. 5. p. 143—307.

———1933, L'Ipermetabolia nei Lepidotteri. — Mem. Ac. Sci. Ist. Bologna (8) 10. p. 115—121.

———1935, Morfologia ed Etologia comparata di Insetti a regime specializzato. VI. La morfologia delle larve minatrici degli Hispini dei gen. Hispa L. e Hispella Chap. — ibid. (9) 2, p. 3—10.

———1936, id. VII. La Morfologia comparata delle larve di alcuni Imenotteri Tentredinidi. — Boll. Ist. Ent. Bologna 8. p. 222—254.

GRIPEKOVEN, H., 1914, Minierende Tendipediden. — Arch. Hydrobiol. 9. p. 129—229.

GRØNLIEN, N., 1925, Mikrolepidoptera fra Voss og Indre-Hardanger samt enkelte andre lokaliteter. — Norsk. Ent. Tidskr. 2. p. 39—52.

———1926, Bladminerare fra Voss og Indre-Hardanger I. — ibid. 2. p. 89—108.

———1927, Eine neue Nepticula von Betula odorata. Nepticula vossensis. — ibid. 2. p. 217—219.

———1932, Notes on Norwegian Nepticulids. — Norsk. Ent. Tidskr. 3. p. 103—114.

———1932, A new Lithocolletid. — ibid. 3. p. 115—116.

GROSCHKE, F., 1939, Die Kleinschmetterlinge der Grafschaft Glatz. — Inaug. Diss. München p. 1—92.

GROSCHKE, F., 1944, Neues über Minierer aus dem Mittelmeergebiet. — Mitt. Münch. ent. Ges. 34. p. 115—124.

GROSSER, W., 1905, Einige Schädlinge unserer Kulturpflanzen. — Jahresb. Schles. Ges. f. vaterl. Kultur 2. Abt. Bot. Zool. vol. 82. p. 3—5.

GRÜNBERG, K., 1909, Lepidoptera. — Brauer, Süsswasserfauna Deutschl. 8. p. 96—159.

DE GRYSE, J. J., 1915, Some modifications of the Hypopharynx in Lepidopterous Larvae. Proc. ent. Soc. Washingt. 17. p. 173—178.

———1916, The Hypermetamorphism of the Lepidopterous Sap feeders. — ibid. 18. p. 164—168.

GUDMANN, F., 1929, Xystophora gudmanni Lars. og dens Biologi. — Faun. & Flor. p. 78 ff.

———1931, Nepticula tengströmi. — ibid. 1931, p. 58—60.

GUSSEW, W. I. & RIMSKY—KORSAKOW, M. N., 1934, Bestimmungsbuch der Frassbilder. Leningrad, 587 pp. (russian).

HAASE, J., 1942, Minenfunde aus der Umgebung von Trautenau, Sudetenland. — Mitt. Münch. ent. Ges. 32. p. 227—236.

HABERLANDT, G., 1918, Mikroskopische Untersuchungen über Zellwandverdauung. — Beitr. z. Allgem. Botanik 1. p. 501—535.

HAGEN, H. A., 1882, Larval mines. — Nature 25. p. 265.

HALIDAY, A. H., 1857, On some remaining blanks in the natural history of the native Diptera. — Proc. Soc. Nat. Hist. Rev. 4. p. 177—192.

———1857, Additional note on the metamorphosis of some species of Diptera, hitherto undescribed, or known but imperfectly. — Proc. Nat. Hist. Rev. 4. p. 192—196.

HAMM, A. H., 1918, Trypetidae from the Oxford district with notes on their time of appearance and food-plants. — Ent. Monthl. Mag. 54. p. 87—91.

———1921, Diptera. — The Natural Hist. of the Oxford District. p. 248—279.

HARDY, J., 1849, On the Primrosa-leaf Miner. — Ann. Mag. Nat. Hist. (2) 4. p. 385—392.

HARLAND, S. C., 1915, Insect pests of Lima Beans in St. Vincent. — Agr. news (Barbados) 14. p. 352—353, 346—347. 363.

HARTIG, F., 1938, Sechs neue Mikrolepidopteren. — Ztschr. Österr. Ent. Ver. 23. p. 82—83, 88—90.

———1939, Sulla Minefauna della Venezia Tridentina. — Archiv. per L'Alto Ad. 34. p. 1—70.

HARTIG, F., 1941, Neue Mikrolepidopteren. — Mitt. Münch. ent. Ges. 31. p. 154—163.

——— & AMSEL, HG., 1939, Contributo alla conoscenza della Fauna Entomologica della Sardegna. — Mem. Soc. ent. Ital. 17. p. 63—84.

HARTIG, R., 1870, Coleophora lutipennella Z. — Ztschr. Forst- & Jagdwesen p. 405.

———1896, Die Tannennadelmotte, Argyresthia fundella. — ibid. 1896, p. 303.

HARTMANN, A., 1870, Die Kleinschmetterlinge der Umgegend Münchens und eines Theiles der bayerischen Alpen. München, 96 pp.

———1880, Die Kleinschmetterlinge des europäischen Faunengebietes. München, 182 pp.

HARUKAWA, C. & YAGI, N., 1918, The serpentine Leaf-miner of the Peach, a species of Lyonetia. — Ber. Ohara-Inst. 1. p. 335 ff. — On the life-history and habits of a Peach leaf-miner (Ornix spec.) — ibid. p. 325 ff.

HASEMANN, L., 1916, Ornix geminatella, the unspotted Tentiform Leaf-miner of apple. — Journ. Agr. Res. 6. p. 289—295.

HAUDER, F., 1896, Verzeichnis der um Kirchdorf im Kremsthale in Oberösterreich gesammelten Microlepidopteren. — Jahresber. Ver. Naturk. Linz 25., p. 1—26.

———1913, Beitrag zur Mikrolepidopteren-Fauna Oberösterreichs. — Linz, 321 pp.

———1914, Raupe und Puppe von Lithocolletis hauderiella Rbl. — Verh. zool. bot. Ges. Wien 1914, p. 153.

———1916, Cemiostoma wailesellum Stt. an Genista germanica. — Ent. Ztschr. Frankfurt 30. Nr. 8.

———1917, Elachista paludum Frey. — Ztschr. Öst. Ent. Ver. 2. Nr. 3.

———1918, id. ibid. 3. Nr. 3.

———1920, Zusammenstellung der seit 30 Jahren nach oberösterreichischen Funden beschriebenen neuen Arten und Formen. — Kranch. ent. Jahrb. 1920. p. 118—128.

———1922, Zucht und Präparation der Nepticuliden. — Ztschr. Österr. Ent. Ver. 7. Nr. 1/2.

——— & MITTERBERGER, K., 1916, Die Zucht von Gracilaria hauderi Rbl. — Verh. zool. bot. Ges. Wien 1916, p. 109—113.

———, ———1924, Nachtrag zur Microlepidopterenfauna Oberösterreichs. — Jahresber. Oberöst. Mus. Ver. 80. p. 267—294.

HEALY, C., 1869, Life-History of Fenusa pumila p. 211, F. ulmi p. 297, F. pygmaea p. 299, Phyllotoma melanopoda p. 138, Ph. tormentillae p. 140, Ph. microcephala p. 176 & 224. — Entomologist 4.

HEDICKE, H., 1918, Mitteilungen über Zeugophora flavicollis, Coleophora saponariella und Agromyza curvipalpis. — Deutsch. ent. Ztschr. 1918, p. 384.

———1918, Agromyza abdominalis. — ibid. p. 425.

——— & HERING, M., 1924, Vorschläge für eine Terminologie der Blattminen. — Deutsch. ent. Ztschr. 1924. p. 185—194.

HEEGER, E., 1851, Beiträge zur Naturgeschichte der Insekten. — SB. Ak. Wiss. Wien 7. p. 209—211.

———1853, id., ibid. 11. p. 42—45.

———1858, Neue Metamorphosen einiger Dipteren. — ibid. 32. p. 295—309.

HEIKERTINGER, F., 1914, Neue oder interessante Psylliodes-Formen der Fauna Niederösterreichs. — Verh. zool. bot. Ges. Wien 1914. p. 95 ff.

———1918, Kleine Mitteilungen zur Biologie der pflanzenfressenden Käfer. Coleopt. Rundschau 7. Nr. 1/2. p. 11.

———1924/26, Resultate fünfzehnjähriger Untersuchungen über die Nahrungspflanzen einheimischer Halticinen. — Ent. Blätt. 20. p. 214—224; 21. p. 10—19, 81—92, 119—131, 155—163; 22, p. 1—9. 49—62.

———1928, Halticinen, Erdflöhe. — Sorauer, Handb. Pflanzenkr. 4th ed. 5. p. 199—212.

———1941, Über das Sammeln von Phyllotreta-Arten. — Koleopt. Rdsch. 27. p. 1—14. — Bestimmungstabelle der paläarktischen Phyllotreta-Arten ibid. p. 15—64, 69—116.

VON HEINEMANN, H., Über Nepticula-Raupen. — Korr. Bl. f. Samml. v. Ins. 1. p. 174 ff.

———1862, Einige Bemerkungen über die Arten der Gattung Nepticula. Wien. ent. Monschr. 1862. Nr. 8.

———1871, Nachtrag zu den Bemerkungen über die Arten der Gattung Nepticula. — Berlin ent. Ztschr. 15. p. 209 ff.

——— & WOCKE, F., 1859/1877. Die Schmetterlinge Deutschlands und der Schweiz. Braunschweig. 4 vol.

HEINRICH, C., 1915, On Acrocercops strigifinitella Cl. — Proc. ent. Soc. Washingt. 17. 1.

———1918, On the Lepidopterous Genus Opostega and its Larval affinities. — Proc. ent. Soc. Washingt. 20. Nr. 2. p. 27—38.

HEINRICH, C., 1926, Revision of the North American Moths of the Subfamilies Laspeyeresiinae and Olethreutinae. — Bull. U. S. Nat. Mus. 132., 207 pp.

———1929, Two new American Coleophoridae. — Proc. ent. Soc. Wash. 31. Nr. 1. p. 18—19.

HENDEL, F., 1909, Über eine neue palaearktische Cordylurinen-Gattung. — Dtsch. Ent. Ztschr. 1909, p. 777.

———1914, Die Gattungen der Bohrfliegen. — Wien. ent. Zeit. 33. p. 73—98.

———1920, Die Palaearktischen Agromyziden. Prodromus einer Monographie. — Arch. Naturg. 84 (1918) A. 7. p. 111—174.

———1922, Nachtrag zu den paläarktischen Agromyziden. — ibid. 88. A. 7. p. 174—176.

———1922, Blattminierende Fliegen (Musciden. — Wien ent. Zeit. 39. p. 65—72.

———1923, Blattminierende Fliegen. — Deutsch. ent. Ztschr. 1923, p. 386—400.

———1923, Neue europäische Melanagromyza-Arten. — Konowia 2. p. 142—145.

———1924, Acht neue europäische Agromyziden. — ibid. 3. p. 140—148.

———1924, Über das Genus Parallelomma Beck. und seine Verwandten in Europa. — Ent. Mitt. 13. p. 82—84.

———1925, Eine neue in Carduus glaucus Baumg. minierende Anthomyidengattung aus den Alpen. — Ztschr. Morph. Ökol. Tiere 4. p. 333—336.

———1925, Neue europäische Minierfliegen. — Konowia 4. p. 301—309.

———1926/28, Blattminenkunde Europas. I. Die Dipterenminen. Wien 100 pp.

———1927, Beiträge zur Systematik der Agromyziden. — Zool. Anz. 69, p. 248—271.

———1927, Beiträge zur Ökologie der Agromyziden. Ztschr. Morph. Ökol. Tiere 7. p. 480—488.

———1927. Trypetidae. — Lindner, Fliegen pal. Reg. 49. 221 pp.

———1928, Zweiflügler oder Diptera. Allgemeiner Teil. — Dahl, Tierwelt Deutschlands 11., 135 pp.

———1928, Über die minierenden europäischen Scaptomyza-Arten und ihre Biologie. — Zool. Anz. 76. p. 289—302.

HENDEL, F., 1931/36, Agromyzidae. — Lindner, Flieg. pal. Reg. 59, 570 pp.

——— & BEIER, M., 1937, Diptera-Fliegen. — Kükenthal, Handb. Zool. 4. II. Insec. 2. p. 1729—1998.

HENNIG, W., Übersicht der bisher bekannten Metamorphosestadien der Ephydriden. — Arb. morph. tax. Ent. Dahlem 10. p. 105—144.

HENRIOT, R., 1933, Brythis pancratii Cyr. — L'Amateur Pap. 6. p. 171—175.

HENSCHEL, G. A., 1895, Die schädlichen Forst- und Obstbauminsekten, ihre Lebensweise und Bekämpfung. Berlin, 3rd ed.

HERING, E., 1891, Ergänzungen und Berichtigungen zu F. O. Büttners Pommerschen Mikrolepidopteren. — Stett. ent. Zeit. 52. p. 135—227.

———1891, Saisondimorphismus und ungelöste Rätsel bei der Gattung Gracilaria. — ibid. p. 84—101.

———1893, Zuträge und Bemerkungen zur Pommerschen Mikrolepidopteren-Fauna. — ibid. 54, p. 80—120.

HERING, M., 1918, Für die Mark neue Kleinschmetterlinge. — Deutsche Ent. Ztschr. 1918, p. 151. — Auf Weiden vorkommende Mikrolepidopteren. — p. 410—411. — Nepticula intimella Z. neu für die Mark. p. 416.

———1919, Für die Mark neue Kleinschmetterlinge. — ibid. 1919, p. 188 ff., 428 ff.

———1919, Zur Biologie und systematischen Stellung von Scythris temperatella Led. — Iris 32. p. 123—130.

———1920/44, Minenstudien: 1. Deutsch. Ent. Ztschr. 1920, p. 133—143. — 2. ibid. 1921, p. 123—147. — 3. ibid. 1923, p. 188—206. — 4. Ztschr. Morph. Ök. Tiere 2. p. 217—250. — 5. Ztschr. wiss. Ins. Biol. 20. p. 125—136, 161—174. — 6. Ztschr. Morph. Ök. Tiere 4. p. 502—539. — 7. ibid. 5. p. 447—488. — 8. Ztschr. angew. Ent. 13. p. 156—198. — 9. Zool. Jahrb. Syst. 55. p. 533—588. — 10. Ztschr. angew. Ent. 17. p. 431—471. — 11. Ztschr. wiss. Ins Biol. 26. p. 93—108, 157—182. — 12. Ztschr. f. Pflanzenkrankh. 41. p. 529—551. — 13. ibid. 42. p. 567—579. — 14. ibid. 44. p. 49—70. — 15. ibid. 45, p. 1—15. — 16. Dtsch. Ent. Ztschr. 1941, p. 10—23. — 17. Mitt. Dtsch. Ent. Ges. 12. p. 56—62. & 13, p. 46. — 18. ibid. 13. p. 116—119.

———1920, Für die Mark neue Kleinschmetterlinge. — Deutsch. Ent. Ztschr. 1920, p. 211—213. — Minen auf Moorpflanzen. p. 432—434.

———1921, Für die Mark neue Kleinschmetterlinge. — ibid. 1921, p. 91—92. — 25 Kleinschmetterlinge und Fliegen neu für die Mark. p. 274 f.

———1922, Blattminen und Blattminierer. — Kranch. ent. Jahrb. Monatliche Sammelanweisungen.

HERING, M., 1922, Drei neue Arten der blattminierenden Agromyziden. — Deutsch. ent. Ztschr. 1922. p. 423—436.

———1923, Durch Insektenlarven erzeugte Blattminen. — P. Schulze, Biologie d. Tiere Deutschlands 43, p. 1—17.

———1924, Zur Kenntnis der Blattminenfauna des Banates. — Ztschr. wiss. Ins. Biol. 29. p. 1—15, 31—41.

———1924/25, Das histologische Bild der von Insektenlarven erzeugten Blattminen. — Mikrokosmos 17. Nr. 4. 5. & 18, Nr. 8. 9.

———1924, Oligophagie bei Blattminierern und ihre Verwendungsmöglichkeit zur Lösung pflanzenverwandtschaftlicher Probleme. — Kranch. Ent. Jahrb. 1924/25, p. 81—88.

———1924, Das Lepidopteren-Genus Cosmopteryx Hb. in seinen paläarktischen Arten. — ibid. p. 109—116.

———1925, Die Oligophagie blattminierender Insektenlarven als Hilfsmittel der Pflanzenverwandtschaftsforschung. — Naturforscher 9. p. 393—397, 454—459.

———1926, Biologie der Schmetterlinge. Berlin, 480 pp.

———1926, Bemerkungen zu einigen Agromyziden der Loewschen Sammlung. — Deutsch. Ent. Ztschr. 1925, p. 376—380.

———1926, Die Oligophagie der blattminierenden Insekten in ihrer Bedeutung für die Klärung phyto-phyletischer Probleme. — Verh. III. Int. Ent. Kongr. Zürich 1925, p. 216—230.

———1926, Eine neue Phytomyzoptera- (Phytomyptera-) Art. — Konowia 5. p. 21—24.

———1926, Beiträge zur Kenntnis der Agromyziden. — Dtsch. Ent. Ztschr. p. 219—224.

———1926, Nachgewiesene parthenogenetische Fortpflanzung bei einer blattminierenden acalyptraten Muscide. — Zool. Anz. 68. p. 223—227.

———1926, Die Ökologie der blattminierenden Insektenlarven. — Zool. Bausteine I. 2. Berlin, 253 pp.

———1927, Galle und Mine. — Kranch. ent. Jahrb. 36. p. 65—79.

———1927, Zwei neue Agromyziden aus dem Naturschutzgebiet von Bellinchen a. Oder. — Dtsch. ent. Ztschr. 1926, p. 331—334. — Eine neue Agromyzide von der Murmanküste. — ibid. p. 157—158. — Das Problem der Oligophagie. — ibid. p. 238—240.

———1927, Agromyzidae. — Dahl, Tierwelt Deutschlands 6. 172 pp.

———1927, Über das Weibchen und die Lebensweise von Vidalia spinifrons (Schroeder). — Stettin. ent. Zeit. 88. p. 110—111.

HERING, M., 1927, Die Minenfauna der Canarischen Inseln. — Zool. Jahrb. Syst. 53. p. 405—486.

———1929, Die Blattminierergattung Pelmatopus Httg. — Ztschr. wiss. Ins. Biol. 24. p. 97—107.

———1929, Von Minierlarven, die nicht in Blättern leben. — Kranch. ent. Jahrb. 1929, 5 pp.

———1929, Synopsis der Blattminen an Ulmus. — ibid. 1930, p. 59—70.

———1930, Beitrag zur Kenntnis der Lebensweise und Morphologie einiger Heteroceren. — Iris 43. p. 165—168.

———1930, Eine neue Lithocolletis als Schädling an Apfelbäumen. — Mitt. Dtsch. Ent. Ges. 1. p. 62—64.

———1930, Eine Agromyziden-Mine aus dem Tertiär. — Deutsch. Ent. Ztschr. 1930. p. 63—64.

———1930, Sammeln und Züchten blattminierender Käfer. — Koleopt. Rdsch. 16. p. 127—137.

———1930, Coleophora flavipennella und Phyllotoma aceris. — Mitt. Dtsch. ent. Ges. 1. p. 97—98.

———1930, Eine Minierfliege als Schädling an Erbsenpflanzen. — Anz. Schädl. Kd. 6. p. 61—64.

———1930, Eine in den Kannen von Nepenthes minierende Phyllocnistis (Lepid.) und ihr Parasit, eine neue Coprodiplosis (Dipt. Cecidom.) — Archiv Hydrobiol. Suppl. 8. p. 50—70.

———1930, Die "grünen Inseln" im verfärbten Herbstlaub. — Kranch. ent. Jahrb. 1931, p. 90—95.

———1931, Ein neuer Parietaria-Minierer, Cosmopteryx parietariae sp. n. — Mitt. Dtsch. Ent. Ges. 2. p. 67—70.

———1931, Eine neue Agromyzide aus Carex-Minen aus der Tschechoslovakei. — ibid. 2. p. 93—96.

———1931, Über die Mine und das Vorkommen von Scolioneura nana Klug in der Mark Brandenburg. — ibid. 2. p. 109—111.

———1931, Minierer an Wasserpflanzen. — Kranch. ent. Jahrb. 1932. p. 69—76.

———1932, Die Blattminen des Dummersdorfer Ufers. — Das Linke Untertraveufer p. 190—205. Lübeck.

———1932, Verfärbungserscheinungen am Blatt unter dem Einfluss minierender Insekten. — Progr. 5. Congr. Int. Ent. Paris p. 58—60.

Hering, M., 1932, Die Minenfauna von Tighina (Bender) in Bessarabien. — Bull. Acad. Roum. Sci. 15. Nr. 1/2. p. 1—22.

———1926/32, Schmetterlinge, Lepidoptera. — Tierw. Mitteleuropas 6. XVIII. p. 1—94. & Suppl. 1. p. 1—545.

———1932, Für die Mark neue Lepidopteren und Dipteren. — Mitt. Dtsch. Ent. Ges. 3. p. 129.

———1932, Beitrag zur Kenntnis der Minenfauna der Riviera. — ibid. 3. p. 137—143.

———1932, Die Farbe der Blattminen und Verfärbungen minierter Blätter. — Kranch. ent. Jahrb. 42 (1933) p. 78—81.

———1932, Ein neuer Loranthus-Minierer (Phyllocnistis voûtei (Lepid.). — Treubia 14. Nr. 1. p. 1—2.

———1933, Zwei neue Microlepidopteren aus Italien. — Boll. Lab. Ent. Bologna 5. p. 104—108.

———1933, Sur quelques Microlépidoptères français peu connus. — L'Amateur de Papill. 6. p. 180—185.

———1933, Recherches de mines aux environs de Paris. Liste des mines d'Insectes observées à Lardy et à Saint-Germain. — ibid. 6. p. 187—212.

———1933, Neue Agromyziden. — Konowia 12. p. 33—40.

———1933, Die palaearktischen Arten der Gattung Leucoptera Hbn. (Cemiostoma Zell.) — Mitt. Zool. Mus. Berlin 19. p. 64—79.

———1933, Die Blattminen an Apfel- und Birnbaum. — Anz. Schädlingskd. 9. p. 149—159.

———1934, Drei für die Mark neue Minierinsekten. — Mitt. Dtsch. Ent. Ges. 4. p. 146—147.

———1934, Probleme der Blattminenforschung. — ibid. 4. p. 130.

———1934, Minen von Shanghai, gesammelt von H. Hoene mit Beschreibung einer neuen Argopistes-Art (Col. Haltic.) von S. Maulick, London. — ibid. 5. p. 23—28.

———1934, Synopsis der Ahornminen. — Kranch. ent. Jahrb. 43. p. 66—74.

———1934, Verfaerbungserscheinungen an Pflanzenteilen unter dem Einfluss minierender Insekten. — Trav. V. Congr. Int. Paris p. 547—556.

———1934, Die Blattminen an Pflaumen- und Kirschbaum. — Anz. Schädl. Kd. 10, p. 51—55.

———1934, Probleme der Minenforschung. Ent. Beihefte Dahlem 1. p. 44—47.

Hering, M., 1934, Neue Gattungsbezeichnungen minierender Tenthrediniden. — Int. Ent. Ztschr. Guben 28. p. 353.

———1934, Elévage et préparation d'Insectes-Mineurs. — L'Amateur de Papill. 6. p. 301—309.

———1934, Les mines des feuilles de Phillyrea avec description d'un nouveau genre et d'une espèce nouvelle de mouche à scie mineuse de feuilles. — L'Amateur de Papillons 7. p. 154—160.

———1934, Monatliche Sammelanweisungen. Minierfliegen (Agromyzidae). — Kranch. ent. Jahrb. 44. p. 17—33.

———1935, Neue märkische Insekten. — Mitt. Dtsch. Ent. Ges. 5. p. 67—68.

———1935, Geäder-Mutationen bei Minierfliegen. — Dtsch. Ent. Zeitschr. 1934, p. 317—323.

———1935, Une nouvelle mine de Rhamnus alaternus. — L'Amateur de Pap. 7. p. 278—280.

———1935, Bestimmungstabelle der Blattminen an Linden. — Kranch. ent. Jahrb. 45 (1936) p. 52—58.

———1935/37, Die Blattminen Mittel- und Nordeuropas einschliesslich Englands. — Neubrandenburg, 631 pp.

———1936, Lycaena oder Chrysophanus astrarche Bergstr.? — Dtsch. Ent. Ztschr. 1936, p. 191—192.

———1936, Blattminen von Spanien. — Eos 11, p. 331—384.

———1936, Subspecies in statu nascendi. — Zool. Anz. 114. p. 266—271.

———1936/37, Agromyziden-Nachlese. Neue Liriomyza- und Phytomyza-Arten. — I. Dtsch. Ent. Ztschr. 1936. p. 73—80. II. Mitt. Dtsch. ent. Ges. 8. p. 76—77.

———1937, Die märkischen Blattwespen der Birke. — Märk. Tierwelt 3. p. 71—76.

———1937, Bohrfliegen. — Kranch. ent. Jahrb. 1937. p. 162—168.

———1940, Lepidopterologisches Wörterbuch. — Stuttgart, 123 pp.

———1941, Zur Lebensweise nichtminierender Minierfliegen. — Mitt. Dtsch. Ent. Ges. 10. p. 67—68.

———1941, Fünf ähnliche deutsche Stenoptilia-Arten. — Mitt. Dtsch. Ent. Ges. 10. p. 35—38.

———1942, Neotropische Buprestiden-Minen. — Arb. phys. angew. Ent. Dahlem 9. p. 241—249.

HERING, M., 1942, Verkannte Coleophoren. — Mitt. Dtsch. Ent. Ges. 11. p. 20—26.

———1942, Ein neuer Spargelfeind. Die Zwergspargelfliege (Ptochomyza asparagi g. n. sp. n.) — Ztschr. Pflanzenkrankh. 52. p. 529—533.

———1942, Eine neue Nepticula der subbimaculella-Gruppe. N. zimmermanni sp. n. von Quercus pubescens Wlld. — Mitt. Dtsch. Ent. Ges. 11. p. 26—28.

———1942, Eine neue stengelaushöhlende Minierfliege. Melanagromyza arnicarum sp. n. — ibid. 11. p. 66—68.

HERING, E. M., 1942, Die Coleophora-Arten an Aster linosyris (L.) Bernh. — Ztschr. Pflanzenkrankh. 52. p. 29—40.

———1942, Genitalpräparate von Macrolepidopteren. — Mitt. Dtsch. Ent. Ges. 11. p. 31—32.

———1943, Eine minierende Gallmücke (Clinodiplosis syringogenea sp. nov.) Eos 19. p. 211—214.

———1943, Neue paläarktische Agromyzidae. Mit einem Anhang: Agromyziden-Funde in Spanien. — Eos 19. p. 51—62.

———1943, Untersuchungen über die Weiden-Neptciulen. I. — Ztschr. Wien. ent. Ges. 28. p. 273—278.

———1944, Aus dem Raritätenkabinett der Minenforschung. — ibid. 29. p. 161—165.

———1943/44, Dipteren-Biologien. I. Mitt. Dtsch. Ent. Ges. 12. p. 16. — II. ibid. 13. p. 42—43.

———1944, Anfertigung von Genitalpräparaten bei Microlepidopteren. — ibid. 13, p. 44—45.

HERING, O., 1926, Die Blattminierergattung Tischeria in ihren palaearktischen Arten. — Kranch. ent. Jahrb. 1926, p. (1—8).

———1926, Die Blattminen der Rosen. — Anz. Schädl. Kd. 2. p. 13—15, 29—32.

HERRICK, W. G., 1912, The Larch Case-Bearer (Coleophora laricella Hb.). — Bull. Corn. Univ. 322. p. 39—54.

HEWETT, C. G., 1917, Insects affecting garden and greenhouse plants. — Rept. Agr. Dept. Canad. Dom. Entom. 1916, p. 34—35.

HEYDEN, C. VON, 1860, Mompha langiella Hb. — Corrsp. Bl. Samml. Ins. 1. p. 22.

HEYDEN, C. VON, 1862, Verwandlungsgeschichte des Trachys minuta und Rhamphus pulicarius. — Berlin. Ent. Zeit. 6. p. 61—63.

——1862, Nepticula fossilis. — Palaeont. 10. p. 77.

HINE, J. S., 1901, Observation on insects. — Ohio Natural. 2. p. 169—170.

HODSON, W. E. H., 1927, The Azalea Leaf Miner, Gracilaria azaleella Brants. — Journ. R. Hortic. Soc. 52. p. 54—59.

HOLLRUNG, M., 1905, Einige Bemerkungen über die Blattminierfliege (Anthomyia conformis) sowie die Trockenfäule der Zuckerrüben. — Centralbl. Biol. II, Abt. 14. p. 43—50.

HOFFMANN, A., 1893, Die Entomologie in der christlichen Legende. — Stettin. ent. Zeit. 54. p. 214—215.

HOFMANN, O., 1860, Über die Naturgeschichte der Psychiden. — Berlin. Ent. Zeit. 4. p. 1 ff.

——1861, Zur Naturgeschichte der Micropterygiden. — Korr. Bl. f. Samml. Ins. 2. p. 116.

——1869, Beiträge zur Naturgeschichte der Coleophoren. — Stettin. ent. Zeit. 1869. p. 107—122, 187—190.

——1871, Naturgeschichte der Cidaria incultaria HS. — ibid. 32. p. 43—44. Naturgeschichte der Heliozela stanneella F.R. — ibid. p. 44—47.

——1893, Beiträge zur Naturgeschichte der Tineinen. — ibid. 54. p. 307—311.

——1898, Elachista Martinii n. sp. — Iris 11. p. 143—146.

HOLMGREN, A. E., 1880, Anthomyia spinaciae Holmgr. — Ent. Tidskr. 1. p. 89.

HOUSER, S. J., 1918, Destructive insects affecting Ohio shade and forest trees. — Bull. Ohio Exp. Sta. 332. p. 161—487.

——1912, The wheat leaf miner. — Bull. Ohio Agr. Exp. Sta. 251. p. 79—86.

——1923, Apple flea-weevil. — Ibid. 372. p. 397—434.

HOWARD, L. O., 1891, An Anthomyid injuring sugar beets. — Insect Life 3. p. 470.

——1895, The beet-leaf Pegomyia (Pegomyia vicina Lintn.). — ibid. 7. p. 379—381.

——1896, The fruit flies or vinegar flies (Drosophila ssp.). — Bull. U. S. Div. Ent. n. s. 4. p. 109—111.

——1898, A peculiar Damage to the Apple. — Bull. U. S. Dept. Agr. n. s. 10. p. 87—88.

HOWARD, L. O., 1898, Danger of importing insect pests. — Yearb. U. S. Dept. Agr. 1897. p. 529—552.

———1901, The Insect Book. — 429 pp.

HUTCHINGS, C. B., 1925, The lilac leaf-miner. — Ann. Rept. Ent. Soc. Ontario 55 (1924) p. 19—23.

JACK, J. G., 1890, Diseases of Chrysanthemums caused by insects. — Garden & Forest 3. p. 439—440.

JACKH, E., 1934, Über einige das nordwestdeutsche Faunengebiet kennzeichnende Microlepidopteren. — Mitt. ent. Ver. Bremen 21. p. 6—10.

———1940, Die Insekten des Naturschutzparkes der Lüneburger Heide. III. Die Kleinschmetterlinge. — Abh. Nat. Ver. Bremen 31, p. 786—806.

———1942, Die Microlepidopterenfauna des rechtsseitigen Mittelrheintales nebst Beschreibung von Borkhausenia magnatella sp. n. — Ztschr. Wien. Ent. Ver. 27. p. 137—141.

JENSEN, A. S., 1932, Studier over Incurvaria koerneriella. — Vidensk. Selsk. Biol. Medd. X. 5. p. 1—49.

IMMS, A. D., 1915, Notes on the beet or mangold fly. — Journ. Board Agr. London 9. p. 881—884.

INCHBALD, P., 1881, Remarks on our Dipt.-Plantminers and the plants they affect. Entomologist 14. p. 41—43.

———1881, Dipt. Plantminers in their perfect states. — ibid. p. 290—292.

———1883, Observations on insect life in 1883. — Entomol. Nr. 244. p. 193—195.

DE JOANNIS, J., 1901, Description d'une nouvelle espèce de Microlépidoptère de France, Zelleria ribesiella. — Bull. ent. Soc. Fr. 1901, p. 391—393.

———1907, Deux espèces nouvelles de Nepticula. — ibid. 1907, p. 326—329.

———1907, Sur une espèce de Microlépidoptères nuisible aux Azalées. — ibid. p. 341—342.

———1918, Observation sur l'hibernation en masse de Lithocolletis populifoliella Tr. — ibid. 1918, p. 150—152.

———1919, Une visite à la collection de Microlépidoptères d'Achille Guenée. — Ann. Soc. ent. Fr. 88. p. 1—40.

———1920, Note sur une petite collection de Microlépidoptères provenant de St. Saens (Seine-Inf.). — Bull. Soc. ent. Fr. 1920. p. 142 ff.

———1921, Les Lithocolletis des Erables. — Ann. Soc. ent. Fr. 89. p. 405—416.

de Joannis, J., 1928, Observations sur Coleophora flavaginella Z. — Bull. Soc. ent. Fr. 1928, p. 208—211.

Johannsen, O. A., Spruce bud worm and spruce leaf miners. — Bull. Maine Agr. Exp. Sta. 210. p. 1—36.

Johnson, C. W., 1902, North American Diptera. — Canad. Ent. 34. p. 240—242.

———1908, The Diptera of the Bahamas, with notes and description of one new species. — Psyche 15. p. 69—80.

———1913, The Dipteran Fauna of Bermuda. — Ann. Soc. ent. America 6. p. 443—452.

Jones, T. H., 1915, Insects affecting vegetable crops in Porto Rico. — Bull. U. S. Agr. Dept. 192. p. 1—11.

———1918, Miscellaneous truck-crop insects in Lousiana. — ibid. 703. 19 pp.

———1923, The Eggplant Leaf-miner, Phthorimaea glochinella Z. — Journ. Agr. Res. Washing. 26. p. 567—570.

Jones, W. W., 1933, The Description and Biology of Nepticula braunella new species, a species of leaf-miner on Prunus ilicifolia Walp. and the variety integrifolia Sarg. — Publ. Univ. Calif. Entom. 6. Nr. 4. p. 49—78.

Jourdheuille, C., 1869/70, Calendrier du Microlépidoptériste. — Ann. Soc. ent. Fr. (4) 9. p. 533—548; 10, p. 111—134, 233—266.

Judeich, J. F. & Nitsche, H., 1885/95, Lehrbuch der mitteleuropäischen Forstinsektenkunde. Berlin, 2 vol.

Kaltenbach, J. H., 1856/69, Die deutschen Phytophagen aus der Klasse der Insekten. — Verh. naturf. Ver. Rheinl. & Westf. 13, p. 165—265; 15, p. 77—192; 16. p. 216—299; 17, p. 203—260; 19, p. 1—106; 21, p. 228—404; 24, p. 21—117; 26, p. 106—224.

———1874, Die Pflanzenfeinde aus der Klasse der Insekten. — Stuttgart, 848 pp.

Karsch, C. A., 1883, Die Insektenwelt. — 702 pp.

Karsch, F., 1887, Orchestes populi F. — Berlin. ent. Zeit. 1887. SB. p. 35—36.

Karl, O., 1926, Minenzuchtergebnisse. — Stettin. ent. Zeit. 87. p. 136—138.

———1928, Muscidae. — Dahl, Tierwelt Deutschlands 13. 232 pp.

———1930, Thalassobionte und thalassophile Diptera Brachycera. — Tierwelt der Nord- u. Ostsee XI. e. 2. p. 1—52.

———1930, Fliegen von der Insel Amrun. — Deutsch. Ent. Ztschr. 1930. p. 193—206.

KAUFMANN, O., 1928, Stand der Rübenfliege im Jahre 1928. — Zuckerrübenbau 1929, Nr. 6. p. 1—8.

———1928, Geschichte des Rübenfliegenbefalls im Schlesien in Jahre 1925. — Arb. Biol. Reichsanst. 16. p. 469—573.

———1929, id. 1928. — ibid. 17. p. 195—224.

——— & BREMER, H., 1927, Bericht über Versuche zur Bekämpfung der Rübenfliege im Jahre 1926. — Zuckerrübenbau 1927, Heft 3. p. 1—5.

KAZIAKIN-VINOGRADOW, B. N., 1932, On the Biology of the Pocket-Moth (Ornix guttea Hw.). Bull. Leningr. Inst. Contr. Farm & For. Pests 3. p. 249—252 (Russian).

KELLER, C., 1917, Zur Biologie von Chrysomela aenea L. und Coleophora fuscedinella Z. — Vierteljahrsschr. naturf. Ges. Zürich 62.

KEMNER, A., 1917, Bjorksäckmalen (Coleophora fuscedinella). — Medd. Centr. Anst. försöks Ent. Adveln. 28. p. 1—28.

———1925, Betflugan (Pegomyia hyoscyami Pnz.) — ibid. 47.

———1926, Clercks Mineralmal (Lyonetia clerkella L.). — ibid. Nr. 49. p. 1—59.

KINNEY, L. F., 1896, The Spinach Leaf-miner. — Bull. Rhode Isl. Agr. Exp. Sta. 41. p. 118 ff.

KIRBY, W. F., 1892, Elementary Text-book of Entomology. — 2d ed. 281 pp.

KIRCHNER, O., 1906, Die Krankheiten und Beschädigungen unserer landwirtschaftlichen Kulturpflanzen. — Stuttgart, 2. Aufl.

KLEINE, R., 1921, Die Rübenblattfliege. — Blätt. f. Zuckerrübenbau 28. p. 136—139.

———1923, Die Runkelfliege (Pegomyia hyoscyami Pnz.) und die landwirtschaftliche Praxis. — ibid. 30. p. 1—23.

———1924/25, Über die Larvenminen einiger Orchestini. — Ztschr. wiss. Ins. Biol. 19. p. 203—207, 251—254; 20, p. 24—27, 44—53.

KLIMESCH, J., 1936, Ein kleiner Beitrag zur Kenntnis der Nepticuliden des Vintschgaues. — Stettin. ent. Zeit. 97. p. 194—211.

———1937, Leucoptera onobrychidella spec. nov. — Ztschr. Österr. Ent. Ver. 22. p. 4—7.

———1937, Nepticula loranthella spec. nov. — ibid. p. 33—35.

———1938, Leucoptera cytisiphagella nov. spec. — ibid. 23. p. 49—51.

———1938, Coleophora perserenella Rbl. — ibid. 23, p. 93—95, 110—112.

Klimetsch, J., 1938, Piccolo contributo alla conoscenza dei Microlepidotteri della Valle Venosta. — Studi Trentini 19. p. 38—51.

———1939, Coleophora asterifoliella nov. sp. — Ztschr. österr. Ent. Ver. 24, p. 2.

———1939, Die Raupe der Coleophora rectilineella F. R. — ibid. 24. p. 39—43.

———1939, Zur Kenntnis der Biologie der Scirtopoda myosotivora M.-R. — ibid. 24, p. 65—69.

———1940, Beschreibung einiger neuer Nepticula-Arten. — ibid. 25, p. 79—81.

———1940, Zur Artberechtigung und Lebensweise der Coleophora prinziella Krone. — ibid. 25. p. 129—134.

———1940, Über eine Nepticuliden-Ausbeute von Triest, sowie Beschreibung der Mine von Leucospilapteryx cupediella. — Ztschr. Wien. Ent. Ver. 25. p. 176—179.

———1941, Eidophasia (Lunakia nov. subg.) alyssella nov. sp. — ibid. 26. p. 129 ff.

———1941, Nepticula Preisseckeri spec. nov. — ibid. 26, p. 162—168.

———1941, Coleophora medelichensis Krone. — L'Amateur de Pap. 9. p. 7—11.

———1942, Über einige Microlepidopteren der alpinen Felsflora. — Ztschr. Wien. Ent. Ver. 27. p. 145—157.

———1942, Über Microlepidopteren-Ausbeuten aus der Gegend von Zaton bei Gravosa. — Mitt. Münch. ent. Ges. 32. p. 347—399.

Knab, F., 1916, Four European Diptera established in North America. — Insec. insc. menstr. 4. p. 1—4.

Knowles, C. H., 1917, The Maize Leaf-miner (Phytomyza sp.) — Pamphl. Fiji Dept. Agr. 27. p. 1—2.

———1919, The Small Leaf Moth of Coconuts in Fiji (Levuana iridescens B.-B. Bull. Fiji Dept. Agr. 12. p. 1—8.

Kollar, V., 1840, A treatise on insects injurious to gardeners, foresters and farmers. — 377 pp.

Konow, F. W., 1901/05, Systematische Zusammenstellung der bisher bekannt gewordenen Chalastogastra. — Teschen, Teil I, II.

Korolkov, D. M. & Savenko, R. F., 1929, The life history of Prays oleellus Fab. and its control in Psyrtskha Soviet Estate (formerly New-Athos). — Subtropiki 1929, Nr. 3/4, 25 pp.

Körting, A., 1931, Über die Entwicklung der grauen Gerstenminierfliege (Hydrellia griseola Fall.) bei verschiedener Ernährung. — Ztschr. Pflanzenkr. 41, p. 321—333.

Kramer, H., 1917, Die Musciden der Oberlausitz. — Abh. naturf. Ges. Görlitz 28. p. 257—352.

Kryger, J. P. & Sønderup, H. P., 1940, Biologiske Iagttagelser over 200 Arten af danske Billelarver. — Ent. Medd. 22. p. 57—136.

Kuntze, R., 1938, Dritter Beitrag zur Kenntnis der Halticinen Polens. — Polsk. Pism. Entom. 16/17, p. 94—125.

Küster, E., 1925, Pathologische Pflanzenanatomie. 3rd ed. Jena, 558 pp.

Kvíčala, B., 1938, Škody Způsobené Minujícím Hmyzem Rostlinám na Kroměřížsku. — Entomol. listy 1. p. 141—156.

Laboulbène, A., 1863, Description et Figure d'une Chenille mineuse des feuilles du Bouleau. — Ann. Soc. ent. Fr. (4) 3. p. 99—104.

———1873, Métamorphose de la Cécidomyie du buis Cecidomyia (Diplosis) buxi. — ibid. (5) 3. p. 313—326.

Lafaury, C., 1879/86, Descriptions de Chenilles de Microlépidoptères inédites ou peu connues. — ibid. (5) 2. p. 73—84; (6) 5. p. 397—422.

Lamborn, W. A., 1914, The Agricultural Pests of the Southern Provinces, Nigeria. — Bull. Ent. Res. 5. p. 197—214.

Landgraf, 1924, Die Fliedermotte (Gracilaria syringella F.). — Die kranke Pflanze 1. p. 173—174.

Larsen, C. S., 1927, Tillaeg til Fortegnelsen over Danmarks Microlepidoptera. — Ent. Medd. 17. p. 1—211.

———1928, Upon some new and little-known species and aberrations of Microlepidoptera from Denmark. — ibid. 18. p. (1—6).

Lasaroff, A. W., 1939, Beitrag zur Biologie und Bekämpfung der im Obstbau schädlichen Sackträgermotten. — Die Gartenbauwissenschaft, 14. p. 77—140.

Lea, A. M., 1908, The Marguerite or Cineraria Fly. — Tasmania Counc. Agr. p. 81—84.

Leach, J. G., 1927, The relation of Insects and Weather to the development of Heart rost of Celery. — Journ. Ser. Minnesota Agr. Exp. Sta. 714.

Le Cerf, F., 1925, Note sur une Gelechide parasite de l'Alfa. — Enc. entomol. B. 3. Lep. 1. p. 11—28.

Lehmann, H., 1934, Luzerneschädlinge, 2. Diptera, Minierfliegen, Gallmücken. — Ztschr. Pflanzenkrankh. 44. p. 331—348.

LEHMANN, H., 1935, Luzerneschädlinge. 4. Blattschädlinge. — ibid. 45. p. 416—431.

LE MARCHAND, S., 1925, Les Nepticula. — L'Amateur Papill. 2. p. 202—208.

———1925, Les Lithocolletis. Les Tischeria. — ibid. p. 237—251.

———1926, Lithocolletis tinctoriella nova species. — ibid. 3. p. 13—16.

———1929, A propos de Cosmopteryx turbidella Rbl. — ibid. 4. p. 201—205.

———1936, Clé ou table analytique pour la détermination des espèces françaises de Lithocolletis. — ibid. 8. p. 83—114.

———1937, Nepticula alaternella nova species du midi de la France. — ibid. 8. p. 233—239.

———1937, Les Cosmopteryx. — ibid. 8. p. 171—175.

VON LENGKERKEN, H., 1922, Eine Mordellistena (Coleopt.) als Schädling an der Orchidee Cattleya labiata f. Trianaei Duch. — Zool. Jahrb. Syst. 44. p. 579—594.

———1928, Lebenserscheinungen der Käfer. — Leipzig, 147 pp.

———1941, Von Käfern erzeugte Pflanzengallen. — Ent. Blätt. 37. p. 121 ff.

———1941, In Blättern minierende Käferlarven. — Biolog. Gen. 15. p. 236—281.

LENGERSDORF, F., 1930, Lycoriidae (Sciaridae). — Lindner, Flieg. pal. Reg. 7. 71 pp.

LEONARDO, G. & SILVESTRI, F., 1922/27, Elenco delle specie di Insetti dannosi e loro parassiti ricordati in Italia fino all'anno 1911. — Portici, pt. 1—3.

LEPRIEUR, C. E., 1861, Essai sur les métamorphoses du Trachys pygmaea. — Ann. Soc. ent. Fr. 1. p. 456—466.

LETZNER, K., 1856, Über Larve und Puppe des Orchestes populi L. — Arb. Schles. Ges. Vaterl. Kult. 1856, p. 98—102.

LEWIN, L., 1892, Über die Ähnlichkeit mancher Blattkrankheiten mit Hautkrankheiten des Menschen. — Wien. mediz. Presse 1892. Nr. 43.

LHOMME, L., 1929, Cosmopteryx turbidella Rbl. en Aveyron. — L'Amateur Pap. 4. p. 205—206.

———1933, Amateur, cherchez des mines. — ibid. 6. p. 175—179.

———1934, Excursion au pays des mines et description d'une sous-espèce nouvelle de Lithocolletis. — ibid. 7. p. 129—1398.

LINDEMANN, H., 1887, Über Agromyza lateralis Macq. und ihre Verwandlungen. — Bull. Soc. Imp. Nat. Moscou 62 (1886) II. p. 9—14.

LINDROTH, C. H., 1931, Die Insektenfauna Islands und ihre Probleme. — Zool. Bidr. Uppsala 13. p. 105—599.

LINNANIEMI, W. M., 1913, Zur Kenntnis der Blattminierer, speziell derjenigen Finnlands. I. — Acta Soc. Flor. Faun. Fenn. Nr. 4.

LINTNER, J. A., 1882, Notice of some Anthomyians mining beet leaves. — Ann. Rept. State N. Y. 1. p. 203—211.

———1882, Leaf-mining Anthomyidae. — Canad. Ent. 14. p. 96—97.

———1883, Leaf-mining Anthomyidae. — Rept. Ent. Soc. Ontario 1882. p. 31.

———1885, Beet-leaf mining Anthomyidae. — Rept. State N. Y. 2. p. 46.

———1888, Phytomyza lateralis. — ibid. 4. p. 73—79, 207.

———1891, Phytomyza chrysanthemi Kow. — ibid. 7. p. 242—246.

———1893, Beet insects. — ibid. 9. p. 374—376.

———1895, The Chrysanthemum Fly. — ibid. 10. p. 509—513.

VON LINSTOW, 1909, Revision der deutschen Psychidengattungen. — Berlin. Ent. Zeit. 54, p. 89—102.

LOCHHEAD, W., 1904, Beet leaf-miner. — Rept. Ent. Soc. Ontario 34. p. 32.

LOTH, N., 1913, Tenthredinidenstudien. I. — Berlin. Ent. Zeit. 58. p. 46—95.

LÜDERS, L., 1900, Beitrag zur Kenntnis der Lepidopterengattung Phyllocnistis Z. — Beitr. Schulber. Realschule St. Pauli. Hamburg.

LUDWIG, F., 1907, Weiteres zur Biologie von Helleborus foetidus. — Ztschr. wiss. Ins. Biol. 3. p. 45—50.

———1908, Noch einige nachträgliche Bemerkungen über die Helleborus-Parasiten. — ibid. 4, p. 102.

LUGINBILL, P. & URBAHNS, T. D., 1916, The spike-horned leaf-miner, an enemy of grains and grasses. — Bull. U. S. Dept. Agr. 432. p. 1—20.

LUNDBLAD, O., 1927, Lilla vinbärsmalen. — Centralanst. Jordbruksf. Flygbl. 121.

———1930, Lilla Vinbärsmalen. — Medd. Centralanst. Jordbruks. 374. 25 pp.

LUNDBLAD, O. & LINDBLOM, A., 1925, Selleriflugan. — Medd. Centr. Anst. Ent. Avdeln. Nr. 45.

LUNDETRAE, O. B., 1940, Beitrag zur Kenntnis der westnorwegischen Schmetterlingsfauna. — Bergens Mus. Arb. 1938, Naturv. R. 10. p. 1—15.

Lüstner, G., 1924, Stärkeres Auftreten der Luzernegallmücke (Dasyneura ignorata Wcht. und der Luzernefliege (Phytomyza affinis Fall.) im Rheingau. — Nachr. Bl. dtsch. Pflanzenschutzdienst 4. Nr. 8.

——1925, Auftreten der Buchenblattminiermotte (Lithocolletis faginella Z.) ibid. 5. Nr. 1.

Malaise, R., 1920, Beiträge zur Kenntnis schwedischer Blattwespen. — Ent. Tidskr. 41. p. 97—128.

——1931, Entomologische Ergebnisse der schwedischen Kamtchatka-Expedition 1920—1922. 35. Tenthredinidae. — Ark. Zool. 23. A. Nr. 8.

Malloch, J. R., 1909, Some Phoridae in Dumbartonshire in 1908. — Ent. Month. Mag. 45, p. 34—36.

——1913, A revision of the species in Agromyza Fall. and Cerodonta Rond. — Ann. Ent. Soc. America 6. p. 269—340.

——1913, A Synopsis of the genera of Agromyzidae, with descriptions of new genera and species. — Proc. U. S. Nat. Mus. 46. p. 127—154.

——1914, Formosan Agromyzidae. — Ann. Mus. Nat. Hung. 12. p. 306—336.

——1914, Notes on North American Agromyzidae. — Ent. News 25. p. 308—314.

——1915, North American Diptera. — Canad. Ent. 47. p. 12—16.

——1915, Flies of the Genus Agromyza, related to Agromyza virens. — Proc. U. S. Nat. Mus. 49. p. 103—108.

——1918, Partial key to the genus Agromyza. — Canad. Ent. 50. p. 315—318.

——1920, Descriptions of new North American Anthomyidae. — Trans. Amer. Ent. Soc. 46. p. 133—186.

——1920, A Synopsis of the North American species of the Genus Pegomyia R.-D. — Bull. Brookl. Ent. Soc. 15. p. 121—127.

——1933/34, Acalyptratae. — Dipt. Patag. & S. Chile 6. fasc. 4/5. p. 177—489.

Mamonov, B. A., 1929, Mining Diptera in the Vicinity of Rostov on the Don. — Bull. Agr. Exp. Sta. North Caucas. Div. Ent. 292.

Marchal, P., 1903, Sur la biologie des Hydrellia. — Bull. Soc. ent. Fr. 1903, p. 236—237.

Marcovitch, S., 1916, Insects attacking weeds in Minnesota. — 16. Rept. State Ent. Minnesota p. 135—152.

MARCU, O., 1928, Fagul şi Orchestes fagi L. in Oltenia. — Ecou de Codru 1. Nr. 3.

MARELLI, C. A., 1926, La Agromyza productora de agallas nodícolas en la lagunilla y sus parásitos e hiperparásitos. — Rev. Soc. Ent. Argent. 1. Nr. 2.

MARGABANDHU, V., 1933, Insect Pests of Oranges in the Northern Circars. — Madras Agr. Journ. 21. p. 60—68.

MARIANI, M., 1932, Novità di Lepidotterologia in Sicilia. — Boll. Soc. ent. Ital. 62, p. 105—110.

———1932, Nota preventiva sulle Cosmopteryx d'Europa. — Boll. Soc. Sci. Nat. Econ. Palermo 1932, p. (1—15).

———1935, Monografia sulle Cosmopteryx d'Europa. — Giorn. Sci. Nat. Econ. 38. Mem. 1. p. 1—54.

———1936, Una nuova specie di Eustaintonia Spul. e studio sulla Morfologia comparata delle Larve del genere. — Boll. Soc. ent. Ital. 68. p. 97—103.

———1939, Un nuovo genere e due nuove specie di Lepidotteri di Sicilia. — Giorn. Sci. Nat. Econ. Palermo 11. Mem. 1. p. 1—7.

———1942, Una infestion di Antispila rivillei St. su vigneti del Palermitano. — Giorn. Sci. Nat. Econ. Palermo 43, p. 1—14.

MARTINI, W., 1898, Antispila petryi nov. spec. — Stettin. ent. Zeit. 1898, p. 398—405.

———1902, Nepticula pulverosella Stt. — Ztschr. f. Ent. N. F. 27. p. 31—33.

———1902, Elachista variabilis n. sp. — Ztschr. f. Ent. N. F. 27. p. 26—30.

———1912, Die Gattung Elachista. — Ent. Ztschr. Frankf. 26. p. 83—84.

———1912, Beiträge zur Kenntnis der Elachista-Raupen. — Iris 1912, p. 92—95.

———1912, Über die Heinemannsche Elachistengruppe D. — ibid. p. 155—188.

———1912, Beiträge zur Kenntnis von Arten der Gattung Elachista. — ibid. p. 208—211.

———1916, Verzeichnis Thüringer Falter aus den Familien der Pyralidae-Micropterygidae. — Iris 30. p. 110—144, 153—186.

MATTEI, G. E., 1926, La variegatura delle foglie è dovuta a batterii? — Rev. Biol. 8. fasc. 1. 23 pp.

———1934, Danni causati alle piante dagli insetti. Le mine fogliari. — Rivist. Ital. Ess. Profum. & Piant. off. p. XXV—XXVI.

MAULIK, S., 1929, New injurious Hispinae. p. 81—94. — Injurious Hispinae from the Solomon Islands. p. 233—239. — Bull. ent. Res. 30.

———1931/33, On the structure of Larvae of Hispine Beetles. — Proc. Zool. Soc. Lond. 1931, p. 1137—1162; 1932, p. 293—322.: 1933, p. 669—680, 935—939.

———1937, Distributional Correlation between Hispine Beetles and their Host-plants. — ibid. 1937. A. p. 129—159.

MEADE, R. H., 1878, Notes on the Anthomyiidae of North America. — Ent. Month. Mag. 14. p. 250—252.

———1883, Annotated list of British Anthomyidae. — ibid. 20. p. 9—14.

———1887, Supplement to the Annotated list of British Anthomyiidae. — ibid. 24. p. 54—58.

MEDER, O., 1930, Über den Stand unserer Kenntnisse von den schleswig-holsteinischen Kleinfaltern. — 4. Wandervers. Deutsch. Ent. Kiel p. 93—98.

———1930, Über die Nährpflanzen der Lyonetia ledi Wck. — Int. ent. Ztschr. Guben 24. p. 163—164.

———1934, Mitteilungen über Kleinfalter der Nordmark. — ibid. 27. p. 489—493.

———1934, Übersicht über die bisher auf den Nordfriesischen Inseln, besonders Amrun, festgestellten Kleinfalter. Schrift. nat. Ver. Schlesw. — Holst. 20. p. 357—364.

———1939, Ein Beitrag zur Kleinfalterfauna Helgolands. — Verh. Ver. naturw. Unterh. Hamburg 27. 14 pp.

VAN DER MEER MOHR, J. C., 1929, Spinnen in Nepenthes Bekers. — Trop. Natuur 1929, 73—77.

MEESS, A., 1923, Die cecidogenen und cecidocolen Lepidopteren. — Zoologica 61. p. 499—584.

MEIGEN, J. W., 1818/38, Systematische Beschreibung der bekannten europäischen zweiflügeligen Insekten. 7 Teile.

DE MEIJERE, J. C. H., 1895, Über zusammengesetzte Stigmen bei Dipterenlarven, nebst einem Beitrag zur Metamorphose von Hydromyza livens. — Tijdschr. Ent. 38. p. 65—100.

———1909, Zur Kenntnis der Metamorphose der Lauxaniinae. — Ztschr. wiss. Ins. Biol. 5. p. 152—155.

———1911, Über in Farnen parasitierende Hymenopteren- und Dipteren-Larven. Tijdschr. Ent. 1911, p. 8—127.

de Meijere, J. C. H., 1916, Beiträge zur Kenntnis der Dipteren-Larven und -Puppen. — Zool. Jahrb. Syst. 40. p. 177—322.

———1922, Zur Kenntnis Javanischer Agromyzinen. — Bijdr. Dierkd. 22. p. 17—24.

———1924, Verzeichnis der holländischen Agromyzinen. — Tijdschr. Ent. 67. p. 119—155.

———1925/49, Die Larven der Agromyzinen. I. Tijdschr. Ent. 68. p. 195—293. II. ibid. 69, p. 227—317. — 1. Nachtrag. ibid. 71. p. 145—178. — 2. ibid. 77. p. 244—290. — 3. ibid. 80. p. 167—243. — 4. ibid. 81. p. 61—116. — 5. ibid. 83. p. 160—188. — 6. ibid. 84. p. 13—30. — 7. ibid. 86, p. 61—76. — 8. ibid. 87. p. 65—74. — 9. ibid. 92. p. 15—33.

———1928, Vierde Supplement op de Nieuwe Naamlijst van Nederlandsche Diptera. — ibid. 71. p. 11—83.

———1938, Phytagromyza buhri n. sp., eine Agromyzide, deren Larve im Stengel der Blütenstände von Galium mollugo lebt. — Ent. Berichten 10. p. 83.

———1940, Über die Larven der in Orchideen minierenden Dipteren. — Tijdschr. Ent. 83. p. 122—127.

———1940, Über Melanagromyza centrosematis n. sp. aus Java nebst Bemerkungen über andere tropische Melanagromyzen. — ibid. 83. p. 128—131.

———1940, Hydromyza livens Fall. en Notiphila brunnipes R.-D. — Ent. Bericht. 10. Nr. 232. p. 220—222.

———1943, Over de Metamorphose van Metopia leucocephala Rossi, Cacoxenus indigator Lw., Palloptera saltuum L., Paranthomyza nitida Mg. en Hydrellia nigripes Zett. — Tijdschr. Ent. 86. p. 57—61.

Meixner, A. & Meyer, 1912/14, Monatliche Sammelanweisungen. Microlepidoptera. — Kranch. ent. Jahrb. 1912—1914.

Melis, A., 1933, Contributo alla conoscenza dello Sphaeroderma rubidum Graells. — Redia 20. p. 189—227.

———1935, Contributo alla conoscenza morfologica et biologica della Phytomyza atricornis Meig. — ibid. 21. p. 205—262.

Mendes, C., 1913, Lepidopteros de S. Fiel. Supplemento. — Broteria 11. p. 15—44.

Menozzi, C., 1930, Insetti dannosi alla Barbabietola. — Boll. Soc. Ent. Ital. 1930, p. 1—98.

Menzel, R., 1929, De Plagen van de The in Nederlandsch-Indie. Batavia 106 pp.

MESNIL, L., 1934, Ophiomyia pinguis Fall. nuisible aux Endives. — Bull. ent. Soc. Fr. 39. p. 131—136.

MEYRICK, E., 1929, A Revised Handbook of British Lepidoptera. — London, 914 pp.

MIALL, L. C. & TAYLOR, T. H., 1907, The Structure and Life-history of the Holly-Fly. — Trans. Ent. Soc. Lond. 1907, p. 259—283.

MICHALK, O., 1937, Über Glottula pancratii Cyr. — Ent. Rundsch. 55. p. 50—53, 74—76.

MIK, J., 1894, Über eine neue Agromyza, deren Larven in den Blüthenknospen von Lilium martagon leben. — Wien. ent. Zeit. 13. p. 284—290.

———1900, Dipterologische Miszellen. — ibid. 1900. p. 148 ff.

MILLER, D., 1916, Contributions to the Diptera Fauna of New Zealand. I. — Trans. N. Zeal. Inst. 49. p. 172—194.

——— & WATT, M. N., 1915, Contribution to the Study of New Zealand Entomology, from an Economical and Biological Standpoint. — ibid. 47. p. 274 ff.

MILLIERE, P., 1856, Histoire de Choreutis dolosana H.-S. — Ann. Soc. ent. Fr. (3) 3. p. 39—49.

———1859/74, Iconographie et Description de Chénilles et Lépidoptères inédits. — Lyon, 3 Pt.

MITTERBERGER, K., 1915, Nepticula sericopeza Z., ein Schädling unserer Ahornfrüchte. Soc. ent. 230 p. 23—25.

———1919, Die Kieferntriebmotte Heringia (Teleia) dodecella L. — ibid. 34. p. 1—2, 7—8, 9—11.

———1914, Nepticula splendidissimella H. S. — Lotos 62, p. 155—161.

———1916, Zur Bestimmung der auf Eichen vorkommenden heimischen Coleophora-Säcke. — Ztschr. österr. Ent. Ver. 1. Nr. 2.

———1918, Die röhrentragenden heimischen Coleophora-Arten. — Kranch. ent. Jahrb. 1918.

———1919, Die Nahrungspflanzen der heimischen Coleophora-Arten. — Arch. Naturg. 83. A. 6. p. 55—75.

———1920, Beiträge zur Kenntnis der Kiefern- und Fichtenfeinde unter den Kleinschmetterlingen. — Kranch. Ent. Jahrb. 1920, p. 127—133.

———1920, Die Lithocolletis-Arten um Steyr in Oberösterreich. — Deutsch. Ent. Ztschr. 1920, p. 331—354.

MITTERBERGER, K., 1922, Lithocolletis alpina Frey. — Kranch. Ent. Jahrb. 1922, p. 101—103.

———1924, Der Kiefernwickler (Cacoecia piceana L.) — Kranch. Ent. Jahrb. 1924/25, p. 105—108.

———1925, Cedestis gysselinella Dp. und farinatella Z. — Soc. ent. 40. p. 13—15, 18—19.

———1926, Der Zwergnadelwickler (Steganoptycha nanana Tr.) — Kranch. ent. Jahrb. 1926. 4 pp.

———1929, Ornix finitimella Z. — ibid. 1929, p. 129—130.

———1930, Die an Achillea vorkommenden Mikro-Lepidopteren-Raupen. — ibid. 1931, p. 154—156.

———1930, Unsere Nepticula-Arten an Weissdorn. — Ent. Rundsch. 47, p. 25—27. 32—35.

———1930/31, Die Nahrungspflanzen der heimischen Nepticula-Arten. — Soc. ent. 45. p. 49—51; 46, p. 6—8, 16—19, 28—30, 53—55, 72—74, 86—89.

———1934, Bucculatrix thoracella Thbg. — Kranch. ent. Jahrb. 43. p. 119—121.

———1934, Die Schmetterlingsfauna Oberösterreichs. — ibid. 44. 7 pp.

———1935, Unsere einheimischen Cemiostoma-Arten. — ibid. 45. p. 94—98.

MJÖBERG, E., 1908, Om en bladminerande skadeinsekt. Ekbladmineraren (Orchestes quercus L.) — Skogsvårdsför. Tidskr. 56. p. 257—263.

MÖBIUS, E., 1913, Septemberfang in Bozen. — Iris 1913, p. 104 ff.

MOKRZECKI, Z., 1928, Rep. Inst. For. Prot. & Entom. Skiernew. 1924—27. — 61 pp.

———1931, Monofagizm i polifagizm u owadow, biologicznie zwiazanych z roslinami. — Choroby Roslin 1. p. 1—8.

MONCREAFF, H., 1869, Orchestes alni in a Branch of Whitethorn. — Entomolog. 4. p. 218.

MOORE, E., 1915, The Potamogetons in relation to pond culture. — Bull. U. S. Bur. Fish. 33. p. 251—291.

MORSTATT, H., 1921, Ein Rüsselkäfer als Sisal-Schädling. — Tropenpflanz. 24. p. 33—37.

MÖSCHLER, B. H., 1861, Die Schmetterlinge der Oberlausitz II. — Neues Lausitz. Mag. 38.

Motaș, C. & Zaharesco, V., 1936, Sur la biologie du Lyonetia clerckella L. et les moeurs de sa chenille. — Bull. Sci. Acad. Roum. 17. 11 pp.

Müller, A., 1872, Dipterous larvae in the fronds of Pteris aquilina. — Ent. Month. Mag. 8. p. 181.

Müller, G. W., 1921, Insektenlarven an Wurzeln von Wasserpflanzen. — Mitt. Naturw. Ver. Neuvorpomm. & Rügen 48. 21 pp.

———1925, Kalk in der Haut der Insekten und die Larve von Sergus cuprarius L. — Ztschr. Morph. Ökol. Tiere 3. p. 542—566.

Müller-Rutz, J., 1920, Aus der Welt der Kleinschmetterlinge mit Beschreibungen neuer Arten und Formen. — Mitt. Entom. Zürich 5.

———1925, Gracilaria azaleella Brants in der Schweiz. — Schweiz. ent. Anz. 4

———1927/32, Die Schmetterlinge der Schweiz. Nachtr. 4. — Mitt. Schweiz. ent. Ges. 13. p. 217—259. — id. 5, ibid. p. 499—533. — id. 6. — ibid. 15. p. 221—266.

———1931, Unica oder fragliche n. sp., auch für die Schweiz neue Arten. — ibid. 15. p. 153—154.

———1934, Über Microlepidopteren. — ibid. 16. p. 118—128.

———1937, Eine neue Elachistide, Scirtopoda myosotivora. — ibid. 17. Heft. 4.

Murtfeldt, M. E., 1874, The larvae of Depressaria dubitella and Gelechia rubensella. — Canad. Ent. 6. p. 222.

Natanson-Grodzinska, N., 1932, Sur la structure du tégument chez les chenilles aquatiques de Cataclysma lemnata L. et son rôle dans la respiration. — Bull. Inst. Cracov. 3. B. 2. p. 203—219.

Natvig, L. R., 1928, Norske Insekter. I. Oslo.

Naumann, J. & Jächk, E., 1931, 2. Beitrag zur Kenntnis der Schmetterlingsfauna des Königsmoores. — Mitt. ent. Ver. Bremen 1930. 9 pp.

Nedeltcheff, N., 1930, Un nouvel ennemi de la vigne en Bulgarie. — La chenille mineuse des feuilles de vigne. — Antispila rivillei Stt. — Ann. Fac. Agr. Sylvicult. Sofia 8. p. 365—368.

Needham, F. G., Frost, S. W. & Tothill, B. H., 1928, Leaf-mining Insects. Baltimore.

Neuhaus, G. H., 1886. Diptera marchica. Berlin, 371 pp.

Nickerl, O., 1908, Die Tineen Böhmens. — Beitr. Ins. Fauna Böhm. herausgeg. Ges. Physiokr. Prag.

Nicolay, S. & Weiss, H. B., 1920, The Genera Pachyschelus and Taphrocerus. — Journ. N. Y. ent. Soc. 28. p. 136—150.

NICOLAY, S. & WEISS, H. B., 1923, The group Trachys in North America. II. The Genus Brachys. — ibid. 31. p. 59—76.

NOEL, P., 1914, Les ennemis du topinambour. — Bull. Lab. Ent. Agr. Seine-Inf. 1914, p. 13 ff.

NOLCKEN, J. H. W. VON, 1868/71, Lepidopterologische Fauna von Estland, Livland und Kurland. Riga.

NÖRDLINGER, H., 1848, Nachtrag zu Ratzeburgs Forstinsekten. — Stettin. ent. Zeit. 9. p. 233—234.

——— 1869, Die kleinen Feinde der Landwirtschaft. 2. Aufl. Stuttgart.

NORDMANN, A. F., 1926, Über die Biologie und Verbreitung von Zelleria ribesiella de Joann. — Notul. Ent. 6. p. 51—55.

——— 1928, Ett Bidrag till fjärilarnas äggläggningsbiologi. — ibid. 8. p. 118 ff.

NORTON, J. B. S. & SYMONS, T. B., 1907, The Parsnip Leaf-miner (Acidia fratria). — Bull. Maryland Agr. Exp. Sta. 115. p. 193.

NOVICKIJ, S., 1928, Lita paniculatella (Rössler i. l.) n. sp. — Sborn. ent. Nar. Mus. Prag 2. p. 137—139.

NÜSSLIN, O., 1913, Leitfaden der Forstinsektenkunde. 2. Aufl. Berlin.

OBENBERGER, J., 1918, Revision der palaearktischen Trachydinen. — Arch. Naturg. 82. A. 11.

——— 1939, Über die von Herrn H. Schmidt in Costa Rica gesammelten Buprestiden. — Arb. morph. tax. Ent. Dahlem 6. p. 301—314.

OERSTADIUS, E., 1918, Ytterligare anteckningar om fjärilar huvudsakligen småfjärilar, i Kronbergs län. — Ent. Tidskr. 39. p. 69—81.

——— 1932, Anmärkningsvärda fynd av småfjärilar. — ibid. 53. p. 108—113.

VON OETTINGEN, H., 1935, Die wichtigsten grasbewohnenden Fliegenlarven Norddeutschlands. — Nachr. Schädingsbekämpf. 2. 8 pp.

O'KANE, W. C., 1920, Injurious Insects. — New York, 414 pp.

O'KANE, W. C. & WEIGEL, C. A., 1921, Experiments with Contact Sprays for Leaf Miners. — Techn. Bull. N. Hampsh. Agr. Exp. Sta. 17.

DE ONG, E. R., 1922, A rice Leaf-miner. — Journ. Econ. Ent. 15. Nr. 6. p. 432.

ORMEROD, E. O., 1880, Tephritis onopordinis, the Celery fly. — Rept. Observ. inj. Ins. 1879, p. 13—14.

——— 1881, Anthomyia betae, the Mangold-fly. — Entomol. 14. p. 165—166.

ORMEROD, E. A., Celery and Parsnip Fly. Tephritis onopordinis Curt. — Rept. Obs. inj. Insects 1883, p. 20—23.

ORMEROD, E. A., 1885, Mangold and Beet Fly, Anthomyia betae Curt. ibid. 1884, p. 63—67.

———1890, A Manual of Injurious Insects. 2d ed. 410 pp.

———1891/93, Mangold or Beet Fly. — Rept. Observ. Injur. Ins. 14, p. 72—73; 15. p. 59—65; 16. p. 83—89.

OSTEN-SACKEN, C. R., 1881, An essay of comparative Chaetotaxy, or the arrangement of characteristic bristles of Diptera. — Mitt. Münch. Ent. Ver. 5. p. 121—138.

PACKARD, A. G., 1888, Guide to the Study of Insects. — 9th ed.

———1890, Insects injurious to Forest and Shade Trees. — Washington.

PANDELLE, L., 1901, Etudes sur les Muscides de France. III. — Rev. ent. Fr. 20. p. 293 ff.

PAPE, H., 1925, Neue und wenig bekannte Insektenschäden bei der Tomate. — Die Gartenwelt 29. Nr. 36/37.

PATTERSON, J. E., 1921, Life History of Recurvaria milleri, the Lodge-pole pine needle-miner. — Journ. Agr. Res. Washingt. 21. p. 127—143.

PERRIS, E., 1873, Résultats de quelques promenades entomologiques. — Ann. Soc. ent Fr. (5) 3. p. 61—98.

———1876, Nouvelles promenades entomologiques. — ibid. 1876. p. 171—244.

PETERS, L. & SCHWARTZ, M., 1912, Krankheiten und Beschädigungen des Tabaks. — Mitt. Biol. Reichsanst. 1912. Heft 13.

PETERSEN, W., 1924, Lepidopteren-Fauna von Estland. — 2d ed. Reval, 588 pp.

———1927/30, Die Blattminierer-Gattungen Lithocolletis und Nepticula. — Stettin. Ent. Zeit. 88. p. 113—174; 91, p. 1—82.

———1929, Die Bedeutung der Monophagie für die Erhaltung der Art. — Beitr. Kd. Estlands 14. p. 180—191.

———1930, Nachträge zur Lepidopterenfauna von Estland mit besonderer Berücksichtigung der Gattungen Lithocolletis und Nepticula. — Korr. Bl. Naturf. Ver. Riga 60. p. 89—101.

———1931, Nahrung und Genotypus. — Ztschr. Morph. Ök. Tiere 20. p. 679—690.

———1938, Wie verhalten sich die monophagen Insekten zu den bisherigen Ergebnissen der Serodiagnostik? — Arch. Fr. Naturg. Meckl. N. F. 13. p. 31—43.

PETRY, A., 1904, Nepticula thuringiaca n. sp. — Stettin. Ent. Zeit. 1904, p. 179—181.

———1904, Beschreibung neuer Microlepidopteren aus Korsika. — ibid. p. 242—254.

———1910, Beiträge zur Kenntnis der heimatlichen Pflanzen- und Tierwelt. — Beil. Schulprogr. Nordhausen 1909/10, 37 pp.

———1911, Gypsophila fastigiata und ihre Bewohner unter den Lepidopteren als Zeugen einer einstigen Periode kontinentalen Klimas. — Dtsch. Ent. Nationalbibl. 2. Nr. 23.

———1912, Über die deutschen an Artemisia lebenden Arten der Gattung Bucculatrix Z. nebst Beschreibung einer neuen Art. — Iris 26, p. 111—115.

———1914, Zwei für Deutschland neue Microlepidopteren. — ibid. 28. p. 23—25.

———1916, Über die Lepidopterenfauna des Kyffhäuser-Gebirges. — Ent. Mitt. 5. Nr. 1—8.

———1923, Über Elachista hedemanni Rbl. — Dtsch. Ent. Ztschr. Iris 37. p. 1—4.

PEYERIMHOFF, P. DE, 1910, Un nouveau type de larves mineuses appartenant au genre Xenostrongylus. — Bull. Soc. ent. Fr. 1910, p. 266—270.

———1921, Ethologie des Brachyleptus et notamment de B. algiricus Grouv. — ibid. 1921, p. 281—285.

———1926, Notes sur la biologie de quelques Coléoptères phytophages du Nord Africain. IV. Ann. Soc. ent. Fr. 95. p. 319—390.

PHILLIPS, W. J., The Corn-leaf blotch miner. — Journ. Agr. Res. 2. p. 15—31.

PICARD, F., 1914, Les insectes de la luzerne. — Progr. Agr. Vitic. 61. p. 555—561.

———1913, Charançons nuisibles à la betterave dans le midi de la France. — Bull. Soc. Etude et Vulg. Zool. Agr. 12. p. 129—137.

PIERCE, F. N. & METCALFE, J. W., 1935, The Genitalia of The Tineid Families of the Lepidoptera of the British Islands. — Oundle, 116 pp.

PISSOT, E., 1891, La Tischeria complanella Hb. — Le Naturaliste 13. p. 236 ff.

———1892, L'Orchestre du Hêtre. — ibid. 14. p. 91—92.

PJATAKOVA, V. L., 1931, To the Biology of the Case bearers (Coleophorae) of Fruit-trees. — Bull. Mlew Hortic. Exp. Sta. Ent. Sect. 45. 24 pp.

POLEJAEFF, W., 1934, Der Kampf ums Dasein bei der Pappelmotte Lithocolletis populifoliella Tr. — Zool. Journ. 13. p. 485—506.

PORTIER, P., 1918, Les Symbiotes. — Paris, 315 pp.

———1931, Les chenilles mineuses et la bactériologie. — L'Amat. Pap. 5. p. 265—270.

POUTIERS, R., 1924, Notes biologiques sur la Mineuse du Platane (Lithocolletis platani Stgr.) et ses parasites. — Riviera Scient. 11. p. 79—84.

PREISSECKER, F., 1923, Neue Microlepidopterenarten in Niederösterreich. — Verh. zool. bot. Ges. Wien 73. p. (76)—(79).

———1937, Niederösterreichische Falter. — ibid. 36/37, p. 418—420, 427—432.

———1942, Zwei neue Nepticula-Arten aus dem Gebiete des heutigen Reichsgaues Wien. — Ztschr. Wien. Ent. Ver. 27. p. 208—211.

PRINZ, 1909, Lithocolletis platani Stgr. aus dem Prater bei Wien. — Verh. zool. bot. Ges. Wien 1909, p. (234).

PROHASKA, K. & HOFFMANN, F., 1929, Die Schmetterlinge Steiermarks. 9. 10. — Mitt. Naturw. Ver. Steiermk. 64/65. p. 164—198, 272—321.

PUPPINI, G., 1930, Contributo alla conoscenza della Anarsia lineatella Z. e appunti sulla Recurvaria nanella Hb. — Boll. Lab. Ent. Bologna 3. p. 182—219.

QUANTZ, B., 1927, Zur Geschichte der Erklärungsversuche der Frassgänge von Lyonetia clerkella L., dem Obstlaub-Minierer. — Ztschr. wiss. Ins. Biol. 22. p. 7—18.

RAGONOT, E. L., 1874/75, Microlépidoptères nouveaux ou peu connus. — Ann. Soc. ent. Fr. (5) 4. p. 579—604; 6. p. 401—422.

REAUMUR, R. A. F. DE, 1734/42, Mémoires pour servir à l'histoire des Insectes. Paris, 6 vol.

REBEL, H., 1892—1938, Zur Lepidopterenfauna der Canaren. — 1. Ann. Hofmus. Wien 7. p. 242—284. — 2. ibid. 9. p. 1—96. — 3. ibid. 11, p. 102—148. — 4. ibid. 13. p. 361—381. — 5. ibid. 21. p. 42—44. — 6. ibid. 24. p. 327—374. — 7. ibid. 31, p. 1—62. — 8. ibid. 49. p. (43)—(68).

———1900, Neue palaearktische Tineen. Iris 12. p. 161 ff.

———1913, Lithocolletis hauderiella n. sp. — Verh. zool. bot. Mus. Wien 1913. p. 45.

———1913, Zur Unterscheidung und Synonymie einiger Arten der Gattung Gracilaria. — Iris 27. p. 168—172.

———1923, Tabakschädlinge auf Sumatra. — Verh. zool. bot. Ges. Wien 73. p. (73)—(75).

REBEL, H., 1939, Die Arthropodenfauna von Madeira nach den Ergebnissen der Reise von Prof. Dr. O. Lundblad Juli—August 1935. — Arkiv Zool. 32. A. Nr. 2, 3, 5.

REH, L., 1920, Insekten-Minen in Blättern. — Verh. nat. Ver. Hamburg (3) 27. p. 30—31.

REICHERT, A., 1917/30, Entomologisches aus Miltitz. — Ber. Schimmel & Co. Miltitz b. Leipzig.

———1933, Die Tenthredinoidea von Leipzig und Umgebung. — SB. naturf. Ges. Leipzig vol. 56—59. 38 pp.

REINIG, W., 1934, Was ist Leben? — Berlin, 419 pp.

REITTER, E., 1908/16, Fauna Germanica. Die Käfer des Deutschen Reiches. Stuttgart, 5 vol.

RENNIE, J., 1846, Insect Architecture. — 420 pp.

REUTER, E., 1895, Coleophora laricella, härjande lärkträd. — Medd. Soc. Flor. Fauna Fenn. 21. p. 44.

———1899/1904, Bidrag till kännedom om Microlepidopterfauna i Ålands och Åbo skärgårdar. — Act. Soc. Faun. Flor. Fenn. 15. Nr. 5, 26. Nr. 1.

REUTTI, C., 1898, Übersicht der Lepidopterenfauna des Grossherzogtums Baden. 2. Aufl. herausg. v. Mees & Spuler. Berlin, 361 pp.

RICCHELLO, A., 1928, Contributo alla conoscenza della Mosca del Carciofo (Agromyza andalusiaca Strobl) e dei suoi parassiti. — Boll. Lab. Ent. Portici 22. p. 81—147.

RICHTER, O., 1915, Über das Erhaltenbleiben des Chlorophylls in herbstlich verfärbten und abgefallenen Blättern durch Tiere. — Ztschr. Pflanzenkrankh. 25.

RICHTER, R., 1938, Was führt die Wege der Minierlarven? — Nat. & Volk 68, p. 343—345.

RICHTER VON BINNENTHAL, F., 1903, Die Rosenschädlinge aus dem Thierreiche, deren wirksame Abwehr und Bekämpfung. Stuttgart.

RIEL, 1910/12, Compte Rendu des excursions mycologiques et entomologiques de la Société Linnéenne de Lyon. — Ann. Soc. Linn. Lyon 57, p. 1—14; 59, p. 61—87.

———1927, Présentation de Diptères. II. Fam. des Agromyzidae. Notes d'élévage. Bull. Soc. Linn. Lyon. 8. Nr. 13. p. 95.

RILEY, C. V., 1885, The Cabbage Oscinis (Oscinis brassicae n. sp.). — Rept. Ent. U. S. Comm. Agr. Rept. 1884, p. 322.

——— & HOWARD, L. O., 1891. An Anthomyiid injuring sugar beets. — Ins. Life 3. p. 470.

RIMSKY-KORSAKOW, M., 1929, Beschädigungen der Bäume und Sträucher in den Parkanlagen des Leningrader Forstinstitutes. — Mitt. Leningr. Forstinst. 37. p. 269—300.

——1929, Notiz über einige Wasserschlupfwespen aus dem Flusse Donetz. — Arb. Naturf. Ges. Charkow 52. p. 311—315.

—— & FUNTIKOW, G., Verzeichnis der Beschädigungen der Laubbäume und Sträucher in dem Versuchsrevier von Siwerskaja. — Mitt. Forstl. Vers. Leningrad 5. p. 21—27.

RIPPER, W. E., 1931, Über blattminierende Tenthrediniden-Larven an Birken. — Ztschr. Pflanzenkrankh. 41. p. 182—191.

RITZEMA BOS, J., 1891, Tierische Schädlinge und Nützlinge. Berlin.

——1891, Beschädigungen von Ulmen durch den Rüsselspringkäfer Orchestes alni L. — Ztschr. Pflanzenkrankh. 1. p. 338 ff.

——1902, Een mineerrupsje van den Appelboom (Cemiostoma scitella). — Tijdschr. Plantenz. 1902, p. 62—63.

——1913, Ziekten, door dieren veroorzaakt. — Medd. Land-, Tuin- & Boschbousch. 6. p. 133—158.

ROBINEAU-DESVOIDY, J. B., 1851, Description de plusieurs espèces de Myodaires dont les larves sont mineuses des feuilles de végétaux. — Rev. Mag. Zool. 3. p. 229—236.

——1851, Description d'Agromyzes et Phytomyzes écloses chez M. le Colonel Goureau. — ibid. p. 393—405.

ROEMER, J. J., 1793, Beschreibung und Abbildung einiger kleiner Nachtvögelchen und ihrer Raupen. — Schrift. Naturf. Fr. Berlin 11. p. 156—165.

RONNA, E., 1923, Molestias parasitarias dos Chrysánthemos. — Egatea Porto Alegre 8. p. 7—11.

ROSS, H. & HEDICKE, H., 1927, Die Pflanzengallen (Cecidien) Mittel- und Nordeuropas. 2. Aufl. Jena, 348 pp.

ROSS, H., 1932, Praktikum der Gallenkunde (Cecidologie). 312 pp.

RÖSSLER, A., 1866, Verzeichnis der Schmetterlinge des Herzogthums Nassau. — Nass. Naturw. Jahrb. 19/20, p. 99—442.

——1877, Grapholitha fuchsiana und Coleophora sarothamni. — Stett. ent. Zeit. 38. p. 75—78.

——1881, Die Schuppenflügler des Kgl. Reg. Bez. Wiesbaden und ihre Entwicklungsgeschichte. — Jahrb. Nass. Ver. Naturk. 33/34, p. 1—392.

DE ROUGEMONT, F. & ESCHER-KÜNDIG, J., 1912, Détails biologiques sur la Phytomyza du Thalictrum. — Bull. Soc. Ent. Suisse 12. p. 82—87.

RÜBSAAMEN, E. H., 1899, Über Gallmücken auf Carex und Iris. — Wien. ent. Zeit. 18. p. 57—76.

——— & HEDICKE, H., 1925/39, Die Zoocecidien, durch Tiere erzeugte Pflanzengallen Deutschlands und ihre Bewohner. II. Die Cecidomyiden (Gallmücken) und ihre Cecidien. Stuttgart, 350 pp.

RUIZ, P. F., 1924, Notes biológicas sobre algunos insectos chilenos. — Revista Chil. Hist. Nat. 28. p. 76—80.

RUMJANZEW, P. D., 1934, Zur Biologie der Pappelmotte (Lithocolletis populifoliella Tr.) unter den Lebenverhältnissen der Stadt Moskau. — Zool. Journ. 13. p. 257—279.

RUSCHKA, F. & THIENEMANN, A., 1913, Zur Kenntnis der Wasser-Hymenopteren. — Ztschr. wiss. Ins. Biol. 9. p. 48—52.

RYDÉN, N. S., 1926, För Sverige nya Agromyzider och deras minor. — Ent. Tidskr. p. 119—132.

———1929, Bidrag till kännedomen om svenska bladminerare. — ibid. 50. p. 1—7.

———1934, id. III. — ibid. 55. p. 149—159.

———1936/40, id. 4. — Opuscula Entomol. 1. p. 29—31. — id. 5. — ibid. 2. p. 42—45. — id. 6. — ibid. 5. p. 15—21.

SAALAS, U., 1935/36, Über Incurvaria pectinea. — Ann. Ent. Fenn. 1. p. 113—137; 2. p. 1—16.

SACHAROV, N., 1915, The injurious insects noticed in the Government of Astrakhan 1912—1914. — Ent. Sta. Astr. Ges. Gartenbau, Gemüse- u. Ackerbau 1915.

SANDERS, J. G., 1912, A Remedy for the Chrysanthemum Leaf-miner. — Journ. Econ. Ent. 5. Nr. 6. p. 472.

SANDERS, J. G. & FRACKER, S. B., 1916, The Chrysanthemum Leaf-miner (Napomyza chrysanthemi Kow.). — Bull. Wiscons. Dept. Agr. 10. p. 56.

SASSCER, E. R., 1918, Important Foreign Insect Pests collected on imported nursery stock in 1917. — Journ. Econ. Ent. 11. p. 125—129.

SÄURICH, P., 1907. Im Gewässer. Leipzig, 173 pp.

———1909, Wiese und Hang. Leipzig, 195 pp.

SCHAUFUSS, C., 1908, C. G. Calvers Käferbuch. Naturgeschichte der Käfer Mitteleuropas.

SCHINER, J. R., 1862/64, Fauna Austriaca. Die Fliegen. Wien, 2 vol.

SCHINER, J. R., 1856, Anmerkungen zu Frauenfelds Beitrag zur Insektengeschichte. — Verh. zool. bot. Ges. Wien 6. p. 215—224.

———1858, Die österreichischen Trypeten. — ibid. 8. p. 635—700.

SCHLECHTENDAHL, D. VON, 1901, Biologische Beobachtungen II. Phytomyza vitalbae Klt. — Allg. Ztschr. Ent. 6. p. 193—196.

SCHMID, A., 1863, Beiträge zur Naturgeschichte der Schmetterlinge. — Berlin. Ent. Zeit. 7. p. 57. ff.

———1899, Raupenkalender, herausg. v. O. Hofmann. Regensburg, 275 pp.

SCHMITT, C., 1918, Insekten als Blattminierer. — Naturw. Wochenschr. N. S. 17. p. 721—724.

SCHNABL, J. & DZIEDZICKI, H., 1911, Die Anthomyiden. — Nov. Act. Leopoldina 95. II. p. 55—358.

SCHNEIDER, C., 1937, Für Württemberg neue Microlepidopteren. — Ent. Rdsch. 54, p. 284; 290—292, 298—299.

SCHNEIDER-ORELLI, O., 1909, Die Miniergänge von Lyonetia clerkella und die Stoffwanderung in Apfelblättern. — Centralbl. Bakt. Parasit. 2. v. 24. p. 158—181.

SCHOEVERS, T. A. C., 1921, Een voor Cattleyas schadelijk kevertje. — Tijdschr. Plantenziekten 27. p. 65—71.

SCHOLTZ, H., 1847/50, Über den Aufenthalt der Dipteren während ihrer ersten Stände. — Ent. Ztschr. Breslau 1—3. p. 1—24; 4. p. 25—34.

SCHØYEN, T. H., 1920, Incurvaria pectinea (Hw.) som skadeinsekt. — Norsk. Ent. Tidskr. 1920. Nr. 1. p. 1—4.

SCHRANK, FRANZ V. PAULA, 1780, Naturgeschichte der Minirraupen in den Fliederblättern. — Neue Abh. Baier. Ak. Philos. 2. p. 385—404.

———1798/1804, Fauna Boica. — Nürnberg, 3 vol.

SCHRÖDER, CH., 1897, Nepticula aurella (F.) — Ill. Wochschr. Ent. 2. p. 113—116.

———1897, Aus der Kleinschmetterlingsgattung Lithocolletis. — ibid. p. 385—388, 625—630.

———1898, Musciden-Minen. — ibid. 3. p. 1—3.

———1903, Über experimentell erzielte Instinktvariationen. — Verh. deutsch. zool. Ges. 1903, p. 158—166.

SCHUMACHER, F., 1919, Minierende Buprestiden. — Deutsch. Ent. Ztschr. 191. SB. p. 215.

SCHÜTTE, L., 1921, Die Metamorphose von Hydromyza livens. — Inaug. Diss. Greifswald.

SCHÜTZE, K. T., 1900/30, Die Kleinschmetterlinge der sächsischen Oberlausitz. — Iris 13. p. 269—287; 14, p. 116—138; 15, p. 1—49; 44, p. 1—82.

———1904, Mitteilungen über einige Kleinschmetterlinge. — ibid. 17. p. 204.

———1930, Sammelt Kleinschmetterlinge! — Ent. Ztschr. Frankf. 44. 13 pp.

———1934, Epiblema tedella Cl. und proximana Hw. — ibid. 48. p. 145—148.

SCHWARTZ, M., 1908, Die Runkelfliege (Anthomyia conformis). — Dtsch. landw. Presse 35. p. 656—657.

SCHWARZ, I., 1929, Untersuchungen an Mikrosporidien minierender Schmetterlingsraupen, den „Symbionten" Portiers. — Ztschr. Morph. Ök. Tiere 13. p. 665—705.

SÉGUY, E., 1923, Diptères Anthomyides. — Fauna de France 6. Paris, 393 pp.

———1932, Etudes sur les Anthomyides. 6. Notes biologiques et taxonomiques sur les mouches de l'oeillet. — Enc. méth. Diptera 6. p. 71—82.

———1932/34, Contributions à l'étude des mouches phytophages de l'Europe occidentale. — ibid. 6. p. 145—194; 7, p. 167—268.

———1934, Diptères (Brachycères). — Faune de France 28. 832 pp.

SÉGUY, E. & PAUCHET, L., 1929, Etude sur le Chylizosoma paridis. — Bull. Soc. Linn. Nord de la France 1929. Nr. 414. p. 47—63.

SEIDEL, J., 1924, Zur Kenntnis der Blattminen der Kreise Reichenbach und Frankenstein in Schlesien. — Jahresh. Ver. Schles. Ins. Kd. 14. p. 62—85.

———1926, Zwei sehr ähnliche Tenthrediniden-Minen (Phyllotoma vagans Fall. und Fenusa dohrni Tischb.) an Alnus. — Ztschr. wiss. Ins. Biol. 21. p. 239—248.

———1926, Beobachtungen an Blattminen. — ibid. 21. p. 33—39, 84—91.

———1931, Blattminen der Oberglogauer Gegend. — Abh. Beuth. Ver. Oberschl. Heimatforsch. 4/5, p. 1—50.

SERVADEI, A., 1930, Contributo alla conoscenza della Hyponomeuta padellus L., cognatellus Hb. e vigintipunctatus Rtz. — Boll. Lab. Ent. Bologna 3. p. 254—301.

SICH, A., 1901, Observations on the Early Stages of Phyllocnistis suffusella Z. — Proc. City of Lond. Ent. & Nat. Hist., 14 pp.

———1908/09, Notes on the Life-history of Nepticula acetosae, Stt. — Entom. Rec. 20. Nr. 11, 21. Nr. 5. 8 pp.

SICH, A., 1910, Larval stages of Chrysopora hermannella Fab. — Proc. South Lond. Ent. & Nat. Hist. Soc. 1909/10, 7 pp.

———1913, Some notes on the Early Stages of Cemiostoma laburnella. — ibid. 25. p. 182—188.

———1924, Larval habits of Teichobia verhuellella Stt. — ibid. 36. p. 65—68.

———1925, A note on Phyllocnistis sorhageniella Lüders. — ibid. 37. p. 97—99.

———1926, The Life-cycle of Coleophora ornatipennella Hb. — ibid. 38. p. 8—9.

SILVESTRI, F., 1907, La Tignola dell'olivo. — Boll. Lab. Zool. Agr. Portici 2. p. 83—184.

———1908, Descrizione e prime notizie biologiche dell'Ecofillembio dell'-Olivo (Oecophyllembius neglectus Silv.) — ibid. p. 195—216.

———1923, Contribuzioni alla conoscenza degli Insetti del Nocciulo. — ibid. 16. p. 211—300.

———1924, Rapporte tra insetti di piante spontanee e piante coltivate. Lotta biologica contre piante dannose. — L'Italia Agr. 70. Nr. 2. p. 93—119.

SIMM, K., 1924, Phytomyza lateralis Fall. — Ein Beitrag zur Kenntnis der Morphologie und Biologie. — Bull. Ac. Polon. Sci. 1924. B. p. 735—752.

SIMMONDS, H. W., 1921, Notes on Levuana iridescens Beth. Bakk. — Circ. Agr. Fiji Dept. 2. Nr. 4. p. 84—85.

SIRRINE, F. A., 1896, The spinach-leaf maggot, or miner. (Pegomyia vicina Lintn.) Bull. N. Y. Agr. Exp. Sta. 99. p. 19—31.

SKALA, H., 1923, Beitrag zur Lepidopterenfauna Mährens und österr. Schlesiens. — Ztschr. österr. Ent. Ver. 8. Nr. 9/10.

———1929, Falter aus dem oberen Mühlviertel. — ibid. 14. Nr. 4.

———1929, Minensuche im oberen Mühlviertel. — Ent. Rdsch. 46. Nr. 2. p. 5—8.

———1930, Zur Verbreitung der Gracilariidae im früheren Österreich—Ungarn. — Societ. Ent. 45. p. 6—8, 12.

———1930, Deutsche Miniermotten und ihre Futterpflanzen. — Kranch. ent. Jahrb. 1931, p. 119—149.

———1930, Minen aus Oberösterreich und Salzburg. — Ent. Ztschr. Frankf. 44.

SKALA, H., 1930, Falter aus dem oberen Mühlviertel. — Ztschr. österr. Ent. Ver. 15. 6 pp.

——1931, Zur Verbreitung der Nepticulidae. — Ent. Anz. 11, p. 206—209, 275—278. 287—288.

——1931, Zur Falterfauna Mährens, Schlesiens und der Slowakei. — ibid. 11. p. 469—472.

——1932, Zur Falterfauna des oberen Mühlviertels. — Kranch. ent. Jahrb. 42. p. 101—107.

——1933, Neue Neptikel. — Ztschr. österr. Ent. Ver. 18. Nr. 3/4.

——1934, Wenig bekannte Neptikel. — Kranch. ent. Jahrb. 43. p. 122—130.

——1934, Die Nepticulidae Oberösterreichs. — Ent. Anz. 13. p. 64—66, 99—101, 118—119, 129—132, 149—152, vol. 14. p. 5—8, 29—32, 41—42, 55—57, 77—78, 85—88, 97—99.

——1934, Zur Verbreitung einiger Miniermotten. — Kranch. ent. Jahrb. 43. p. 131.

——1935, Zur Verbreitung einiger Blattminen in Oberösterreich. — Entom. Anz. 15, p. 153—155, 177—179.

——1935, Zur Minenfauna Oberösterreichs. — Ztschr. österr. Ent. Ver. 20. Nr. 9.

——1936, Neue oberösterreichische Minen. — ibid. 21. p. 55—56.

——1936, Minen aus Mittel- und Südeuropa. — ibid. 21. p. 78—79; 22, p. 10—11, 19—20.

——1937, Einiges über Falter-Minen aus dem Mediterrangebiet. — ibid. 22. p. 109—112; 23, p. 8—10, 30—31, 43—45.

——1939, Miner in deutschen Landen. — ibid. 24. p. 27—30, 43—45, 62—64, 91—95, 111—112, 125—128, 138—144, 149—152.

——1940, In der Welser Heide beobachtete Falter. — ibid. 25. p. 143—144, 159—160, 179—180, 186—188.

——1941, Neues über Miner. — ibid. 26, p. 55—57, 77—80, 123—125.

——1941, Minen aus der Normandie. — ibid. 26, p. 148—149.

——1941, Einige Minen aus Südbayern und Nürnberg. — Mitt. Münch. ent. Ges. 31. p. 702—709.

——1942, Falter aus Mähren und Schlesien. — Ztschr. Wien. Ent. Ver. 27. p. 274—277, 289—294.

——1943, Nepticula groschkei spec. nov. — ibid. 28. p. 86.

SKALA, H., 1943, Kerfe aus Oberdonau und anderes. — ibid. 28. p. 101—104.

———1944, Beitrag zur Falterfauna Mährens und Schlesiens. — Entom. listy 7. p. 111.

SMITH, J. B., 1900, Insects of New Jersey. — Rept. New Jersey Agr. St. 27 suppl. — 755 pp.

SMITH, J. B. & DICKERSON, E. L., 1912, The spinach, or beet-leaf maggot (Pegomyia vicina Lintn.) — Rept. Ent. N. Jers. Agr. Exp. Sta. 32. p. 444—453.

SMULYAN, M. T., 1915, The marguerite fly or Chrysanthemum leaf-miner. — Rept. Agr. Exp. Sta. Massach. 27. p. 21—52.

SOFFNER, 1942, Eine neue Nepticula-Art. — Mitt. Dtsch. ent. Ges. 11. p. 56—61.

SØNDERUP, H. P. S., 1921, Et interessant Fund. — Flora og Fauna 1921, p. 28—31.

———1922, Et par interessante Miner. — ibid. 1922. p. 84.

———1923, En sjaelden Bille. — ibid. 1923, p. 131—132.

———1928, Meddelelser om Smaasommerfugle. — ibid. 1928, p. 11—14.

———1934, Nye danske Nepticulider. — ibid. p. 130—132. — To interessante Fund. — 110—112. — Smaasommerfuglen Tischeria dodonaea St. — p. 123.

———1934, Ny dansk Smaasommerfugl. — ibid. p. 123—124. — De springende Knopper endnu en Gang. — p. 124—125. — Didactylota kinkerella Sn. — p. 140—141.

———1937, Phyllotoma Slaegten. — ibid. 1937, p. 10—12.

———1939, Miner og Minedannare. — Naturens Verden 1939, p. 445—457.

———1939, Beretning om mine entom. Fund i 1938. — Flora og Fauna 1939, p. 150—154.

———1941, Insektfund fra 1939—40. — ibid. 1941, p. 63—68.

SORAUER, P. & REH, L., 1925/28, Handbuch der Pflanzenkrankheiten. Die tierischen Schädlinge, 4. Aufl. vol. 4, 5.

SORHAGEN, L., 1881/85, Aus meinem entomologischen Tagebuche. — Berlin. Ent. Ztschr. 1881, p. 17—34; 1882, p. 129—158; 1885, p. 81—108.

———1883, Beitrag zur Auffindung und Bestimmung der Mikrolepidopteren-Raupen. — ibid. 27. p. 1—8.

———1885/86, Die Mikrolepidopteren der Hamburger Torfmoore. — Ent. Nachr. 11. p. 257—264, 273—285, 337—341.

SORHAGEN, L., 1886, Die Kleinschmetterlinge der Mark Brandenburg. — Berlin, 368 pp.

———1900, Restitutio in integrum. — Ill. Ztschr. Ent. 5. Nr. 8.

———1900, Beitrag zur Kenntnis der Gattung Lithocolletis. — ibid. 5. p. 211—213, 232—233, 248—251.

———1901, Grabowiana, Nachtrag zu den Kleinschmetterlingen der Mark Brandenburg. — ibid. 6. Nr. 16, 17. p. 241 ff.

———1905, Ornix sauberiella n. sp. — Ztschr. wiss. Ins. Biol. 1. p. 256.

———1922, Beiträge zur Biologie europäischer Nepticula-Arten. Herausg. von E. Strand. — Arch. Naturg. 1922. Abt. A. 3. p. 1—60.

SPEISER, P., 1905, Ergänzung zu Czwalinas "Neuen Verzeichnis der Fliegen Ost- und Westpreussens", IV. — Ztschr. wiss. Ins. Biol. 1. p. 465—467.

———1905, Die Minierfliege des Leberblümchens. — Schr. Phys. — ökon. Ges. Königsberg 1905, p. 194—196.

SPRINGENSGUTH, W., 1935, Physiologische und ökologische Untersuchungen über extraflorale Nektarien und die sie besuchenden Insekten. — SB. & Abh. naturf. Ges. Rostock (3) 5. p. 31—110.

SPULER, A., 1910, Die Schmetterlings Europas. 3 vol., Stuttgart.

STACKELBERG, A. A., 1932, Verzeichnis der schädlichen Insekten der paläarktischen Region. I. Schädlinge der Landwirtschaft. — Bull. Plant. Prot. I. Entom. 500 pp.

STAINTON, H. T., 1853, An introduction to the study of the Nepticulae. — Zoolog. 11, p. 3952—3960.

———1854, Insecta Britannica, Tineina. — London, 313 pp.

———1854/57, Entomological Botany. — Zoologist 12. p. 4468—4472, 4506—4510; 13, p. 4553—4557, 4680—4684, 4771—4777, 4842—4846, 4890—4894; 14, p. 5009—5012, 5135—5139, 5196—5199, 5238—5241; 15, p. 5398—5402.

———1855, Note sur les moeurs d'une chenille, qui vit dans l'intérieur des feuilles de la vigne. — Ann. Soc. ent. France (3) 3. p. 211—213.

———1855, Leaves of Vicia sepium containing larvae and pupae of Lithocolletis Bremiella. — Proc. ent. Soc. Lond. 1855, p. 108.

———1855, On some of the Difficulties of Entomological Students, as exemplified by recent Experience in the Genus Elachista. — Trans. Ent. Soc. London. (2) 3. p. 84—89.

———1855/74, The Entomologist's Annual. London, 20 vol.

Stainton, H. T., 1855/73, The Natural History of the Tineina. London, 13 vol.

———1856, Descriptions of three species of Indian Micro-Lepidoptera. — Trans. Ent. Soc. Lond. (2) 4. p. 301—340.

———1856/59, A Manual of British Butterflies and Moths. — London, 2 vol.

———1856, Entomologist's Weekly Intelligencer vol. 1.

———1856, A new Nepticula larva. — Substitute 1856/57, p. 56.

———1857, Entomologist's Weekly Intelligencer. vol. 2.

———1858, On the aberrant species hitherto placed in the genus Elachista. — Trans. Ent. Soc. Lond. (2) 4. p. 261—271.

———1858, Synopsis of the Genus Elachista. — ibid. p. 292—338.

———1857/58, The Entomologist's Weekly Intelligencer. vol. 3, 4.

———1858, Notes on the habits of Lithocolletis helianthemi. — Proc. Ent. Soc. Lond. 1858, p. 79.

———1859, The Entomologist's Weekly Intelligencer 5, 6.

———1860, id. 7, 8.

———1861, id. 9.

———1862, Notes on mining larvae of the genus Micropteryx. — Proc. Ent. Soc. Lond. 1861, p. 10; 1862, p. 60—61; Zoologist 20, p. 7910—7911.

———1862, Note respecting the Micropteryx bred from hazel-leaves by Kaltenbach. — Proc. Ent. Soc. Lond. 1862, p. 97—98; Zoologist 20. p. 8260.

———1862, Ten new species of Nepticula. — Proc. Ent. Soc. Lond. p. 101.

———1862/63, On the generic characters afforded by the habits of various leaf-mining Micro-Lepidopterous larvae. — Trans. ent. Soc. Lond. (3) 1. p. 604—610. — Proc. Ent. Soc. 1863, p. 165.

———1863, Note on larva of Incurvaria pectinata. — Proc. Ent. Soc. Lond. 1863, p. 149.

———1863, Sur les chenilles mineuses appartenant au genre Micropteryx. — Ann. Soc. ent. France (4) 3. p. 293—296.

———1863, Observations on American Tineina. — Proc. Ent. Soc. Philad. 2. p. 130—132.

———1863, Descriptions of Nine Exotic Species of the Genus Gracilaria. — Trans. Ent. Soc. Lond. (3) 1. p. 291—300.

STAINTON, H. T., 1864, On the European Species of the Genus Cosmopteryx. — ibid. 1. p. 637—659.

———1867, The Tineina of Syria and Asia Minor. — London, 84 pp.

———1869, The Tineina of Southern Europe. — London, 370 pp.

STANGE, G., 1891, Biologische Notizen über einige Mikrolepidopteren. — Stett. Ent. Zeit. 52. p. 132.

———1899, Die Tineinen der Umgegend von Friedland in Mecklenburg. — Wiss. Beil. Schulprogr. Gymnas. Friedland 706.

———1912, Nachträge zur Schmetterlingsfauna Friedlands. — ibid. Nr. 962.

———1916, Beitrag zur Hallenser Lepidopteren-Fauna. — Mitt. Ent. Ges. Halle 10. p. 50—59.

STÄGER, R., 1917, Beobachtungen an der Raupe von Coleophora gryphipennella Bché. Ztschr. wiss. Ins. Biol. 13. p. 204—207, 235—237.

STARKE, H., 1942, Die Minierfliegen (Agromyziden) und deren Substrate. — Abh. Naturf. Ges. Görlitz 33. p. 74—80.

———1941, Zweiter Nachtrag zu den Schmetterlingen der sächsischen Oberlausitz Iris 55. p. 127—133.

STARÝ, B., 1930, Über minierende Schmetterlinge Mährens und Schlesiens. I. — Sborn. Klubu Brne 13.

———1932, Über minierende Insekten Mährens und Schlesiens. — Act. Soc. Sci. Nat. Morav. 6. p. 125—142.

———1938, Bemerkungen zur Parasitierung der in der Landwirtschaft schädlichen Insekten in der Tschechoslowakei in den letzten Jahren. VII. Int. Kongr. Berlin, Zusammenf. Vortr. p. 143—146.

STEIN, F., 1867, Eine der Gerste schädliche Fliege (Hydrellia griseola). — Berlin. Ent. Zeit. 11. p. 395—397.

STEIN, P., 1906, Die mir bekannten europäischen Pegomyia-Arten. — Wien. ent. Zeit. 25. p. 47—107.

———1908, Analytische Übersicht aller mir bekannten breitstirnigen Anthomyiden-Männchen mit Ausschluss der Gattungen Lispa und Fucellia. — ibid. 27. Heft. 1.

———1915, Die Anthomyiden Europas. — Arch. Naturg. 1915 A. 10.

STEPHENS, J. F., 1829, A systematic catalogue of British Insects. 416 pp.

STERNECK, J. & ZIMMERMANN, F., 1933, Prodromus der Schmetterlingsfauna Böhmens. II. Microlepidoptera. Karlsbad 168 pp.

STOCK, M. N., 1936, A contribution to the knowledge of the Puparia of Anthomyidae. — Tijdschr. Ent. 79. p. 94—168.

STONE, J. L., 1899, Ennemies of the beet crop. — Bull. Corn. Univ. Agr. Exp. Sta. 166. p. 425.

STRICKLAND, E. H., 1920, The Cotton-wood Leaf-miner. — Canad. Ent. 52.

STRITT, W., 1935, Zwei für Deutschland neue Blattwespen sowie einige bemerkenswerte Blattwespenfunde. — Mitt. Bad. Landesver. Naturk. N. F. 3. p. 161—163.

———1935, Die Blatt-, Halm- und Holzwespen Badens. — ibid. p. 184—190.

———1936, Kleine Mitteilungen über Blattwespen. — Arb. morph. tax. Ent. Dahlem 3. p. 54—60.

———1944, Der Blattminierer Fenusella recta Thoms. — Arb. physiol. angew. Ent. 11. p. 25—31.

STROBL, G., 1900, Dipterenfauna von Bosnien, Herzegowina und Dalmatien. — Wiss. Mitt. Bosn. 7. p. 552—670.

STRONG, S. & VAN RENSSELAER, L., 1897, Two insect pests. — Gard. & Forests 10. p. 278.

STURTEVANT, A. H., 1916, Notes on North American Drosophilidae with descriptions of 23 new species. — Ann. Ent. Soc. America 9. p. 323—343.

SUBRAMANIAM, P. V., 1920, Life-history of the Buprestid Leaf-miner (Trachys bicolor), a Pest on Butea frondosa in Mysore. — Journ. Bomb. Nat. Hist. Soc. 27, p. 178—179.

SUIRE, J., 1926, Sur quelques insectes de l'Herault. — Feuill. Natural. 47. Nr. 31.

———1927, Notes biologiques sur quelques Coleophores de la région de Montpellier. — Bull. Soc. ent. Fr. 1927, p. 45—48; 1928, p. 135—140.

———1930, Une espèce nouvelle de Coleophore Coleophora joannisella n. sp. et ses premiers états. — ibid. 1930. p. 25—29.

———1935, Remarques sur l'évolution des chenilles du genre Coleophora. — L'Amateur Pap. 7. p. 191—208.

SURANYI, P., 1942, Aknazo rovarlarvak. — Kül. Potf. Termesz. Közlön. 1942, X/XII.

———1942, Über die minierenden Insektenlarven Ungarns. — Fol. Ent. Hung. 7. p. 1—64.

SWAMMERDAM, J. J., 1737/38, Bybel der natuur. — Amsterdam, 2 vol., 910 pp.

SWEZEY, O. H., 1917, A leaf-mining cranefly in Hawaii. — Proc. Haw. Ent. Soc. 3. p. 87.

——1923, The Leaf-Miners of Pipturus (Lepidoptera). — ibid. 5. p. 293—296.

TASCHENBERG, E. L., 1879/80, Praktische Insektenkunde. Bremen, 4 vol.

——1871, Entomologie für Gärtner und Gartenfreunde. — 585 pp.

TAYLOR, T. H., 1928, The Watercress Stem-miner. — Ent. Month. Mag. 64, p. 126—128.

——1918, Oviposition in the Celery Fly. — Ann. Appl. Ent. 5. p. 60—61.

TEMPEL, W., 1926, Auftreten eines Stachelkäfers (Mordellistena beyrodti v. Leng.) an Orchideen. — Die kranke Pflanze 3. p. 161—164.

——1921, Zur Futteralmottenbekämpfung. — ibid. 6. p. 21—23.

TENGSTRÖM, J. M., 1847, Bidrag till Finlands fjärilfauna. — Act. Soc. Faun. & Flor. Fenn. p. 69—164.

——1859, Anmärkningar och tilläg till Finlands småfjärilar. — ibid. (2) 4. p. 155—226.

THEOBALD, F. V., 1904, Grubs of Phytomyzidae in Marguerite and Cineraria leaves. — Rept. Eco. Zool. Lond. 2. p. 159.

——1904, The Apple Leaf-miner (Lyonetia clerkella L.). — ibid. p. 37—41.

——1905, The Larch Coleophora or Leaf-miner. — Two Leaf-miners: I. The Lilac Leaf-miner. II. The Laburnum Leaf-miner. — ibid. 1905, p. 111—125.

——1907, The Apple Leaf-blister Moth (Ornix petiolella Sta.). The wild Crab Leaf-miner (Nepticula malella St.) p. 27—30. The Pear Leaf-blister Moth (Cemiostoma scitella Z.) p. 67. — The Iris Leaf-miner (Agromyza iraeos Duf.). p. 129. — ibid. 1907.

——1922, The Apple and Plum Case-bearer (Coleophora nigricella) and its Treatment. — Journ. Pomology & Hort. Sci. 1922.

THIENEMANN, A., 1908, Die Metamorphose der Chironomiden (Zuckmücken). — Verh. nat. Ver. Rheinl. Westf. 65. 12 pp.

——1916, Über Wasser-Hymenopteren. — Ztschr. wiss. Ins. Biol. 12. p. 49—54.

——1932, Die Tierwelt der Nepenthes-Kannen. — Arch. Hydrobiol. Suppl. 11. vol. 3. p. 1—54.

THOMSEN, M., 1921, Bøgeloppen i Skoven og Frugthaven. — Ent. Medd. Copenhag. 13. p. 336—337.

TOLL, S. GRAF, 1932, Einige Worte über die heimischen Arten der Gattung Nepticula Z. — Bull. ent. Pol. 11. p. 22—51.

———1934, Nepticula heringi sp. nova. — Ann. Mus. Zool. Pol. 11. p. 1—3.

———1934, Noch über die heimischen Arten der Gattung Nepticula Z. — Bull. ent. Pol. 13. p. 61—84.

———1936, Beitrag zur Kleinschmetterlingsfauna der Wojewodeschaften Poznan und Pomorze. — Bull. Ent. Pologne 14/15, p. 227—261.

———1936, Untersuchung der Genitalien bei Pyrausta purpuralis und P. ostrinalis Hb. nebst Beschreibung 11 neuer Microlepidopteren-Arten. — Ann. Mus. Zool. Pol. 11. p. 403—413.

———1938, Zwei neue Microlepidopteren-Arten aus Podolien. — ibid. 13. p. 77—80.

———1941, Die Genitalien der europäischen "Hyponomeuta". — Ztschr. Wien. Ent. Ver. 26. p. 170—176.

———1942/44, Studien über die Genitalien einiger Coleophoriden. — Ztschr. Wien. Ent. Ver. 27. p. 78—84. — id. II. ibid. 27. p. 213—216. — id. III. — Veröff. Deutsch. Kolon. & Überseemus. 3. p. 288—299. — IV. — Ztschr. Wien. ent. Ges. 28. p. 89—94. — id. V. ibid. 28. p. 225—231. — id. VI. Larvivarie bei Coleophora leucapennella Hb. — Mitt. Deutsch. ent. Ges. 10. p. 55—56. — id. VII. Coleophora marianii spec. nov. — ibid. 11. p. 89—90. — id. VIII. ibid. 13, p. 27—36.

———1947, Some new Nepticulidae of Poland. — Document. Physiogr. Polon. 6. p. 1—37.

TOWER, D. G., 1915, A curious feeding habit of Chrysopa rufilabris Burm. — Journ. Econ. Ent. 8. p. 431.

TOWNSEND, C. H. T., 1892, The North American genera of Calyptrate Muscidae. — Trans. Amer. Ent. Soc. 19, p. 133—144, 273—284, 290—294.

TRÄGÅRDH, I, 1909, Zur Kenntnis von Phytomyza xylostei Kltb. — Ztschr. wiss. Ins. Biol. 5. p. 301—304.

———1910, Lärkträdsmalen (Coleophora laricella Hb.) p. 258—264. — Clerks minerarmal (Lyonetia clerkella L.) p. 266—271. — Ent. Tidskr. 32. H. 4.

———1910, Contribution towards the metamorphosis and biology of Orchestes populi, fagi and quercus. — Ark. Zool. 6. Nr. 7. 25 pp.

———1911, Notes on the earlier stages of Orchestes fagi 1. — Ent. Tidskr. 32. p. 73—78.

———1911, Den randiga jordloppan (Phyllotreta nemorum L.) — Uppsatser prakt. Ent. 21. p. 95—101.

TRÄGÅRDH, I., 1911, Syrénmalen (Gracilaria syringella F.). — ibid. p. 31—37.

———1911, Om biologien och utvecklingshistorien hos Cedestis gysselinella Dp., en barrminerare. — Medd. Centr. Anst. Ent. Avd. 9. p. (53).

———1913, Contributions towards the Comporative Morphology of the trophi of the Lepidopterous Leaf-miners. — Ark. Zool. 8. Nr. 9.

———1913, Om Nepticula sericopeza Z. ett skadedjur på lönnens frukter. — Skogv. Tidskr. 1913. H. 4.

———1913, Hvarpå beror bladminerarnes förmåga att om hosten konservera klorofyllet i bladen? — Ent. Tidskr. 34. p. 179—190.

———1914, Bladminerare. — Medd. Centr. Anst. försöksv. Ent. Avd. 16. 29 pp.

———1915, Bidrag till kännedomen om tallens och grenens fienden bland småfjärilarna. — Medd. Stat. Skogsförsökanst. 12.

———1923, Skogsentomologiska Bidrag II. — ibid. 20. p. 401—424.

TRIMBLE, F. M., 1924, The Azalea Leaf-miner. — Ent. News 35. p. 275—279.

TROUVELOT, B., 1925, Sur les causes de disparition d'une invasion de Pegomyia hyoscyami Panz. survenue en 1924 dans le departement de l'oise. — Feuill. Natural. 46. p. 54—56.

TRYON, H., 1925, The Bean Fly (Agromyza phaseoli). — Queensl. Agr. Journ. 24. p. 243—244.

TULLGREN, A., 1905, Studier och iakttagelser rörande Skadeinsekter. Diptera. 15. — Medd. Landbruksstyr. 111. p. 1—55.

———1905, Om fluglarver på spenat. — Ent. Tidskr. 26. p. 172—176.

———1907, Våra frukttträds fiender bland insekterna. — ibid. 28. p. 201—222.

———1908, Svensk insektfauna. Hymenoptera 4. — ibid. 29. p. 141—220.

———1913, Skadedjur i Sverige 1911. — Uppsats. prakt. Ent. 22. p. 43—134.

———1915, Drivhusvaxternas fiender. 1. Skadeinsekter på Chrysanthemum. — Tragården, Stockholm, p. 9—12.

———1918, Lökmalen (Acrolepia assectella Z.). — Medd. Centr. Anst. försöksv. 167 Ent. avd. Nr. 30.

———1927, Azaleamalen (Gracilaria azaleella Brants). — Centralanst. Jordbruksf. Flygbl. 118. 7 pp.

TUOMINEN, O., 1933, Ein neues Minensubstrat. — Notulae Ent. 13. p. 76—77.

TURATI, E., 1923, Cinque anni di ricerche nell'Appenino modenese. — Atti Soc. Ital. Sci. Nat. 62. p. 1—74.

TURATI, E., 1929, Eteroceri di Tripolitania. — Boll. Lab. Zool. Gen. Agr. 23. p. 98—128.

———1930, Lepidotteri di Rodi. — Mem. Soc. ent. Ital. 9. p. 196—214.

TURNER, H. J., 1906, Notes on Coleophora. — Ent. Rec. 18.

TUTT, J. W., 1898, On the British species of Lithocolletis of the spinicolella-group. — Ent. Rec. 10. p. 164—168.

———1899, A Natural History of the British Lepidoptera. I. 560 pp.

UFFELN, K., 1930, Die sogenannten Kleinschmetterlinge Westfalens. — Abh. Westf. Prov. Mus. Naturk. 1. p. 1—98.

UHMANN, E., 1934, Hispinen-Minen aus Costa-Rica. — Arb. phys. ang. Ent. Dahlem 1. p. 272—277.

———1935, Neue Hispinen aus Costa-Rica. — Ent. Blätt. 31. p. 103—106.

———1937/44, Hispinen-Minen aus Costa Rica. — Arb. phys. angew. Ent. Dahlem 4. p. 61—66 & 11. p. 59—61.

ULMER, G., 1911, Unsere Wasserinsekten. Leipzig.

———1932, Gliederfüssler, Arthropoda. — In Brohmer, Tierbestimmungsbuch, 2. Aufl. p. 37—144.

URBAN, C., 1922, Zur Biologie der Zeugophora flavicollis Mrsh. — Deutsch. Ent. Zeitschr. 1922, p. 405—408.

———1922, Zur Lebensweise des Tanysphyrus lemnae Payk. — Ent. Blätt. 18. p. 73—75.

———1926, Über das Leben und die Larve von Hydronomus alismatis Mrsh. — Ent. Blätt. 22. p. 109—113.

———1929/30, Beiträge zur Naturgeschichte einiger Rüsselkäfer. — Ent. Blätt. 25. p. 65—79 & 26, p. 97—179.

VAN DEN BRUEL, 1933, Contribution à l'étude des Mouches de la Chicorée Witloof. Bull. Inst. Agr. Gembloux 1933. p. 17.

VELITCHKEVITCH, A. J., 1921, On the biology of flies mining the leaves of Graminaceae in the Novgorod government. — Trav 2d All-Russ. Ent. Phytopath. Meet. Petrogr. 1920, p. 129—135 (Russ.).

VENTURI, F., 1934/35, Contributo alla conoscenza dell'Entomofauna del frumento. — Boll. Lab. Entom. Bologna 6. p. 231—238 & 8, p. 1—26.

———1935/40, Contributo alla conoscenza dell'Entomofauna delle Graminacee coltivate e spontanee. III. — Boll. Inst. Bologna 8. p. 141—149. — IV. ibid. 9. p. 1—22. — VI. Redia Firenze 26. p. 27—70.

VIMMER, A., 1909, Anatomické poznámky o larvách Blepharoptera serrata L. a Pegomyia conformis Fall. — Časopis České Spol. Ent. 6. p. 109—114.

——1913, Ergänzungen zu dem Aufsatze: Zur Kenntnis der Phytomyza xylostei Klt. Ztschr. wiss. Ins. Biol. 9. p. 19.

——1918, O Topografii tracheální soustavy Larev hmyzu dvojkřídlého. — Act. Soc. Ent. Bohem. 15. p. 15—42.

——1922, Příspěvky k metamorfose některých českých Anthomyií. — Časopis Česk. Spol. Ent. 18. H. 1/2.

——1923, id. II. ibid. 19. H. 3—6.

——1925, Larvy a Kukly. — Prag, 348 pp.

——1926, O nových Dipterohyponomech v Čechach. — Sborn. Čsl. Spol. Ent. 3—4. p. 66—69.

——1927, O tracheální soustavě larev Schizophor. — Rozpravy II. Třídy Česk. Ak. 35. Čislo 31.

——1927, Mouchy z náplavů říčnich. — Časopis Čsl. Spol. Ent. 1927. Č. 1—2. p. 6—10.

——1927, Dipterohyponomy z okolí orlíka. — ibid. Č. 3—4, p. 63—69.

——1928, O novém druhu rodu Napomyza Hal. — Sborn. Entom. Nár. Mus. Praze 4. p. 117.

——1928, O larvách a pupách Dipter, nových pro československou republiku. — ibid. 4. p. 119—124.

——1930, Ústní ústrojí Larev několika Trypaneid. — Časopis Spol. Ent. 26, 5—6. p. 102—107.

——1930, Několik poznámek k anatomii a fysiologii larev hmyzu dvojkřídlého. ibid. p. 79—92.

——1930, Muší Rody v Československé Republice. — Prag, 379 pp.

——1931, O Larvách Muších které škodí v Československu rostlinám vyhlodáváním podkopěnek čili hyponomů. — Arch. Přír. Čech. 18. p. 1—159. Prag.

VLACH, V., 1942, Coleophora klimeschi spec. nov. — Časopis Spol. Ent. 39. p. 18—21.

VOGEL, J. F., 1921, De Beukenspringkever (Orchestes fagi L.) — Tijdschr. Plantenz. 27. p. 129—131.

VOIGT, G., 1921, Beiträge zur Kenntnis der Minen und ihrer Erreger, sowie Beobachtungen über das Vorkommen von Minen in Rheingrau und benachbarten rheinischen Gebieten. — Jahrb. Nass. Ver. Naturk. 80. p. 24—73.

VOIGT, G., 1930, Chortophila brunnescens Zett. als Schädling kultivierter Caryophyllaceen. — Ztschr. Pflanzenkrankh. 40. p. 265—269.

———1930, Über Ceutorrhynchus contractus Mrsh. als Schädling kultivierter Cruciferen, besonders des Goldlacks nebst Bemerkungen zur Phänologie und Gradation kaltbrütiger Insekten. — ibid. p. 492—505.

———1931, Bemerkungen über die Rebenminiermotte, Antispila rivillei. Stt. — Anz. Schädlingskd. 7. p. 90—93.

———1931, Beobachtungen über den Befall der Kultur- und Zierpflanzen durch blattminierende Insekten. — Ztschr. Pflanzenkrankh. 41. p. 192—202.

———1932, Beiträge zum Xenophagie-Problem und zur Standpflanzenkunde. Über den Befall sekundärer Substrate durch Blattminierer. — ibid. 42. p. 513—541.

———1932, Galle oder Blattmine? Beitrag zur Histologie der vergallten Mine und anderer Sedum-Minen von Apion sedi Germ. — Anz. Schädlingskd. 8. p. 135—143.

———1933, Über Eichenminen, besonders über die Beschädigung der Eichen durch die Mine der Eichentrieb-Miniermotte Coriscium brogniardellum Z. — Mitt. Deutsch. dendrol. Ges. 45. p. 238—243.

———1934, Frühstadien und Frühbekämpfung der Gespinstmotten im Lichte neuerer Arbeiten. — Anz. Schädlingskd. 10. p. 37—42.

———1946, The preservation of the living chlorophyll in dying leaves contaminated by Bird's Dropping. — Trans. Norf. & Now. Nat. Soc. 1945, p. 169—171.

VORBRODT, K. & MÜLLER-RUTZ, J. P., 1911/14, Die Schmetterlinge der Schweiz. — Bern, 2 vol.

VOSS, E., 1934, Über einen blattminierenden tropischen Rüssler. — Arb. phys. angew. Ent. Dahlem 1. p. 267—271.

———1943, Zur Kenntnis zweier blattminierenden Rüssler. — Ent. Blätt. 39. p. 16—18.

VOÛTE, A. D., De Djeroek-Mineerrups (Phyllocnistis citrella St.) — Landbouw 10, p. 138 ff.

———1932, Een nieuw geval van bestrijding van een insectenplaag met behulp van een inheemschen parasiet. — Tijdschr. Ent. 75. Suppl. p. 128—135.

———1935, Der Einfluss von Ageniaspis sp. auf ihren Wirt Phyllocnistis citrella Stt. unter verschiedenen (mikro-) klimatischen Verhältnissen. — Arch. Néerl. Zool. 1. p. 354—372.

WAHLGREN, E., 1915, Svensk Insektfauna 10. II. Småfjärilar. Microlepidoptera. — Ent. Tidskr. 36. p. 99—181.

WAHLGREN, E., 1927, Svensk Insektfauna. 11. Diptera. Cyclorapha Schizophora. Fam. 21—26. Stockholm, p. 323—416.

WALKER, H. C., 1915, The box Leaf-miner. — Journ. Econ. Ent. 8. p. 306.

WALLENGREN, H. D. B., Skandinaviens med ögenlock försedda Tineider. — Ent. Tidskr. 1881 2. p. 124—136.

———1883, Skandinaviens arter af Tineidgruppen Lithocolletidae. — ibid. 4. p. 195—212.

———1883, Skandinaviens Micropterygider. — ibid. p. 213—216.

WALSHE, B. W., 1947, Feeding mechanism of chironomous larvae. — Nature 160 p. 474.

WALSINGHAM, LORD, 1891, Micro-Lepidoptera collected near Cannes, 1890. — Ent. Month. Mag. (2) 2. p. 137—152.

———1907, Descriptions of new North American Tineid moths, with a generic table of the family Blastobasidae. — Proc. U. S. N. Mus. 33. p. 197—228.

WASHBURN, F. L., 1905, The Diptera of Minnesota. — Bull. Univ. Minnesota Agr. Exp. Sta. 93. p. 19—168.

WATERS, E. G. R., 1924, Tineina in the Oxford District, 1912—1923. — Ent. Month. Mag. 60. p. 93—103.

———1925, Cotoneaster frigida Wallich as a food-plant of Micro-Lepidoptera. p. 18—19. — Micro-Lepidoptera in North Devon and Somerset. p. 45—77. Three Additions to the British List of Tineinae. p. 82—89. — Ent. Month. Mag. 61.

———1925, Tineina in the Oxford District. — ibid. 61. p. 188—196.

———1926, Microlepidoptera. — The Nat. Hist. Oxford Distr. p. 230—247.

———1926, Micro-Lepidoptera in South Devon, August 1925. — Entomol. 59. p. 158—161.

———1927, Tineina in the Oxford District. — Ent. Month. Mag. 63. p. 69—75, 99—102.

———1927, Notes on the Coleophorinae. — ibid. p. 182—184.

———1928, Nepticula suberivora Stt. in the Isle of Wight. — ibid. 64. p. 1—4.

———1928, Further Notes on the Coleophorinae. p. 76—78. — A new Nepticula from North Wales. p. 105—106. — Tineina in the Oxford District. p. 172—178. — Notes on the Nepticulidae. p. 219—224. — Nepticula albifasciella Hein., its early stages and its occurence in Britain. p. 248—251. — Ent. Month. Mag. 64.

WATERS, E. G. R., 1929, Notes on some species of Lithocolletis, with an Addition to the British List. — ibid. 65. p. 163—170.

———1929, A List of the Micro-Lepidoptera of the Oxford District. — Proc. & Rpt. Ashmol. Nat. Hist. Soc. 1928. p. 1—72.

WATSON, J. R., Beet leaf-miner. (Pegomyia vicina). — Bull. Florida Agr. Exp. Sta. 134. p. 57—58.

WATT, M. N., 1916, Contributions to the Entomology of New-Zealand. Nr. 8. Parectopa citharoda Meyr. — Trans. New-Zeal. Inst. 48. p. 407—413.

———1921, A new Tortricid Moth. — New Zeal. Journ. Sci. Techn. 4. p. 257—258.

———1920/24, The Leaf-mining Insects of New-Zealand. — I. Transact. Proc. N. Zeal. Inst. 52. p. 439—466. — II. ibid. 53, p. 197—219. — III. ibid. 54, p. 465—489. — IV. ibid. 55. p. 327—340. — V. ibid. 55. p. 674—687.

WEBER, H., 1939, Vergleichend-funktionsanatomische Untersuchungen an atypischen Beissmandibeln von Insekten mit besonderer Berücksichtigung der Phyllotoma-Larve. — Biol. Zentralbl. 59. p. 541—566.

WEBER, P., 1936, Über Microlepidopteren. — Mitt. Schweiz. ent. Ges. 16. p. 666—672.

———1938, Zwei neue Arten der Mikrolepidopterengattung Nepticula und ein Beitrag zur Kenntnis von Nepticula stelviana Wocke. — ibid. 17. p. 211—215.

WEBSTER, F. M., 1898, Notes and observations on several species of Diptera. — Canad. Ent. 30. p. 18—19.

———1900, Some species of Diptera inhabiting or frequenting the wheat fields of the Middle West. — Canad. Ent. 32. p. 212—213.

WEBSTER, F. M. & PARKS, T. H., 1913, The Serpentine Leaf-miner. — Journ. Agr. Res. 1. p. 59—87.

WEIGEL, C. A., Insects injurious to ornamental Greenhouse plants. — U. S. Dept. Agr. Farmers' Bull. 1362.

WEIGEL, C. A. & MIDDLETON, W., 1926, Insect Ennemies of the Flower Garden. — ibid. 1495.

WEISE, J., 1893, Naturgeschichte der Insekten Deutschlands. Coleoptera 6. Berlin.

WEISS, H. B., 1915, Insect importations into N. Jersey during the spring of 1915. — Canad. Ent. 47. p. 326—328.

———1915, Notes on the occurence of some economic insects not heretofore recorded from New Jersey. — Psyche 22. p. 105—106.

WEISS, H. B., 1916, Monarthropalpus buxi Lab. in New Jersey. — ibid. 23. p. 154—156.

———1916, Diptera. (Foreign pests recently established in N. Jersey. — Journ. Econ. Ent. 9. p. 213—214.

———1917, Notes on several insects not heretofore recorded from New Jersey. — ibid. 10. p. 224.

———1918, The control of imported pests recently found in N. Jersey. — ibid. 11. p. 122—125.

———1918, The boxwood leaf-miner (Monarthropalpus buxi Lab.). — Circ. N. Jers. Agr. Exp. Sta. 100. p. 9—11.

WEISS, H. B. & BECKWITH, C. S., 1921, Notes on Coptodisca kalmiella Dietz., a Leaf-miner of Kalmia angustifolia. — Canad. Ent. 33. p. 44—45.

WEISS, H. B., & LOTT, R. B., 1921, Notes on Orchestes rufipes Lec. in New Jersey. — Psyche 28. p. 152—155.

WEISS, H. B. & NICOLAI, A. S., 1919, Notes on Zeugophora scutellaris Suffr., a European Poplar Leaf-miner, in New Jersey. — Ent. News 30. p. 124—127.

WELCH, P. S., 1917, Further Studies on Hydromyza confluens Loew. — Ann. Ent. Soc. Amer. 10. Nr. 1.

WELLHOUSE, W. H., 1922, The insect fauna of the genus Crataegus. — Mem. Cornell Un. 56. p. 1041—1136.

WERTH, 1925, Die Miniermotte. — Nachrichtenbl. Deutsch. Pflanzenschutzdienst 5. p. 6 ff.

WESTWOOD, J. O., 1883, The marguerite leaf-mining Musca. — Garden chron. (2) 19. p. 593.

WILKE, S., 1922, Die Runkel- oder Rübenfliege (Pegomyia hyoscyami Pnz.) — Nachr. Bl. Deutsch. Pflanzenschutzd. 1922.

———1924, Die graue Gerstenminierfliege (Hydrellia griseola Fall.) Deutsch. Ent. Ztschr. 1924. p. 172—179.

———1927, Neue und wenig bekannte Orchideen-Schädlinge. — Gartenwelt 31, p. 119—122.

WHITEHEAD, C., 1887, Reports on insects injurious to root and certain crops in Great Britain. — Priv. Counc. Off. p. 81—86.

———1890, Third Annual report on insects and fungi injurious to the crops of the farm, the orchard and the garden. 74 pp.

———1891, Ann. Report of the Intelligence Department on injurious insects and fungi. — 31 pp.

WHITEHEAD, C, 1893, Report on insects and fungi injurious to crops. — 60 pp.

WILD, W., 1915, Descriptions of a New Coleophora Moth and Note on a Synonym. — Ent. News 26. p. 320.

WILLISTON, S. W., 1908, Manual of North American Diptera. 3d ed. 405 pp.

VON WINNING, E., 1938, Versuch einer Monographie von Tortrix pronubana Hb. — Ztschr. angew. Ent. 25. p. 215—276.

———1941, Zur Biologie von Phthorimaea operculella Z. als Kartoffelschädling. — Arb. phys. angew. Ent. Dahlem 8. p. 112—128.

WOCKE, F., 1871, Lepidopterologisches. — Stettin. Ent. Zeit. 32. p. 426—430.

WOIT, M., 1925, Umgestaltungen an Blattgeweben infolge des Wundreizes. Langensalza. 26 pp.

———1925, Über Wundreaktionen an Blättern und den anatomischen Bau der Blattminen. — Jahrb. Deutsch. Dendrol. Ges. 35. p. 163 ff.

WOOD, J. H., 1890, Nepticula torminalis, a species new to science. — Ent. Month. Mag. (2) 1. p. 209.

———1891, Micropteryx sangii, a new species from birch. — ibid. (2) 2. p. 100—101.

———1892, Notes on the earlier stages of the Nepticula, with a view to their better recognition at this period of their life. — ibid. 29. p. 197—201, 268—274; vol. 30 (1894), p. 1—4, 43—50, 93—98, 150—154.

———1894, Nepticula confusella, a new birch-mining species. — ibid. p. 272.

———1900, On the larvae, habits and structure of Lithocolletis concomitella Bankes, and its nearest allies. — ibid. 36. p. 30—34, 69—75, 102—106.

WÖRZ, A., 1937, Nepticula hahniella spec. nov. m. — Ent. Rdsch. 54. p. 290—292.

WULP, J. M. VAN DER, 1867, Eenige Noord-Amerikaansche Diptera. — Tijdschr. Ent. 10.

YUASA, H., 1922, A classification on the larvae of the Tenthredinidae. — Ill. Biol. Monogr. 7. pt. 4. 172 pp.

ZACHER, F., 1921, Tierische Schädlinge an Heil- und Giftpflanzen. — Ber. Deutsch. Pharmazeut. Ges. 31. p. 53—65.

———1922, Die Feinde der Syringen. — Gartenwelt 26. Nr. 21.

ZELLER, P. C., 1846, Die Arten der Blattminierergattung Lithocolletis. — Linnaea Ent. 1. p. 166—261.

ZELLER, P. C., 1847, Die Argyresthien. — ibid. 2. p. 234—302, 585—588.

———1847, Die Gracilarien. — ibid. p. 303—383.

———1848, Die Gattungen der mit Augendeckeln versehenen blattminierenden Schaben. — ibid. 3. p. 248—344.

———1849, Beitrag zur Kenntnis der Coleophoren. — ibid. 4. p. 191—416.

———1851, Drei Schabengattungen: Incurvaria, Micropteryx und Nemophora. — ibid. 5. p. 301—362.

———1868, Beitrag zur Kenntnis der Lepidopteren-Fauna der Umgegend von Raibl in Oberkärnthen und Preth. — Verh. zool. bot. Ges. Wien 1868, p. 563—628.

———1872/75, Beiträge zur Kenntnis der nordamerikanischen Nachtfalter Verh. zool. bot. Ges. Wien 22. p. 447—566; 23, p. 201—334; 25, p. 207—360.

ZERNY, H. & BEIER, M., 1936, Lepidoptera. — Kükenthal und Krumbach, Handb. Zool. 4. II. Insecta 2. p. 1554—1728.

ZETTERSTEDT, J. W., 1814/25, Diptera Skandinaviae. Lund, 14 pt.

ZIMMERMANN, A., 1903, Über einige auf den Plantagen von Ost- und West-Usambara gemachte Beobachtungen. — Ber. Land- u. Forstwirtsch. Deutsch. Ostafr. 1. p. 351—381.

ZIMMERMANN, F. & HERING, E. M., 1940, Eine neue Nepticulide aus Deutschland. Zimmermannia liebwerdella (Zimmermann). — Mitt. Zool. Mus. Berlin 24. p. 264—266.

ZIMMERMANN, F., 1944, Zur Kenntnis der Verbreitung der Nepticuliden in den Reichsgauen Wien und Niederdonau. — Ztschr. Wien. ent. Ges. 29, p. 3—6, 60—64, 78—91, 107—122.

ZIRNGIEBEL, A., 1902, Feinde des Hopfens. Berlin.

ZIRNGIEBEL, L., Die Legewerkzeuge der Blattwespen. — Beitr. nat. Forsch. SW. Deutschl. 3. p. 39—65.

ZUMPT, F., 1931, Die Koleopterenfauna des Steppenheide-Biotops von Bellinchen und Oderberg (Fauna marchica). — Inaug. Diss. Berlin.

GENERAL INDEX

(+ = accompanied by figure)

A

B

C

F

G

H

I

J

K

L

M

O

P

Q

R

S

MIX
Papier aus verantwortungsvollen Quellen
Paper from responsible sources
FSC® C105338

FSC
www.fsc.org

If you have any concerns about our products,
you can contact us on
ProductSafety@springernature.com

In case Publisher is established outside the EU,
the EU authorized representative is:
Springer Nature Customer Service Center GmbH
Europaplatz 3, 69115 Heidelberg, Germany

Printed by Libri Plureos GmbH
in Hamburg, Germany